Bob -- from the first development of plant transformation systems in the 1980's to today's adoption of biotech crops by millions of farmers around the globe, it's been a pleasure to witness the phenomenal progress we've made in improving corn as well as the positive impact biotechnolo

agriculture!

However, what is most exciting to me -- and most important -- is the opportunity to double yields in major crops by the year 2030 (off a year 2000 base) to ensure food and energy for a growing planet.

I appreciate your involvement and support and I encourage you to help bring the exciting yield potential of 2030 to life by continuing to advocate for the advancement of science in agriculture!

All the best. Robb Fraley

Biotechnology in Agriculture and Forestry

Volume 63

Biotechnology in Agriculture and Forestry

Further volumes can be found at springer.com.

Volume 30: Somatic Embryogenesis and Synthetic Seed I (1995)
Volume 31: Somatic Embryogenesis and Synthetic Seed II (1995)
Volume 32: Cryopreservation of Plant Germplasm I (1995)
Volume 33: Medicinal and Aromatic Plants VIII (1995)
Volume 34: Plant Protoplasts and Genetic Engineering VI (1995)
Volume 35: Trees IV (1996)
Volume 36: Somaclonal Variation in Crop Improvement II (1996)
Volume 37: Medicinal and Aromatic Plants IX (1996)
Volume 38: Plant Protoplasts and Genetic Engineering VII (1996)
Volume 39: High-Tech and Microprogation V (1997)
Volume 40: High-Tech and Microprogation VI (1997)
Volume 41: Medicinal and Aromatic Plants X (1998)
Volume 42: Cotton (1998)
Volume 43: Medicinal and Aromatic Plants XI (1999)
Volume 44: Transgenic Trees (1999)
Volume 45: Transgenic Medicinal Plants (1999)
Volume 46: Transgenic Crops 1I (1999)
Volume 47: Transgenic Crops II (2001)
Volume 48: Transgenic Crops III (2001)
Volumes 1–48 were edited by Y.P.S. Bajaj†
Volume 49: Somatic Hybridization in Crop Improvement II (2001)
T. Nagata and Y.P.S. Bajaj (Eds.)
Volume 50: Cryopreservation of Plant Germplasm II (2002)
L.E. Towill and Y.P.S. Bajaj (Eds.)
Volume 51: Medicinal and Aromatic Plants XII (2002)
T. Nagata and Y. Ebizuka (Eds.)
Volume 52: Brassicas and Legumes: From Genome Structure to Breeding (2003)
T. Nagata and S. Tabata (Eds.)
Volume 53: Tobacco BY-2 Cells (2004)
T. Nagata, S. Hasezawa, and D. Inzé (Eds.)
Volume 54: Brassica (2004)
E.C. Pua and C.J. Douglas (Eds.)
Volume 55: Molecular Marker Systems in Plant Breeding and Crop Improvement (2005)
H. Lörz and G.Wenzel (Eds.)
Volume 56: Haploids in Crop Improvement II (2005)
C.E. Palmer,W.A. Keller, and K.J. Kasha (Eds.)
Volume 57: Plant Metabolomics (2006)
K. Saito, R.A. Dixon, and L.Willmitzer (Eds.)
Volume 58: Tobacco BY-2 Cells: From Cellular Dynamics to Omics (2006)
T. Nagata, K. Matsuoka, and D. Inzé (Eds.)
Volume 59: Transgenic Crops IV (2007)
E.C. Pua and M.R. Davey (Eds.)
Volume 60: Transgenic Crops V (2007)
E.C. Pua and M.R. Davey (Eds.)
Volume 61: Transgenic Crops VI (2007)
E.C. Pua and M.R. Davey (Eds.)
Volume 62: Rice Biology in the Genomics Era (2008)
H.-Y. Hirano, A. Hirai, Y. Sano and T. Sasaki (Eds.)
Volume 63: Molecular Genetic Approaches to Maize Improvement (2009)
A.L. Kriz and B.A. Larkins (Eds.)

Alan L. Kriz • Brian A. Larkins
Editors

Molecular Genetic Approaches to Maize Improvement

Springer

Prof. Dr. Alan L. Kriz
BASF Plant Science, LLC
26 Davis Drive
Research Triangle Park, NC 27709
USA
alkriz@comcast.net

Prof. Dr. Brian A. Larkins
Department of Plant Sciences
University of Arizona
Marley 822C
P.O. Box 210036
Tucson, AZ 85721-0036
USA
larkins@ag.arizona.edu

ISBN 978-3-540-68919-5 e-ISBN 978-3-540-68922-5

Biotechnology in Agriculture and Forestry ISSN 0934-943X

Library of Congress Control Number: 2008930153

Cover Design: WMXDesign GmbH, Heidelberg, Germany

Printed on acid-free paper

9 8 7 6 5 4 3 2 1

springer.com

Preface

During the past decade, we have seen tremendous progress in maize improvement through the application of molecular genetics and biotechnology. Today, more than half of the US corn acreage is genetically engineered, and further development of this technology will have far-reaching effects on corn production throughout the world. This volume provides an overview of our current knowledge of maize molecular genetics, how it is being used to improve the crop, and future possibilities for crop enhancement. First, we consider the technical approaches for introducing novel genes into the maize genome, regeneration of transformed cells into plants, and creation of transgenic lines for field production. We then consider a number of genetically engineered traits that are currently, or soon will be, in commercial production. Next, we consider how molecular genetic techniques are being used to identify genes and characterize their function, and how these procedures are utilized to accelerate the development of elite maize germplasm. Characterization of the maize genome at the DNA and chromosomal level is providing insight into its structure and evolution, and this information is creating technical approaches that help us to understand and control gene expression. Ultimately, these approaches may lead to the understanding of a fundamentally important feature of maize, hybrid vigor. Currently, more than 70% of corn production is used for food and feed; hence, knowledge of the biochemical genetics of starch, protein and lipid production is fundamental for improving the nutritional and food-making properties of corn. We also consider chloroplasts, as they provide energy for the cell and the precursors for starch, protein, and lipid synthesis. Finally, we consider a new and growing use for corn as biomass for energy production. Thus, this volume provides an in-depth review of the foundation for maize biotechnology.

Alan L. Kriz and Brian A. Larkins

Gales Ferry and Tucson
June 2008

Contents

Part I Introduction

Part II Transgenic Traits

Contributors

W. Brad Barbazuk
Donald Danforth Plant Science Center, 975 N Warson Road, St. Louis, MO 63132, USA
e-mail: bbarbazuk@danforthcenter.org

Fred E. Below
Department of Crop Sciences, University of Illinois at Urbana-Champaign, 389 Edward R. Madigan Laboratory, 1201 W. Gregory Drive, Urbana, IL 61801, USA

Arnold J. Bendich
Biology Department, University of Washington, Box 355325, Seattle, WA 98195-5325, USA
e-mail: bendich@u.washington.edu

Edward S. Buckler
USDA-ARS, US Plant, Soil and Nutrition Laboratory, Ithaca, NY 14853, USA
Institute for Genomic Diversity, 159 Biotechnology, Cornell University, Ithaca, NY 14853, USA
e-mail: esb33@cornell.edu

Ming-Tang Chang
BASF Plant Science LLC, ExSeed Genetics, 2901 South Loop Dr. Bldg 3 Suite 3800, Ames, IA 50010, USA
Current address: 5333 Cervantes Dr., Ames, IA 50014-6914, USA
e-mail: mtcchang@hotmail.com

Bruce M. Chassy
40 NSRC, 1101 West Peabody, University of Illinois, Urbana, IL 61801, USA
e-mail: bchassy@uiuc.edu

Edward H. Coe, Jr
University of Missouri, Curtis Hall, Columbia, MO 65211-7020, USA
e-mail: coee@missouri.edu

R. Kelly Dawe
Department of Plant Biology, Miller Plant Sciences Bldg, University of Georgia, Athens, GA 30602, USA
e-mail: kelly@plantbio.uga.edu

Jill Deikman
Monsanto Company, 1920 Fifth Street, Davis, CA 95616, USA
e-mail: jill.deikman@monsanto.com

Elhan S. Ersoz
Institute for Genomic Diversity, Cornell University, Ithaca, NY 14853, USA

Robert T. Fraley
Monsanto, 800 North Lindbergh Blvd. (A3N), St. Louis, MO 63167, USA
e-mail: janis.braun@monsanto.com

David W. Galbraith
BIO5 Institute for Collaborative Bioresearch and Department of Plant Sciences, University of Arizona, Thomas W. Keating Bioresearch Building, 1657 E. Helen St., Tucson, AZ 85721, USA
e-mail: galbraith@arizona.edu

Thomas Greene
Dow AgroSciences, 9330 Zionsville Road, Indianapolis, IN 46268, USA
e-mail: twgreene@dow.com

Ling Guo
Center for Plant Genomics, Iowa State University, Ames, IA 50011, USA
Current address: Pioneer Hi-Bred Inc., 7100 NW 62nd Avenue, P O Box 1014, Johnston, IA 50131-1014, USA
e-mail: lingguo1@gmail.com

L. Curtis Hannah
University of Florida, Program in Plant Molecular and Cellular Biology and Department of Horticultural Sciences, P.O. Box 110690, 2211 Fifield Hall, Gainesville, FL 32611, USA
e-mail: hannah@mail.ifas.ufl.edu

Graham Head
Monsanto Company, 800 North Lindbergh Blvd, St Louis, MO 63167, USA
e-mail: graham.p.head@monsanto.com

David R. Holding
Department of Plant Sciences, University of Arizona, 303 Forbes Building, P.O. Box 210036, Tucson, AZ 85721, USA

Elizabeth E. Hood
Arkansas State University, P O Box 2760, State University, AR 72467, USA
e-mail: ehood@astate.edu

John A. Howard
Applied Biotechnology Institute, Building 36, Cal Poly State University, San Luis Obispo, CA 93407, USA

An-Ping Hsia
Center for Plant Genomics, Iowa State University, Ames, IA 50011, USA
e-mail: hsia@iastate.edu

Todd J. Jones
BASF Plant Science, LLC, 26 Davis Dr., Research Triangle Park, NC 27519, USA
e-mail: todd.jones@basf.com

Michael R. Kerns
Monsanto Company, 1920 Fifth Street, Davis, CA 95616, USA

Youngmi Kim
Department of Agricultural and Biological Engineering, Purdue University, 225 South University Street, West Lafayette, IN 47907, USA

Alan L. Kriz
BASF Plant Science, LLC, 26 Davis Drive, Research Triangle Park, NC 27709, USA
e-mail: alkriz@comcast.net

Michael Ladisch
Laboratory of Renewable Resources Engineering, Purdue University, Potter Engineering Center, 500 Central Drive, West Lafayette, IN 47907, USA
Department of Agricultural and Biological Engineering, Purdue University, 225 South University Street, West Lafayette, IN 47907, USA
Weldon School of Biomedical Engineering, Purdue University, 206 S. Intramural Drive, West Lafayette, IN 47907, USA
e-mail: ladisch@purdue.edu

Brian A. Larkins
Department of Plant Sciences, University of Arizona, Marley 822C, P.O. Box 210036, Tucson, AZ 85721-0036, USA
e-mail: larkins@ag.arizona.edu

Joachim Messing
Waksman Institute of Microbiology, Rutgers University, 190 Frelinghuysen Road, Piscataway, NJ 08854-8020, USA
e-mail: messing@waksman.rutgers.edu

Rita-Ann Monde
Agronomy Department, Purdue University, 915 West State St., West Lafayette, IN 47907, USA

Stephen Moose
Department of Crop Sciences and Institute for Genomic Biology, University of Illinois at Urbana-Champaign, 389 Edward R. Madigan Laboratory, 1201 W. Gregory Drive, Urbana, IL 61801, USA
e-mail: smoose@uiuc.edu

Nathan Mosier
Laboratory of Renewable Resources Engineering, Purdue University, Potter Engineering Center, 500 Central Drive, West Lafayette, IN 47907, USA
Department of Agricultural and Biological Engineering, Purdue University, 225 South University Street, West Lafayette, IN 47907, USA

John Mullet
Department of Biochemistry and Biophysics, Texas A&M University, College Station, TX 77843, USA
e-mail: jmullet@tamu.edu

Delene J. Oldenburg
Biology Department, University of Washington, Box 355325, Seattle, WA 98195-5325, USA
e-mail: delene@u.washington.edu

Victor Raboy
USDA-ARS, Small Grains and Potato Germplasm Research Unit, 1691 South 2700 West, Aberdeen, ID 83210, USA
e-mail: victor.raboy@ars.usda.gov

Martin M. Sachs
Maize Genetics Cooperation Stock Center, US Department of Agriculture, Agricultural Research Service, Soybean/Maize Germplasm, Pathology and Genetics Research Unit, Department of Crop Sciences, University of Illinois, Urbana, IL 61801, USA
e-mail: msachs@uiuc.edu

Patrick S. Schnable
Center for Plant Genomics, Iowa State University, Ames, IA 50011, USA
e-mail: schnable@iastate.edu

Steven H. Schwartz
Monsanto Company, 1920 Fifth Street, Davis, CA 95616, USA

Stefan Schwietzke
Laboratory of Renewable Resources Engineering, Purdue University, Potter Engineering Center, 500 Central Drive, West Lafayette, IN 47907, USA

A. Mark Settles
Horticultural Sciences Department and Plant Molecular and Cellular Biology Program, University of Florida, Gainesville, FL 32611-0690, USA
e-mail: settles@ufl.edu

Dale L. Val
Monsanto Company, 1920 Fifth Street, Davis, CA 95616, USA

Dannette Ward
Monsanto Company, 800 North Lindbergh Blvd, St Louis, MO 63167, USA

Clifford F. Weil
Agronomy Department, Purdue University, 915 West State St., West Lafayette, IN 47907, USA
e-mail: cweil@purdue.edu

Eduardo Ximenes
Department of Agricultural and Biological Engineering, Purdue University, 225 South University Street, West Lafayette, IN 47907, USA

Jianming Yu
Department of Agronomy, Kansas State University, Manhatten, KS 66506, USA

Part I
Introduction

Chapter 1
Molecular Genetic Approaches to Maize Improvement – an Introduction

Robert T. Fraley

In the following chapters prominent scientists will discuss the recent genetic improvements in maize that have brought us to this point, as well as the potential for future improvement. A review of historical improvements is an instructive base from which to launch this discussion.

Let's look back to 1944, when much like today, the United States was planting as many acres of maize as possible and trying to get as much yield as possible from every acre. According to the US Department of Agriculture, about 85 million acres in the US were planted to maize in 1944 – about the same acreage as 2007. The total production of maize in the US in 1944 was 2.3 billion bushels. In 2007, it was 13.1 billion bushels – that's an increase of 470%. The average yield in 1944 was 33 bushels per acre, production typical of that era. That figure has jumped by 360% to 151 bushels per acre.

Given this stark contrast, the question we must ask ourselves is, how did we come so far so fast? What are some of the factors contributing to higher maize yields? Will these trends continue?

From a historical perspective, the most striking advance in maize production in the last century was the introduction of hybrids. By inbreeding and crossing the resultant lines, maize breeders were able to produce hybrid offspring with higher yields through the phenomenon known as heterosis or "hybrid vigor." In order to harness the power of hybrid vigor, breeders initially began "double-crossing," or crossing two inbred parents at the same time as two other inbred parents. Resulting hybrids are then crossed together to produce a hybrid with the characteristics of all four parents. This method was used into the 1960s, by which time inbreds had improved to the point where single cross hybrids became standard in US maize production and have been widely adopted around the world.

Along with advances in breeding, the steady rate of yield gain in maize benefited from improvements in nutrient management, tillage practices and chemical

Robert T. Fraley
Monsanto, 800 North Lindbergh Blvd. (A3N), St. Louis, MO 63167, USA
e-mail: janis.braun@monsanto.com

A.L. Kriz, B.A. Larkins (eds.), *Molecular Genetic Approaches to Maize Improvement*
Biotechnology in Agriculture and Forestry, Vol. 63

methods for weed and insect management. The agricultural sector built upon these developments by establishing commercial soil and plant testing laboratories and producing new farm implements for no-till systems. Crop consultant enterprises developed to help implement these more information-intensive crop and soil management practices (Duvick and Cassman 1999). The historically strong connection between public institutions (universities, extension services, USDA, etc) and private industry (seed, crop chemical and equipment producers) served as a template for the integration of different production advances, enabling the industry to leverage these technical advances into rapid yield gains. More recent advancements in precision planting equipment and variable rate nutrient delivery ensure these productivity increases will continue.

The application of biotechnology techniques has facilitated rapid progress in breeding using molecular markers. In the last two decades, multiple generations of DNA detection technology were introduced: restriction fragment length polymorphisms (RFLPs), simple sequence repeats (SSRs), amplified fragment length polymorphisms (AFLP) and single nucleotide polymorphisms (SNPs). These technologies moved quickly from theory to practical applications in breeding. The parallel development of computing capability within this time frame also facilitated the ability to fully exploit molecular markers. In addition, development of laboratory automation technology to allow high-throughput genotyping provided the scalability necessary to make use of molecular markers an integral part of commercial maize breeding programs. Routine application of molecular markers in maize breeding programs should lead, by some estimates, to at least a doubling of the rate of genetic gain compared to conventional breeding programs without markers. In addition, the use of molecular markers has greatly facilitated the introgression of transgenic traits into commercial maize germplasm, which has allowed the rapid adoption of transgenic traits in highly adapted germplasm around the world.

The development of maize transformation technology in the early 1990s rapidly accelerated the application of transgenic approaches to improvement of maize production. The first insect-protected maize plants containing genes to confer resistance to European corn borer and other lepidopteran pests were commercialized in the mid-1990s. Since then, adoption of first-generation biotechnology traits, for insect protection and herbicide resistance in maize, has been rapid, not only in North America, but in other maize-growing areas around the world. In particular, the use of maize containing three stacked genes or multiple traits for lepidopteran insect control, corn rootworm control and herbicide tolerance for more efficient weed control has increased dramatically. By 2007, transgenic maize was planted on more than 30 million hectares in 16 countries. In 2008 more than 80 percent of U.S. maize crops contain at least one biotech traits. In addition to higher yields, these transgenic maize hybrids contribute to reduction in greenhouse gases and pesticide use. While the adoption of first generation traits in maize has been rapid, the next generation, currently in development, holds even more promise. These traits are designed to help maize continue to grow under drought conditions, more efficiently use nitrogen, produce even higher yields, enhance protection against insects and other pests, and improve grain quality for food, animal feed and biofuels.

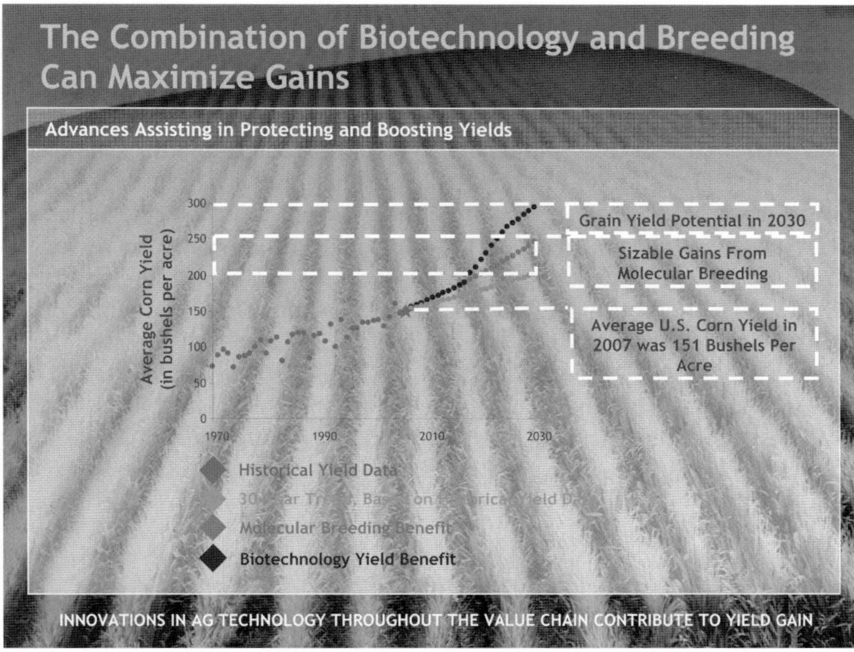

Fig. 1.1

Together with improved farming practices, maize breeding and biotechnology have allowed us to continuously improve productivity to the current US average of 151 bushels per acre. At the same time, maize production has become increasingly sustainable in terms of land, fertilizer and water use. A doubling of yield per acre is a doubling of land use efficiency. Farmers are producing twice as much maize on the same amount of land, and, in the past three decades, sustainability metrics in maize production have improved across the board. For instance, nitrogen fertilizer application rates per acre of maize have held constant while yields have increased by 50%.

New technologies continue to increase our understanding of maize. The complete DNA sequence of the maize genome, along with more comprehensive transcriptome, proteome and metabolome information, will continue to drive innovations in molecular breeding and biotechnology. These additional layers of information help to further unravel the complexities of how genes and gene networks function to produce productive maize plants. This knowledge will lead to improved predictions and capabilities to assemble native gene variation through molecular breeding as well as more optimal gene selection and regulation in the development of future biotechnology products.

In the chapters that follow, leading scientists discuss the recent genetic improvements in maize that have brought us to this point, as well as highlight the immense potential for future improvement. One theme clearly emerges from the text: The pace of new innovation continues to accelerate ... from the development of more powerful breeding tools that are increasing the yield gains ... to the rapid discovery of new transgenes that will further boost yields, mitigate production risks, promote more sustainable production practices and improve grain composition and nutritional content. The effective and rapid integration of these innovations and tools is breathtaking, and this promises to continue to both further increase maize yields in the US and to extend these gains rapidly to other maize production areas around the world, including Asia and Africa. At the same time advances in maize serve as an important roadmap for increasing yields in other crops, including rice and wheat.

In a world with increasing global need and expectation for food and energy security, the pace of technology and yield advances in maize takes on even greater importance. From the chapters in this book, it is clear that we have the genetic tools based on genomics-based breeding and second-generation biotech traits, together with the continued gains from improved agricultural practices and production systems, for the industry to once again double maize yields to 300 bushels per acre.

References

Brookes G, Barfoot, P (2006) Global Impact of Biotech Crops: Socio-Economic and Environmental Effects in the First Ten Years of Commercial Use. AgBioForum, 9(3) 139–151

Duvick DN, Cassman KG (1999) Post-green-revolution trends in yield potential of temperate maize in the north-central United States. Crop Sci 29:1622–1630

Chapter 2
Maize Tissue Culture and Transformation: The First 20 Years

Todd J. Jones

2.1 Introduction

Maize is one of the world's three most widely cultivated crops (along with wheat and rice) and is arguably the most economically important cereal crop on a worldwide basis. That status is only likely to become more apparent in the next decade. Demand for maize is projected to increase by 50% to over 800 million tons per year by the year 2020 and will surpass both rice and wheat in global demand (Pingali and Pandey 2001). Biotechnology is expected to play an increasingly important role in maize genetic improvement to meet this expanding demand.

The genetic marvel that is modern maize belies its humble origins. While still a topic of some debate, the preponderance of evidence clearly suggests that modern maize (*Zea mays* L.) was domesticated from teosinte (*Zea mays* ssp. *parviglumis*) approximately 9000 years ago (Doebley 1990; Doebley and Stec 1991; Matsuoka et al. 2002). Teosinte is an innocuous annual plant with a two-rowed spikelet that produces 6–12 hard cupule-encased kernels, and a growth habit consisting of many, long, lateral branches that are topped by a male inflorescence. Maize, on the other hand, has one central stalk which terminates in a male inflorescence, the tassel, and several modified short axillary shoots that terminate in the female inflorescence, or ear. Unlike the simple teosinte inflorescence, modern maize is capable of producing a large multi-rowed ear that typically produces hundreds of kernels. So different is teosinte from maize that the evolutionary relationship between the two is not readily apparent and is still disputed. This remarkable and rapid transformation of teosinte into domesticated maize was certainly facilitated by Native Americans. Their skillful selection of the useful traits from the existing variation within teosinte created the basis for maize to become the centerpiece of a major agricultural revolution, a revolution that continues today.

Todd J. Jones
BASF Plant Science, LLC, 26 Davis Dr., Research Triangle Park, NC 27709, USA
e-mail: todd.jones@basf.com

A.L. Kriz, B.A. Larkins (eds.), *Molecular Genetic Approaches to Maize Improvement*
Biotechnology in Agriculture and Forestry, Vol. 63

Similarly, but with considerably more rapidity, maize transformation, the technology that enables the introduction of foreign genes into maize, has evolved from conception to commercial reality within the past two decades. The current state of the art utilizes elite germplasm, *Agrobacterium tumefaciens*, standard binary vectors and constructs capable of excising unwanted DNA to efficiently transform maize zygotic embryos and embryogenic cultures (Ishida et al. 1996, 2007; Zhao et al. 2001; Frame et al. 2002; Zhang et al. 2003;). This chapter looks at the development of maize transformation over the past 20 years, with an emphasis on the two key parameters that are required for achieving commercial maize transformation: a reliable system for plant regeneration and an efficient gene delivery mechanism to introduce foreign DNA into maize cells.

2.2 Maize Plant Regeneration Systems

A robust plant transformation system requires an efficient delivery mechanism for DNA transfer into a plant cell and a means to regenerate a fertile plant from the transformed cell. Plant regeneration from tissue culture can be via organogenesis (de novo shoot and root formation) or somatic embryogenesis (de novo formation of embryos from somatic cells). Somatic embryogenesis has the advantage of producing a bipolar structure that can, theoretically, be germinated and regenerated in one step. Somatic embryogenesis also has the capacity to be scaled up rapidly, which can be problematic with many organogenic systems. This chapter will begin with an historical perspective of somatic embryogenesis in maize, as it remains the principal means of plant regeneration, and will then look at alternative organogenic systems.

2.2.1 Somatic Embryogenesis in Maize

The first report of plant regeneration from maize tissue cultures was in 1975, from immature embryos of the inbred line A188 (Green and Phillips 1975). Regeneration of plants in that report was from a compact and organized culture that is now referred to as "type I" callus, a developmentally distinct form of embryogenic regeneration. Subsequent studies of regeneration from immature embryo-derived maize tissue cultures demonstrated unequivocally that plant regeneration was from scutellum cells via somatic embryogenesis (Lu et al. 1982, 1983; Green 1982). These early reports also described regeneration to be from a compact, type-I-like tissue culture, similar to that described by Green and Phillips (1975). In 1982, Green described a "friable embryogenic callus" that was distinctly different, phenotypically, from the compact cultures described in the other reports (Green 1982). This form of embryogenic callus was fast growing, easily dispersed and consisted of many reduced somatic embryos. While highly regenerable, this friable callus type was not easily maintained for extended periods; the embryogenic potential appeared to be reduced with successive subcultures. The problem of "maintainability" of the friable callus phenotype was overcome by Armstrong and Green who were able to produce long-term friable cultures by culturing immature embryos of A188 on N6 medium (Chu 1975) supplemented with 6 mM proline. While proline was certainly a significant addition to the medium composition, the form and amount of reduced nitrogen were also determined to be important for the maintenance of the friable callus phenotype. In this system, N6 salts were also distinctly superior for maintaining a friable embryogenic response compared to MS media (Murashige and Skoog 1962). Media composition and additives are by no means the only critical components for inducing type II somatic embryogenesis in maize. Other factors include the developmental stage of the immature embryo, the type and amount of auxin, subculture timing and, very importantly, genotype. Armstrong and Green (1985) were only able to get a friable embryogenic response from the inbred genotype A188, while seven other inbreds proved recalcitrant to tissue culture. This genotype specificity for embryogenic response has, in large part, been overcome with improvements in tissue culture methodology and genetics (more on this later). The Armstrong and Green paper (1985) was also notable for firmly establishing the terms "type I" and "type II" callus in the maize lexicon, differentiating the compact, complex culture phenotype (type I) from the highly friable, embryogenic cultures (type II). For over 20 years, somatic embryogenic systems based on these early, seminal studies have remained the method of choice for tissue culture, regeneration and transformation of maize.

As mentioned above, somatic embryogenesis has, historically, been considered highly genotype-dependent. Most of the original reports of embryogenesis from maize used the inbred line A188 or hybrids with A188 as one of the parents as the donor for source material (Green and Phillips 1975; Green 1982, 1983; Armstrong and Green 1985). One important stage in maize transformation was the development of Hi-II, a highly type II embryogenic, transformable and publicly available genotype derived from a cross of A188 $\times$ B73 (Armstrong et al. 1991). Armstrong et al. (1991) made selections from the cross and identified a "Parent A" and "Parent B", which, when crossed, produced embryos that produced type II cultures with high frequency. Hi-II has been used extensively in both industrial and academic transformation laboratories. Due to the consistent tissue culture response and ease of transformation, Hi-II has proven to be a stellar "lab rat" for maize transformation experiments and transgenic plant analysis. However, Hi-II is not inbred, it has a high proportion of the less-than-desired A188 genotype and the progeny segregate for phenotypic characteristics. Consequently, Hi-II is not the ideal choice for validating genes with a subtle phenotype or for commercial transformation. Nevertheless, Hi-II has an important place in the history of maize transformation as an extremely prolific genotype bred exclusively for maize embryogenic response and transformation.

It quickly became obvious that the embryogenic response in maize was conditioned by a small number of genes, and they could be transferred from a responsive genotype to a non-responsive but agronomically elite genotype by simple crossing. Rosati et al. (1994) determined that the regeneration capacity could be nearly doubled after only two rounds of selection from a population based on the double-cross (A188 × W64A) × (A634 × B79). The realization that tissue culture response and regeneration in maize is conditioned by a small number of genes led to efforts to identify the genetic basis and possibly the genes that were responsible for the embryogenic response in maize. Armstrong et al. (1992) introgressed A188 into a B73 background and selected for type II tissue culture response from immature embryos. Using RFLP markers, they detected A188 remnant segments on chromosomes 1, 2, 3 and 9 that were correlated with tissue culture response. These quantitative trait loci (QTL) were further tested in an A188 × Mo17 F2 population. Segments on chromosomes 1 and 9 were found to be significant in the A × Mo population, while the region on chromosome 3 could not be adequately tested. Other similar studies evaluated the genetic basis of the androgenic response (Cowen et al. 1992; Wan et al. 1992) from cultured maize anthers and shoot regeneration from maize meristem cultures (Li et al. 2004). All of these studies found the responses to be conditioned by multiple QTLs, some potentially overlapping with those described by Armstrong et al. and some that were unique for the system described. In particular, the regions of chromosome 1, 2 and 3 identified by Armstrong et al. were also found to be involved in the androgenic response by Wan et al. (1992). Similarly, Cowen et al. (1992), in their study of embryo-like structure formation from maize microspores, identified the same region on chromosome 1 as the other two studies and also implicated the region on chromosome 9 identified by Armstrong et al. (1992). Interestingly, both Armstrong et al. (1992) and Cowen et al. (1992) used B73 as the recalcitrant parent in the cross, which may explain the co-identification of the region of chromosome 9 as being involved in embryogenic regeneration in their studies.

Similar studies have also been carried out to identify the genomic regions associated with shoot regeneration from organogenic systems, but there has been little overlap in the regions identified in those studies with those identified for the embryogenic response in maize. For instance, Li et al., (2004) identified a unique region on linkage group seven that conditioned nearly 20% of the shoot regeneration response from cultured maize meristems.

More recently, Krakowsky et al. (2006) have mapped QTLs responsible for callus initiation and totipotency in maize by evaluating type I culture response in a set of recombinant inbred lines (RILs) derived from a cross of H99 × Mo17. The RILs were evaluated over 2 years and, in the means across years, nine QTL on seven chromosomes were detected with main effects. Again, regions on chromosomes 1, 2, 3 and 9 appeared to be involved in callus formation and/or regeneration. In addition to those chromosomal regions, Krakowsky et al. (2006) also detected a unique QTL with main effects on chromosome 5. Interestingly, these QTL also happen to be closely linked to mapped alleles for viviparous mutants (vp1–vp14 and w3), suggesting that abscisic acid (ABA) may be involved in maize somatic embryogenesis. Krakowsky et al. (2006) acknowledge that a definite connection between ABA and embryogenic response has not been established and a functional relationship between ABA and the initiation of somatic embryogenesis is not obvious. However, it is an interesting observation and deserves further exploration.

Inevitably, the identification of QTL for tissue culture response would be extended to also identifying QTL for transformation competence. Lowe et al., (2006) identified markers associated with tissue culture, and by extension transformability, in a cross between Hi-II and the elite, stiff-stalk inbred line FBLL. The markers they identified were located on chromosomes 1S, 1L, 3L, 6L and 10L. The allele associated with culturability was, in every case, provided by a chromosomal segment from the A188 parent of Hi-II. They used the markers in a molecular assisted backcrossing program to introgress transformability into FBLL for three backcross generations and selfed promising lines to produce transformable FBLL-MAB inbred lines. When these lines were tested for transformation efficiency, using either NPTII or glyphosate, they transformed with an average frequency of 3.5%. The inbred FBLL failed to produce any transgenic events under similar conditions (Lowe et al. 2006). Hybrids between three FBLL-MAB lines and two male testers were also evaluated on multiple locations. A slight yield decrease (2–5%) was associated with the FBLL-MAB parents when compared to the FBLL hybrids (Lowe et al. 2006). Nevertheless, this study clearly demonstrated that transformability can be bred into maize inbreds to produce uniform lines with significantly improved transformation efficiency.

While immature embryos remain the popular choice as the starting material for generating embryogenic cultures, alternative explant sources with embryogenic potential have been described. Immature reproductive meristems (tassel and ear) have been utilized to produce regenerable embryogenic cultures (Pareddy and Petolino 1990; Songstad et al. 1992). Mature embryos have been used as a source for embryogenic tissue cultures from several elite inbred lines of maize (Huang et al. 2004). When using mature embryos as an explant, regeneration frequencies of over 30% were reported when a small amount of 6-benzylaminopurine (BA at 0.2 mg/L) and silver nitrate (10 mg/L) was added to the subculture medium. Al-Abed et al. (2006) recently described regeneration from maize "split-seed" explants, derived from germinating mature seeds. The split-seed exposes three distinct embryonic regions capable of regeneration: the scutellum, the coleoptilar-ring and the shoot apical meristem. Regeneration was determined to be via three distinct pathways – callus formation, somatic embryogenesis or direct organogenesis – depending upon the medium composition and the specific region of the split-seed explant. Regeneration was achieved from all inbreds and hybrids tested at relatively high frequencies (28 shoots/explant). Somatic embryogenesis from nodal sections of freshly germinated mature seeds has also recently been described (Sidorov et al. 2006). The use of mature seeds, or explants derived from mature seeds, as an explant source has the advantage of eliminating the dependency on fresh greenhouse- (or field-)grown plants for a year-round, consistent supply of immature embryos.

2.2.2 *Organogenesis in Maize*

Somatic embryogenesis has been the major plant regeneration system for the past two decades, but it is by no means the only system available. Plant regeneration via organogenesis has also been reported for maize, but the amount of research into organogenic systems has been considerably less than that devoted to embryogenesis.

One organogenic system, however, has been explored in some detail and has been successfully used for maize transformation. Regeneration of maize plants from multiplying shoot meristems derived from immature embryos was first described by Jones and Reiter, (1992). Coleoptilar stage immature embryos were cultured on media containing 2,4-D (1.0 mg/L) and a small amount of 6-benzylaminopurine (BAP, 0.05– 0.10 mg/L). This combination of auxin and cytokinin appears to interrupt embryo development at the coleoptilar stage and causes the nascent shoot apical meristem to proliferate as a large, meristematic "field". These meristematic fields had the potential to continually produce new shoots, which were readily rooted on media containing an auxin, such as NAA (Jones and Reiter 1992). A common consequence of many organogenic regeneration systems, multiplying meristems included, is the production of plants with a multicellular origin. If used for transformation, this can lead to the production of chimeric plants, plants that are composed of a combination of transformed and non-transformed cells. In the system described by Lowe et al. (1995), initial transgenic events produced directly from bombarded meristems of coleoptilar-stage immature embryos indeed tended to be chimeric and rarely produced germline events. They overcame this limitation by employing a meristem multiplication step to the apical meristems of chimeric events and were able to select fully transformed plants that transmitted transgenes to the subsequent generation (Lowe et al. 1995). Shoot multiplication from seedling meristems and subsequent organogenic regeneration was also demonstrated by Zhong et al. (1992). They utilized 7-day-old seedlings cultured for 4 weeks in the dark on medium containing 0.5 mg/L 2,4-D and 2.0 BAP. Multiplying meristems were cultured in the light and produced compact, "shoot-tip cultures" that were capable of being regenerated on media without 2,4-D. Zhong et al. (1996) utilized this system to produce transgenic events by particle bombardment. As germline events can only be produced from transgenic events in the L2 layer, they optimized conditions for subepidermal biolistic delivery of DNA into shoot meristem cultures. They determined that 2-month-old shoot tip clumps were the preferred target. Bombardment conditions required a low density of particles (75 μg/shot of 1.0 μm diameter gold or tungsten) accelerated at 1550 p.s.i. More recently, Sairam et al. (2003) demonstrated that maize apical meristems are competent for T-DNA transfer from *Agrobacterium* and possible stable transformation, although the molecular evidence for transformation was scant. In essence, the shoot multiplication system involves converting the single shoot apical meristem of maize into a proliferating culture of adventitious shoot meristems, each with the potential to develop into a new maize shoot that can subsequently be rooted and grown to maturity.

2.3 Maize Transformation Systems

Dicot transformation has been facilitated by the logical use of *Agrobacterium tumefaciens* and its highly evolved DNA delivery mechanism (Herrera-Estrella et al. 1983; Zupan and Zambryski 1995). However, as most monocots are not a natural host for *Agrobacterium*, the initial focus for maize transformation was on naked DNA delivery systems and the development of regeneration systems from transformation-competent cells.

2.3.1 Protoplast Transformation

In the late 1980s, a significant amount of time and effort was spent developing regeneration systems from maize protoplasts (Rhodes et al. 1988a; Shillito et al. 1989). Coupling regeneration via protoplast culture with existing methodology to deliver DNA across cell membranes seemed to be a logical approach. DNA could, in fact, be delivered to maize protoplasts with high efficiency, and protoplast transformation was routinely used for transient transformation experiments. Nevertheless, regeneration of plants from transformed protoplasts remained problematic and idiosyncratic. Specific genotypes were developed with high protoplast regeneration capacity (Mórocz et al. 1990), but these never gained popular acceptance. Regeneration of a transgenic maize plant event via protoplast transformation was first achieved in 1988, but the plants were not fertile (Rhodes et al. 1988b). Time-consuming, genotype-dependent and fussy transformation of maize via plant protoplasts was finally achieved (Golovkin et al. 1993; Shillito et al. 1994), but alternative technologies soon emerged that proved more alluring. An interesting sidebar to protoplast transformation was to deliver DNA molecules to intact, cell-wall-encased maize cells via electroporation (D'Halluin et al. 1992). While initially promising, this technique proved to be technically challenging and, ultimately, inefficient and was never widely adopted.

2.3.2 Particle Bombardment (Biolistics)

The development of particle bombardment technology ("biolistics") was a significant development in crop transformation that eliminated the need to transfer DNA into naked plant protoplasts (Klein et al. 1987, 1988). With the ability to deliver DNA across intact plant cell walls, protoplast regeneration became moot and alternative plant regeneration systems could be used for stable transformation. Coupled with an effective selectable marker for corn cells and a facile means of plant regeneration, particle bombardment proved to be the key to the development of a reproducible maize transformation system. Within months of each other, two reports were published that described maize transformation using particle bombardment to introduce foreign DNA into embryogenic callus or suspension cultures and confirm transgene transmission to the next generation (Gordon-Kamm et al. 1990; Fromm et al. 1990). Particle bombardment transformation of maize was rapidly adopted by many laboratories and used to target other regenerable maize tissues, including immature zygotic embryos (Koziel et al. 1993), type I callus (Wan et al. 1995) and proliferating apical meristems (Lowe et al. 1995; Zhong et al. 1996). Many parameters needed to be optimized, typically empirically, in order to achieve transformation success, including bombardment conditions, DNA concentration and precipitation conditions and tissue pre-treatments. Of particular importance was the discovery that an osmotic pre-treatment, applied prior to particle bombardment, was beneficial to DNA delivery and, ultimately, plant regeneration (Vain et al. 1993). During the early 1990s, particle bombardment was the principal means of maize transformation, both in industry and academia, and the first transgenic maize products to be commercialized were produced using this methodology. In fact, the first 20 maize events to be deregulated by APHIS were produced using particle bombardment and the first *Agrobacterium*-mediated maize event was not deregulated until 2003 (APHIS 2007).

One advantage of particle bombardment is that it can be used to simply and efficiently co-transform genes of interest (GOI) along with a selectable marker gene. Multiple DNAs can simply be mixed, matched and bombarded into plant cells, greatly simplifying the introduction of multiple genes into plants (Chen et al. 1998; Wu et al. 2002). This method eliminates the need for complex vector design and the often tedious and difficult vector cloning steps necessary for *Agrobacterium* transformation. Additionally, large and cumbersome vectors are frequently not easily transformed or maintained in *Agrobacteria* and that alone can be a limitation. In the study by Wu et al. (2002), they demonstrated that independent, non-selected genes regularly integrate into the same locus with the selectable marker and the genes are expressed independently of one another. In fact, they were able to co-introduce nine transgenes and have all of them expressed; additionally, the non-selected transgenes were co-localized with the selectable marker 70% of the time (Wu et al. 2002). Taking this approach one step further, particle bombardment was recently employed to introduce functioning artificial mini-chromosomes into maize, enabling the potential to transform even larger numbers of genes (Yu et al. 2007). Being able to introduce multiple transgenes simultaneously means that it is now possible to manipulate metabolic pathways and produce multimeric proteins in transgenic plants.

A possible disadvantage of particle bombardment is the potential occurrence of multi-copy, complex transgene integrations that may be unstable and prone to silencing. While this may have been the case with early particle bombardment events, and it seems to have been accepted as the prevailing wisdom, it is not necessarily borne out by the facts. Particle bombardment methodology has been refined and, with the use of linear DNA fragments and tightly titrated DNA quantity, clean and simple integration patterns can be regularly achieved (Fu et al. 2000; Altpeter et al. 2004). Altpeter et al. (2004) compared transgenic events of ryegrass produced by *Agrobacterium* and particle bombardment. They found that both methods produced a majority of events with simple integration patterns and stable expression. Gene silencing was observed in lines with five or more copies of the transgene, but these were only about 20% of the lines produced by particle bombardment (Altpeter et al. 2004).

I should note that particle bombardment is not the only technology available to deliver DNA directly across cell walls. Whiskers™ technology utilizes silicon carbide microfibers to penetrate the plant cell wall while intact plant cells (suspension cultures or immature embryos) are immersed in a DNA-containing solution

(Kaeppler et al. 1990; Frame et al. 1994). The cells, whiskers and DNA solution are gently vortexed or mixed. The DNA enters the cells, presumably via "holes" in the cell wall and plasmalemma created by the whiskers. Cellular damage can result from excessive exposure to the whiskers, and the whiskers themselves, being similar in size and shape to asbestos fibers, have the potential to be a respiratory hazard. Consequently, this technique has not been widely adopted. Nevertheless, in skilled hands, whiskers technology has proven to be an efficient transformation system and is used routinely by Dow AgroSciences in their research program. Dow (then Dow Elanco) licensed the technology from Garst Seeds in 1997.

Due to their inherent flexibility and high efficiency, direct DNA delivery methods, and particle bombardment in particular, remain popular and useful maize transformation techniques.

2.3.3 **Agrobacterium-*Mediated Transformation***

The alternative methodology to direct DNA transformation is to utilize disarmed *Agrobacterium tumefaciens* as a vector for plant transformation. Routinely used for dicot transformation since the mid-1980s, transformation of monocotyledonous species by *Agrobacteria* proved to be a difficult challenge. Although there was tantalizing early evidence that particular monocot species could be transformed by some *Agrobacteria*, notably *Asparagus* (Hernalsteens et al. 1984) and *Narcissus* (Hooykaas-Van Slogteren et al. 1984), it was not obvious that the non-host cereal crops could also be transformed. The first credible report of stable *Agrobacterium*-mediated transformation of a monocot crop species was in 1993 when Chan et al. (1993) reported successful regeneration of transgenic rice plants from inoculated immature embryos. They did not regenerate many plants and provided molecular analysis of the progeny from only one individual transformant; nevertheless the data for stable transformation were persuasive and convincing. Shortly after that report, a group from Japan Tobacco Inc. reported an efficient and reproducible system for *Agrobacterium*-mediated rice transformation (Hiei et al. 1994). This seminal paper not only provided data from hundreds of independent events from rice genotypes, but also described the use of a "super-binary" vector derived from pTiBo542 in *Agrobacterium* strain LBA4404 (Hood et al. 1986; Komari 1990). The so-called super-binary vector developed by T. Komari, pTOK233, is a small T-DNA plasmid that contained an extra copy of *virB*, *virC* and *virG*. The super-binary vector proved to be particularly useful when transforming more recalcitrant varieties of rice, such as cv. Koshihikari. Other factors regarded as critical for successful transformation included: the choice of starting material (embryogenic callus was most preferred), tissue culture conditions, co-cultivation conditions and media composition (Hiei et al. 1994).

Using their success in rice transformation as a starting point, the researchers at Japan Tobacco were soon able to develop an efficient and reproducible *Agrobacterium*-mediated system for maize, using maize inbred A188, immature embryos and LBA4404 containing the super-binary vectors pSB131 or pTOK233 (Ishida et al. 1996). They generated 120 phosphinothricin (PPT)-resistant plants and analyzed 33 T0 plants by Southern hybridization. Nineteen of the 33 plants contained a single copy of the *bar* gene for tolerance to phosphinothricin and none contained more than three copies. The progeny from 40 selfed plants were also analyzed for transgene transmission and Mendelian inheritance of the T-DNA. Twenty-eight of the lines had an expected 3:1 segregation ratio, indicative of single-locus integration, and Southern analyses of the T-DNA boundaries were similar to those observed in *Agrobacterium*-mediated dicot transformation (Ishida et al. 1996). Similar to rice, the Japan Tobacco researchers found that other factors were of critical importance to transformation. For maize, these included: the choice of starting material (immature embryos were preferred), genotype and the concentration of *Agrobacterium* and selectable marker. Interestingly, they also noted that optimizing conditions for screenable marker expression (GUS expression) did not correlate with stable transformation, indicating that DNA integration, and not DNA delivery, was the limiting factor. The two reports from Japan Tobacco clearly demonstrated that *Agrobacterium*-mediated T-DNA transfer occurs via a similar mechanism in both dicots and monocots. *Agrobacterium* transformation of cereal crops had become a reality.

2.4 Selectable Marker Systems

2.4.1 Herbicidal Selectable Markers

Selectable marker genes are commonly used to identify transformed cells from non-transformed during plant transformation. In early dicot transformation systems, antibiotic selectable markers, neomycin phosphotransferase II (NPTII) or hygromycin phosphotransferase (HPT) conferring resistance to kanamycin or hygromycin, respectively, were routinely employed. Maize cells, on the other hand, tended to be difficult to select using antibiotic selection. In fact, the first report of fertile transgenic maize plants used herbicidal selection to select the transformed cell lines (Gordon-Kamm et al. 1990; Fromm et al. 1990). The fertile transgenic plants produced by Fromm et al. (1990) used a mutant form of the gene that encodes acetolactate synthase (ALS) that conferred resistance to chlorsulfuron, a member of the sulfonylurea class of herbicides. While not commonly utilized as a selectable marker for maize, the mutant ALS enzyme (also known by its synonym acetohydroxy acid synthase, or AHAS) has proven itself to be a useful and efficient marker. At BASF Plant Science, we prefer to use a mutant form of the AHAS protein that confers specific resistance to the imidazalinone class of herbicides as opposed to the sulfonylureas (Peng et al. 2006). When coupled with the appropriate herbicidal chemistry, AHAS provides a tight selection over a wide range of maize germplasm. We have used AHAS and imazethapyr selection effectively for inbred maize lines from various heterotic groups, with virtually no escapes and a high percentage of single copy events (>60%).

The more commonly used herbicidal selectable marker for maize is the *bar* gene encoding phosphinothricin acetyltransferase (PAT). PAT confers resistance to the herbicidal compounds L-phosphinothricin or bialophos via acetylation (De Block et al. 1987). The *bar* gene was first used successfully for maize by Fromm et al. (1990) and Gordon-Kamm et al. (1990) to select for stable transgenic events and transgenic fertile plants, respectively. *Bar* was quickly adopted by numerous laboratories developing maize transformation. It has been used successfully to transform A188, A188 hybrids, Hi-II and numerous inbred genotypes (Gordon-Kamm et al. 1990; Zhao et al. 2001; Huang et al. 2005; Ishida et al. 2007). Since the first report of reproducible maize transformation, *bar* selection has become the selection marker of choice.

Other herbicidal selection agents have also been utilized successfully, including glyphosate selection using a gene encoding a resistant form of the EPSP synthase (Howe et al. 1992). I should also note that, contrary to my earlier statement, antibiotic selection has also been used successfully in maize. In particular, npt II and the antibiotic kanamycin have been utilized successfully for some genotypes (Lowe et al. 2006).

2.4.2 Alternative Non-antibiotic, Non-herbicidal Selectable Markers

While antibiotic and herbicidal selectable markers have proven laboratory utility, the general public has voiced concerns regarding having transgenic crop plants containing antibiotic- or herbicide-resistant genes. Consequently, alternative, less contentious selectable marker systems have been developed. One such marker system is based on the fact that plant cells cannot metabolize mannose as a carbon source. Mannose is readily taken up by plant cells, where it is rapidly converted into mannose-6-phosphate by the action of hexokinase. Mannose-6-phosphate is not capable of being utilized in plant cells due to the lack of an effective phosphomannose isomerase (PMI), and buildup of mannose-6-phosphate ultimately interferes with glycolysis. Introduction of *pmi*, the gene encoding phosphomannose isomerase (PMI), into plant cells confers the capability to convert mannose-6-phosphate into fructose-6-phosphate, thereby providing a growth advantage to transformed cells over non-transformed cells, a so-called positive selection system (Joersbo and Okkels 1996; Joersbo et al. 1998). Mannose selection has been successfully applied to produce fertile transgenic maize plants (Negrotto et al. 2000). More recently, it was discovered that plant cells are able to absorb D-amino acids, but have a limited capacity to metabolize them (Erikson et al. 2004). In addition, certain D-amino acids, D-serine and D-alamine in particular, appear to be especially toxic to plant cells at even low concentrations (Erikson et al. 2004). Most eukaryotes, including yeast, insects, reptiles, birds and mammals contain a D-amino acid oxidase (DAAO) that can metabolize D-amino acids. This enzyme is apparently missing in plants; no plant DAAO activity has ever been reported. The introduction of the gene encoding DAAO (*dao1*) from the yeast *Rhodoturula gracilis* into plants has been demonstrated to be an effective selectable marker for plants exposed to toxic levels of D-ser and D-ala (Erikson et al. 2004). In addition, a unique attribute of DAAO is that the deamination of the non-toxic D-amino acids D-isoleucine and D-valine produces a toxic keto acid intermediate, 3-methyl-2-oxopentanoate and 3-methyl-2-oxobutanoate respectively. Consequently, plant cells expressing *dao1* can be selected against if grown in the presence of D-isoleucine or D-valine, providing a powerful counter-selection to eliminate transgenic cells when desired. A second D-amino acid catabolizing enzyme, D-serine ammonia lyase (DSD), specific for D-serine and D-threonine, has also been used as a selectable marker for plant cells (Erikson et al. 2005). The gene is encoded by the *dsdA* gene in *E. coli*. In *E. coli*, DSD is used to specifically metabolize D-serine, which it uses as a carbon and nitrogen source. Introduction of the *E. coli dsdA* gene into plant cells also conferred the ability to metabolize and detoxify D-serine and D-alanine in *Arabidopsis* (Erikson et al. 2005). At BASF Plant Science, we have successfully implemented D-amino acid selection using *dao1* and *dsdA* for maize, a system we refer to as SELDA™ (Lai et al. 2007). In addition to selection on D-serine and D-alanine, we have also demonstrated efficient counter-selection of *dao1*-containing maize on D-isoleucine and D-valine. SELDA™ selection has been used to generate fertile transgenic maize plants from A188 and A188 hybrids, Hi-II and numerous inbreds from different heterotic groups. SELDA™ is a reliable, consistent selectable marker system that is comparable to our herbicidal selection system based on a mutant AHAS and imidazilinone herbicides. SELDA™ is also compatible with IMI selection, as cells selected on imazethapyr can subsequently be transformed and selected using D-serine or D-alanine, or vice versa. SELDA™ provides an efficient, versatile, non-antibiotic, non-herbicidal selectable marker for maize.

2.5 Marker-Free Transformation

2.5.1 Co-transformation and Transgene Segregation

Regardless of the selectable marker chosen for maize transformation, there are reasons to consider the means to produce selectable marker-free transformation events. Having a gene for herbicide tolerance may not be desired in the final product, or it may be useful to eliminate the selectable marker to allow for future re-transformation. Whatever the case, strategies to remove the selectable marker have been developed and successfully implemented in maize. Co-transformation, by having your gene of interest on a separate DNA molecule from the selectable marker, has been an obvious and simple solution. In the case of co-transformation, the marker gene may integrate at a completely separate locus from the gene of interest, allowing for segregation in the subsequent generation. Co-transformation is easily achieved via particle bombardment, and other direct DNA transformation techniques, where separate, unlinked DNA molecules are simply bombarded into the plant cells. The selectable marker and gene of interest often integrate into different chromosomal locations and are separable in the subsequent generation (Socher et al. 1986). *Agrobacteria* often contain more than one T-DNA and, consequently, have evolved a natural mechanism for co-transformation (Depicker et al. 1985; Deblock and Debrouwer 1991). Komari et al. (1996) exploited this phenomenon to develop an efficient, super-binary "2 T-DNA" *Agrobacterium* system capable of delivering the selectable marker and an unlinked gene of interest. They demonstrated co-transformation and unlinked transgene segregation in tobacco and rice. The independent loci were then able to be segregated in the next generation. Miller et al. (2002) demonstrated that a 2 T-DNA approach can be used to achieve co-transformation and transgene segregation in maize. In their system, the *gus* screenable marker gene segregated independently from the *bar* marker gene in over 70% of the events (Miller et al. 2002).

An alternative to co-transformation is to excise the marker gene using recombinase systems that recognize specific DNA sequences for recombination. For a comprehensive review of marker excision strategies, see the review by Hare and Chua (2002). I will present a brief summary of molecular means of marker excision and the application of such systems for maize transformation. Three such systems have been utilized in plants: the CRE/loxP system from bacteriophage P1 (Dale and Ow 1991; Russell et al. 1992) and the FLP/FRT or R/RS from yeast (Lyznik et al. 1996; Sugita et al. 2000). In each case the recombinase, CRE, FLP or R, cleaves DNA at its recognition site, loxP, FRT or RS, with subsequent ligation of the cleaved DNA. When the recognition sites are added to a transgene such that they flank the selectable marker in opposite orientation, the recombinase cleaves the DNA on both sides of the selectable and effectively excises it from the site of integration. In practice, the recombinase is often introduced into the transgenic plant targeted for marker excision by sexual crossing with a recombinase-expressing plant. Marker excision occurs in the F1 generation and the marker-free plants are resolved in the subsequent generation. While demonstrated to be effective in maize (Lyznik et al. 1996; Zhang et al. 2003), this method obviously extends the timeline by at least one generation. A faster alternative would be to express the recombinase protein in a controlled fashion so that it is expressed only at the proper time to mediate marker excision in germline cells in the T0 generation, for instance using a chemically inducible promoter that is effective in the L2 layer. This has been demonstrated to work in *Arabidopsis* using a β-estradiol-inducible hybrid-promoter-driving CRE (Zuo et al. 2001) and a similar strategy could prove to be effective in maize. All of the above recombinase/recognition site systems have the unfortunate consequence of leaving a DNA "footprint" following marker excision and ligation. This extraneous DNA, while not necessarily deleterious, means that the recombinase systems are not entirely "clean". One possible method to produce perfectly recombined marker-excision events would be to use homing endonucleases which recognize rare sequences and leave no footprint upon DNA cleavage and ligation. One such endonuclease, I-SCE1, has shown promise for mediating neat excision of selectable markers in tobacco (Siebert and Puchta 2002) and has recently been used to mediate homologous recombination in maize (D'Halluin et al. 2007).

Of course, the logical way to produce marker-free transgenic plants is to not use selectable markers at all. Marker-free transformation has been demonstrated to be an effective strategy for potato (de Vetten et al. 2003). Potato explants were transformed with *A. tumefaciens* strain AGL0 with a granule-bound starch synthase antisense construct and screened using a PCR-based strategy on DNA isolated from pooled regenerated plants. de Vetten et al. were able to identify transgenic plants, containing the gene of interest only, in 4.5% of the harvested shoots (de Vetten et al. 2003). Depending upon the efficiency of the transformation system employed, simply selecting for regenerating plants that contain only the transgene of interest is a viable approach to producing marker-free transgenic plants.

2.6 Future Prospects: Bigger and Better

Much of the maize transformation research over the past two decades has been focused on defining conditions to make the maize transformation process more efficient and genotype-independent. Over the next decade, research will be increasingly focused on enhancing the quality of transformation events. Anticipated quality enhancements include: reliable gene targeting; directed integration of transgenes into a desired genetic locus via homologous recombination; cleaner, well-defined integration events; and, most significantly, the means to transform increasingly large DNA molecules, including those that encode genes for whole metabolic pathways. Interestingly, these are some of the same improvements anticipated by Chuck Armstrong in a similar review written nearly 10 years ago (Armstrong 1999). He was, indeed, prescient, and technology improvement in those areas, while steady, has been incremental.

2.6.1 Homologous Recombination and Targeted Integration

In higher plants, transgene integration typically occurs randomly at breaks in chromosomes via a process of illegitimate recombination (or non-homologous end-joining, NHEJ). In NHEJ, DNA ends are ligated to unrelated, or minimally related, sequences. This random integration into the genome leads to so-called position effects, differences in gene expression due to varying chromosomal locations of the integration events. This, in turn, means that many different independent transformation events need to be analyzed to identify the preferred transgenic event. Homologous recombination, and by extension gene targeting, occurs when long stretches of homologous sequence align during the recombination process. Homologous recombination is enhanced by the occurrence of double-strand breaks (DSBs) in the chromosomes, elegantly demonstrated in plants by Puchta in 1999. Holger Puchta used transient expression of the rare-cutting homing endonuclease I-Sce1 to create a DSB at an introduced recognition site in tobacco. Homologous recombination was increased 100-fold by the creation of DSB at the I-Sce1 target site (Puchta 1999). As mentioned above, the I-Sce1 endonuclease has also been used to effect marker excision in tobacco (Siebert and Puchta 2002) and, very recently, it was used to mediate high-fidelity DNA targeting in maize (D'Halluin et al. 2007). D'Halluin's team used *Agrobacterium*-mediated transformation and particle bombardment to produce numerous events with precise integration of a targeted sequence insertion into a pre-engineered I-Sce1 site. The use of an I-Sce1 gene that had been codon optimized for expression in maize appeared to be an important factor (D'Halluin et al. 2007). Interestingly, both *Agrobacterium*-mediated and particle bombardment methods produced precisely targeted events with a similar high frequency (D'Halluin et al. 2007). This indicates that, in their system, homologous recombination may be the preferred method of DNA integration and can thus be used to produce higher quality events via particle bombardment than can be done using the standard methodology. While this is an exciting prospect, much work remains to be done to validate and improve the technology. For homologous recombination to become a feasible transformation approach in plants, techniques need to be refined for the creation of DSBs at specific, desired genomic sites and to routinely shift the frequency of transgene integration from NHEJ to HR. For instance, it may be possible to molecularly block enzymes involved in illegitimate recombination to further enhance the occurrence of homologous recombination. Ultimately, improved precision in transgene integration and gene targeting would greatly simplify the event sorting process by minimizing expression variability.

An alternative method for introducing DSBs, and hence homologous recombination, is the use of zinc finger nucleases designed to cleave specific nucleic acid sequences. Zinc finger nucleases (ZFNs) contain an endonuclease domain, typically from the *Fok*I restriction enzyme, to introduce a DSB coupled with three to four DNA binding zinc fingers, each designed to recognize a specific 3-base sequence (Kumar et al. 2006). Each zinc finger is stabilized by a zinc ion, hence the name. Different DNA recognition sites can be targeted by mixing and matching zinc fingers with various specificities. ZFNs can, therefore, be theoretically targeted to virtually any site in the genome. Wright et al. (2005) demonstrated the feasibility of this approach by targeting a partially deleted, inactive GUS:NPT II fusion gene in a specific site into tobacco protoplasts that contained the missing sequence from the fusion gene. Homologous recombination restored the function of the GUS:NPT II genes. In this set of experiments, targeted integration was observed 10% of the time, an increase in homologous recombination of four to five orders of magnitude (Wright et al. 2005) over the typical occurrence of HR. Demonstration of ZFN-mediated homologous recombination in maize has yet to be demonstrated, but this technique appears to have great potential for enhancing the precision of maize transformation.

2.6.2 High Molecular Weight DNA Transformation

In addition to increased precision of transformation, there is a need to be able to reliably and reproducibly transform maize with DNA constructs of increasingly high molecular weight and complexity. Few second generation traits will be as simple and easily identified as herbicide or insect tolerance. It is highly likely that transgenic traits for yield, for instance, will be multigenic and may, in fact, require coordinated expression of partial or whole biochemical pathways. In the mid 1990s, Carol Hamilton and her laboratory developed BIBAC (binary bacterial artificial chromosome) vectors that were *Agrobacterium*-transformation-competent and capable of delivering over 150 kb of DNA to tobacco and tomato (Hamilton et al. 1996; Frary and Hamilton 2001). As might be expected, the transformation frequency with BIBACs was inversely proportional to the size of the T-DNA; 150 kb inserts were two- to three-fold more difficult to transform than a 30-kb T-DNA, and up to an order of magnitude more difficult than a standard binary vector (Frary and Hamilton 2001). Interestingly, they also noted that addition of a helper plasmid containing *vir*G, the transcriptional activator of the other *vir* genes, was an absolute requirement for large DNA transformation in tomato (Frary and Hamilton 2001). BIBAC transformation has not been reproducibly demonstrated in maize and one limitation of the technology may be the instability of large (>100 kb) DNA vectors in *Agrobacterium* (Song et al. 2003). Delivery of BAC vectors by particle bombardment may be a simpler and more direct approach. BAC clones of ~45 kb from sorghum have been successfully bombarded into maize cells (Song et al. 2004) and recently a maize BAC of over 100 kb was successfully transformed into rice (Phan et al. 2006). Ultimately, engineered artificial plant chromosomes are likely to be the desired vector for the introduction of large DNA segments into maize. As artificial chromosomes containing a functioning centromere are capable of being replicated and maintained independently in cells, they would provide a versatile, and theoretically predictable, platform for DNA introduction. The transformation of maize with engineered minichromosomes, derived from supernumerary B maize chromosomes by telomere truncation, has recently been demonstrated by Jim Birchler's laboratory (Yu et al. 2007). This ground-breaking work clearly demonstrated GUS gene expression from minichromosomes and meiotic transmission of minichromosomes to progeny (Yu et al. 2007). In addition they demonstrated that it is possible to add successive transgenes to the minichromosomes using the CRE/*lox* recombination system (Yu et al. 2007). Hence, minichromosomes have been demonstrated to be heritable, recombination-competent platforms for gene expression in maize. This opens the door to a variety of large-DNA applications in maize, from multiple gene stacks, to biochemical pathways, to functional genomics.

The next decade of maize transformation is poised to be as exciting, inventive and productive as the previous two. Maize transformation technology continues to evolve rapidly to meet the needs of the maize biotechnology community and the growing demand for improved maize varieties.

References

Al-Abed D, Sairam R, Talla R, Goldman S (2006) Split-seed: a new tool for maize researchers. Planta 223:1355–1360

Altpeter F, Fang Y-D, Xu J, Ma XR (2004) Comparison of transgene expression stability after *Agrobacterium*-mediated or biolistic gene transfer into perennial ryegrass. In: Hopkins A, Wang ZY, Mian R, Sledge M, Barker R (eds) Molecular breeding of forage and turf. Kluwer, Dordrecht, pp 255–260

APHIS (2007) Petitions of nonregulated status granted or pending by APHIS as of 3 July 2007. http://www.aphis.usda.gov/brs/not_reg.html

Armstrong C (1999) The first decade of maize transformation: a review and future perspective. Maydica 44:101–109

Armstrong C, Romero-Severson J, Hodges T (1992) Improved tissue culture response of an elite maize inbred through backcross breeding, and identification of chromosomal regions important for regeneration by R.F.L.P. analysis. Theor Appl Genet 84:755–762

Armstrong CL, Green CE (1985) Establishment and maintenance of friable, embryogenic maize callus and the involvement of L-proline. Planta 164:207–214

Armstrong CL, Green CE, Phillips RL (1991) Development and availability of germplasm with high type II culture formation response. Maize Genet Coop Newsl 65:92–93

Chan M-T, Chang H-H, Ho S-L, Tong W-F, Yu S-M (1993) *Agrobacterium*-mediated production of transgenic rice plants expressing a chimeric α-amylase promoter/β-glucoronidase gene. Plant Mol Biol 22:491–506

Chen L, Marmey P, Taylor NJ, Brizard J, Expinoza C, D'Cruz P, Huet H, Zhang S, de Kochko A, Beachy RN, Fauquet CM (1998) Expression and inheritance of multiple transgenes in rice plants. Nat Biotechnol 16:1060–1064

Chu CC (1975) Establishment of an efficient medium for anther culture of rice through comparative experiments on the nitrogen sources. Sci Sinica 18:659–668

Cowen NM, Johnson CD, Armstrong K, Miller M, Woosley A, Pescitelli S, Skokut M, Belmar S, Petolino JF (1992) Mapping genes controlling in vitro androgenesis in maize using RFLP analysis. Theor Appl Genet 84:720–724

Dale EC, Ow DW (1991) Gene transfer with the subsequent removal of the selection gene from the host genome. Proc Natl Acad Sci USA 88:10558–10562

De Block M, Debrouwer D (1991) Two T-DNA's co-transformed into *Brassica napus* by a double *Agrobacterium tumefaciens* infection are mainly integrated at the same locus. Theor Appl Genet 82:257–263.

De Block M, Botterman J, Vandewiele M, Dockx J, Thoen C, Gossele V, Rao Movva N, Thompson C, Van Montague M, Leemans J (1987) Engineering herbicide resistance in plants by expression of a detoxifying enzyme. EMBO J 6:2513–2518

Depicker A, Herman L, Jacobs A, Schell J Van Montagu M (1985) Frequencies of simultaneous transformation with different T-DNAs and their relevance to the *Agrobacterium*/plant cell interaction. Mol Gen Genet 201:477–484

de Vetten N, Wolters A-M, Raemakers K, van der Meer I, ter Stege R, Heeres E, Heeres P, Visser R (2003) A transformation method for obtaining marker-free plants of a cross-pollinating and vegetatively propagated crop. Nat Biotechnol 21:439–442

D'Halluin K, Bonne K, Bossut M, De Beuckeleer M, Leemans J (1992) Transgenic maize plants by tissue electroporation. Plant Cell 4:1495–1505

D'Halluin K, Vanderstaeten C, Stals E, Cornelissen M, Ruiter R (2007) Homologous recombination: a basis for targeted genome optimization in crop species such as maize. Plant Biotechnol J 6:93–102

Doebley J (1990) Molecular evidence and the evolution of maize. Econ Bot 44:6–27

Doebley J, Stec A (1991) Genetic analysis of the morphological differences between maize and teosinte. Genetics 129:285–295

Erikson O, Hertzberg M, Näsholm T (2004) A conditional marker gene allowing both positive and negative selection in plants. Nat Biotechnol 4:455–458

Erikson O, Hertzberg M, Näsholm T (2005) The *dsdA* gene from *Escherichia coli* provides a novel selectable marker for plant transformation. Plant Mol Biol 57:425–433

Frame BR, Drayton PR, Bagnall SV, Lewnau CJ, Bullock WP, Wilson HM, Dunwell JM, Thompson JA, Wang K (1994) Production of fertile transgenic maize plants by silicon carbide whisker-mediated transformation. Plant J 6:941–948

Frame BR, Shou H, Chikwamba RK, Zhang Z, Xiang C, Fonger TM, Pegg, SE, Li BC, Nettleton DS, Pei DQ, Wang K (2002) *Agrobacterium tumefaciens*-mediated transformation of maize embryos using a standard binary vector system. Plant Physiol 129:13–22

Frary A, Hamilton CH (2001) Efficiency and stability of high molecular weight DNA transformation: an analysis in tomato. Transgenic Res 10:121–132

Fromm ME, Morrish F, Armstrong C, Williams R, Thomas J, Klein TM (1990) Inheritance and expression of chimeric genes in the progeny of transgenic maize plants. Bio/Technol 8: 833–839

Fu XD, Duc LT, Fontana S, Bong BB, Tinjuangjun P, Dudhakar D, Twyman RM, Christou P, Kohli A (2000) Linear transgene constructs lacking vector backbone sequences generate low-copy-number transgenic plants with simple integration patterns. Transgenic Res 9:11–19

Golovkin MV, Abraham S, Morocz S, Bottka A, Feher D, Dudits D (1993) Production of transgenic maize plants by direct DNA uptake into embryogenic protoplasts. Plant Sci 90:41–52

Gordon-Kamm WJ, Spencer TM, Mangano ML, Adams TR, Daines RJ, Start WG, Obrien JV, Chambers SA, Adams WR, Willetts NB, Rice TB, Mackey CJ, Krueger RW, Kausch AP, Lemaux PG (1990) Transformation of maize cells and regeneration of fertile transgenic plants. Plant Cell 2:603–618

Green CE (1982) Somatic embryogenesis and plant regeneration from the friable callus of *Zea mays*. In: Fujiwara A (ed) Plant tissue culture, Maruzen, Tokyo, pp 107–108

Green CE (1983) New developments in plant tissue culture and regeneration. In: Hollaender A, Laskin AI, Rogers P (eds) Basic biology of new developments in biotechnology. Plenum, New York, pp 195–209

Green CE, Phillips RL (1975) Plant regeneration from tissue cultures of maize. Crop Sci 15:417–420

Hamilton CM, Frary A, Lewis C, Tanksley SD (1996) Stable transfer of intact high molecular weight DNA into plant chromosomes. Proc Natl Acad Sci USA 93:9975–9979

Hare PD, Chua N-H (2002) Excision of selectable marker genes from transgenic plants. Nat Biotechnol 20:575–580

Hernalsteens J-P, Thia-Toong L, Schell J, Van Montagu M (1984) An *Agrobacterium*-transformed cell culture from the monocot *Asparagus officinalis*. EMBO J 3:2029–3041

Herrera-Estrella L, Depicker A, Van Montague M, Schell J (1983) Expression of chimaeric genes transferred into plant cells using a T-plasmid-derived vector. Nature 303:209–213

Hiei Y, Ohta S, Komari T, Kumashiro T (1994) Efficient transformation of rice (*Oryza sativa* L.) mediated by *Agrobacterium* and sequence analysis of the boundaries of the T-DNA. Plant J 6:271–282

Hood EE, Helmer GL, Fraley RT, Chilton M-D (1986) The hypervirulence of *Agrobacterium tumefaciens* A281 is encoded in a region of pTiBo542 outside of T-DNA. J Bacteriol 168:1291–1301

Hooykaas-Van Slogteren GMS, Hooykaas PJJ, Schilperoot RA (1984) Expression of Ti plasmid genes in monocotyledonous plants infected with *Agrobacterium tumefaciens*. Nature 311:763–764

Howe AR, Tamayo R, Brown S, Armstrong C, Fromm M, Hart J, Padgette S, Parker G, Horsch R (1992) Development of glyphosate as a selectable for the production of fertile transgenic corn plants. In Vitro Cell Dev Biol-Plant 28:124A

Huang X-Q, Wie Z-L (2004) High-frequency plant regeneration through callus initiation from mature embryos of maize (*Zea mays* L.). Plant Cell Rep 22:793–800

Huang X-Q, Wie Z-L (2005) Successful *Agrobacterium*-mediated genetic transformation of maize elite inbred lines. Plant Cell Tissue Organ Cult 83:187–200

Ishida Y, Saito H, Ohta S, Hiei Y, Komari T, Kumashiro T (1996) High efficiency transformation of maize (*Zea mays* L.) mediated by *Agrobacterium tumefaciens*. Nat Biotechnol 14:745–750

Ishida Y, Hiei Y, Komari T (2007) *Agrobacterium*-mediated transformation of maize. Nat Protocols 2:1614–1621

Joersbo M, Okkels FT (1996) A novel principle for selection of transgenic plant cells: positive selection. Plant Cell Rep 16:219–221

Joersbo M, Donaldson I, Kreiberg J, Petersen SG, Brunstedt J, Okkels FT (1998) Analysis of mannose selection used for transformation of sugar beet. Mol Breed 4:111–117

Jones TJ, Reiter KL (1992) The establishment of proliferative shoot meristem cultures from immature embryos of maize. Proc 34th Annual Maize Genetics Conference, Asilomar

Kaeppler HF, Gu W, Somers CA, Rines HW, Cockburn AF (1990) Silicon carbide fiber-mediated DNA delivery into plant cells. Plant Cell Rep 9:415–418

Klein TM, Wolf ED, Wu R, Sanford JC (1987) High-velocity microprojectiles for delivery of nucleic acids into living cells. Nature 327:70–73

Klein TM, Fromm ME, Weissinger A, Tomes D, Schaaf S, Sletten M, Sanford JC (1988) Transfer of foreign genes into intact maize cells with high-velocity microprojectiles. Proc Natl Acad Sci USA 85:4305–4309

Komari T (1990) Transformation of cultured cells of *Chenopodium quinoa* by binary vectors that carry a fragment of DNA from the virulence region of pTiBo542. Plant Cell Rep 9: 303–306

Komari T, Hiei Y, Saito Y, Murai N, Kumashiro T (1996) Vectors carrying two separate T-DNAs for co-transformation of higher plants mediated by *Agrobacterium tumefaciens* and segregations of transformants free from selection markers. Plant J 10:165–174

Koziel MG, Beland GL, Bowman C, Carozzi NB, Crenshaw R, Crossland L, Dawson J, Desai N, Hill M, Kadwell S, Launis K, Lewis K, Maddox K, McPherson K, Meghil MR, Merlin E, Rhodes R, Warren GW, Wright M, Evola SV (1993) Field performance of elite transgenic maize plants expressing an insecticidal protein derived from *Bacillus thuringiensis*. Bio/Technol 11:194–200

Krakowsky MD, Lee M, Garay L, Woodman-Clikeman W, Long MJ, Sharapova N, Frame B, Wang K (2006) Quantit.ative trait loci for callus initiation and totipotency in maize (*Zea mays* L.). Theor Appl Genet 113:821–830

Kumar S, Allen GC, Thompson WF (2006) Gene targeting in plants: fingers on the move. Trends Plant Sci 11:159–161

Lai F-M, Mankin L, Mei K, Jones T, Song H-S (2007) Selection system for maize. Int Patent Appl WO2007014844

Li W, Sun G, Liu J, Masilamany P, Taylor JH, Yan W, Kasha KJ, Pauls KP (2004) Inheritance of plant regeneration from maize (*Zea mays* L.) shoot meristem cultures derived from germinated seeds and the identification of associated RAPD and SSR markers. Theor Appl Genet 104:681–687

Lowe BA, May MM, Kumpf JM, Rout J, Warner D, Johnson R, Armstrong CL, Spencer MT, Chomet PS (2006) Molecular assisted breeding for transformability in maize. Mol Breed 18:229–239

Lowe K, Bowen B, Hoerster G, Ross M, Bond D, Pierce D, Gordon-Kamm B (1995) Germline transformation of maize following manipulation of chimeric shoot meristems. Bio/Technol 13:677–681

Lu C, Vasil IK, Ozias-Akins P (1982) Somatic embryogenesis in *Zea mays* L. Theor Appl Genet 62:109–112

Lu C, Vasil V, Vasil IK (1983) Improved efficiency of somatic embryogenesis and plant regeneration in tissue cultures of maize (*Zea mays* L.). Theor Appl Genet 66:285–289

Lyznik LA, Rao KV, Hodges TK (1996) FLP-mediated recombination of FRT sites in the maize genome. Nucleic Acids Res 24:3784–3789

Matsuoka Y, Vigouroux Y, Goodman MM, Sanchez GJ, Buckler ES, Doebley JF (2002) A single domestication of maize shown by multilocus microsatellite genotyping. Proc Natl Acad Sci 99:6080–6084

Miller M, Tagliani L, Wang N, Berka B, Bidney D, Zhao Z-Y (2002) High efficiency transgene segregation in co-transformed maize plants using an *Agrobacterium tumefaciens* 2 T-DNA binary system. Transgenic Res 11:381–396

Mórocz S, Donn G, Németh J, Dudits D (1990) An improved system to obtain fertile regenerants via maize protoplasts isolated from a highly embryogenic suspension culture. Theor Appl Genet 80:721–726

Murashige T, Skoog F (1962) A revised medium for rapid growth and bioassays with tobacco tissue cultures. Plant Physiol 15:473–497

Negrotto D, Jolley M, Beer S, Wenck AR, Hansen G (2000) The use of phosphomannose-isomerase as a selectable marker to recover transgenic maize plants (*Zea mays* L.) via *Agrobacterium* transformation. Plant Cell Rep 19:798–803

Pareddy DR, Petolino JF (1990) Somatic embryogenesis and plant regeneration from immature inflorescences of several elite inbreds of maize. Plant Sci 67:211–219

Peng J, Mrabet P, Bagley C, Wang Z, Mankin L, Singh B, Lai F-M, Mei K, Zhong H (2006) Improved methods for the production of stably transformed, fertile *Zea mays* plants. World Intellectual Property Organization, Patent Appl WO 2006136596

Phan BH, Jin W, Topp CN, Zhong CX, Jiang J, Dawe RK, Parrott WA (2006) Transformation of rice with long DNA-segments consisting of random genomic DNA or centromere-specific DNA. Transgenic Res 16:341–351

Pingali PL, Pandey S (2001) World maize needs meeting: technological opportunities and priorities for the public sector. In: Pingali P (ed) CIMMYT 1999–2000. World maize facts and trends. Meeting world maize needs: technological opportunities and priorities for the public sector. CIMMYT, Mexico City, pp 1–3

Puchta H (1999) DSB-induced recombination between ectopic homologous sequences in somatic plant cells. Genetics 152:1173–1181

Rhodes CA, Lowe KS, Ruby KL (1988a) Plant regeneration from protoplasts isolated from embryogenic maize cell cultures. Bio/Technol 6:56–60

Rhodes CA, Pierce DA, Mettler IJ, Mascarenhas D, Detmer JJ (1988b) Genetically transformed maize plants from protoplasts. Science 240:204–207

Rosati C, Landi P, Tuberosa R (1994) Recurrent selection for regeneration capacity from immature embryo-derived calli in maize. Crop Sci 34:343–347

Russell SH, Hoopes JL, Odell JT (1992) Directed excision of a transgene from the plant genome. Mol Gen Genet 234:49–59

Sairam RV, Parani M, Franklin G, Lifeng Z, Smith B, MacDougall J, Wilber C, Sheikhi H, Kashikar N, Meeker K, Al-Abed D, Berry K, Vierling R, Goldman SL (2003) Shoot meristem: an ideal explant for *Zea mays* L. transformation. Genome 46:323–329

Shillito RD, Carswell GK, Johnson CM, DiMaio JJ, Harms CT (1989) Regeneration of fertile plants from protoplasts of elite inbred maize. Bio/Technol 7:581–587

Shillito R, Carswell G, Harms C, Boyce C, Chang Y-F (1994) *Zea mays* plants and transgenic *Zea mays* plants regenerated from protoplasts or protoplast-derived cells. US Patent 5,350,689

Sidorov V, Gilbertson L, Addae P, Duncan D (2006) *Agrobacterium*-mediated transformation of seedling-derived maize callus. Plant Cell Rep 25:320–328

Siebert R, Puchta H (2002) Efficient repair of genomic double-strand breaks by homologous recombination between directly repeated sequences in the plant genome. Plant Cell 14:1121–1131

Socher RJ, Shillito RD, Saul MW, Paszkowski J, Potrykus I (1986) Co-transformation of unlinked foreign genes into plants by direct gene transfer. Bio/Technol 4:1093–1096

Song J, Bradeen JM, Naess SK, Helgeson JP, Jiang J (2003) BIBAC and TAC clones containing potato genomic DNA fragments larger than 100 kb are not stable in *Agrobacterium*. Theor Appl Genet 107:958–964

Song R, Segal G, Messing J (2004) Expression of the sorghum 10-member kafirin gene cluster in maize endosperm. Nucleic Acids Res 32:e189

Songstad DD, Peterson WL, Armstrong CL (1992) Establishment of friable embryogenic (type II) callus from immature tassels of *Zea mays* (Poaceae). Am J Bot 79:761–764

Sugita K, Kasahara T, Matsunaga E, Ebinuma H (2000) A transformation vector for the production of marker-free transgenic plants containing a single copy transgene at high frequency. Plant J 22:461–469

Vain P, McMullen MD, Finer JJ (1993) Osmotic treatment enhances particle bombardment-mediated transient and stable transformation of maize. Plant Cell Rep 12:81–88

Wan Y, Rocheford TR, Widholm JM (1992) RFLP analysis to identify putative chromosomal regions involved in the anther culture response and callus formation in maize. Theor Appl Genet 85:360–365

Wan Y, Widholm JM, Lemaux PG (1995) Type I callus as a bombardment target for generating fertile transgenic maize (*Zea mays* L.). Planta 196:7–14

Wright DA, Townsend JA, Winfrey RJ, Jr, Irwin PA, Rajagopal J, Lonosky PM, Hall BD, Jondle MD, Voytas DF (2005) High-frequency homologous recombination in plants mediated by zinc-finger nucleases. Plant J 44:693–705

Wu L, Nandi S, Chen L, Rodriguez RL, Huang N (2002) Expression and inheritance of nine transgenes in rice. Transgenic Res 11:533–541

Yu W, Han F, Gao Z, Vega JM, Birchler JA (2007) Construction and behavior of engineered minichromosomes in maize. Proc Natl Acad Sci USA 104:8924–8929

Zhang W, Subbarao S, Addae P, Shen A, Armstrong C, Paschke V, Gilbertson L (2003) *Cre/lox* mediated marker gene excision in transgenic maize (*Zea mays* L.) plants. Theor Appl Genet 107:1157–1168

Zhao Z-Y, Gu W, Cai T, Tagliani L, Hondred D, Bond D, Schroeder S, Rudert M, Pierce D (2001) High throughput genetic transformation of maize mediated by *Agrobacterium tumefaciens*. Mol Breed 8:323–333

Zhong H, Srinivasan C, Sticklen MB (1992) In-vitro morphogenesis of corn *Zea mays* L. I. Differentiation of multiple shoot clumps and somatic embryos from shoot tips. Planta 187:483–489

Zhong H, Zhang S, Sun B, Warkentin D, Sticklen MB (1996) The competence of maize shoot meristems for integrative transformation and expression of transgenes. Plant Physiol 110:1097–1107

Zuo J, Niu Q-W, Moeller SG, Chua N-H (2001) Chemical-regulated, site-specific DNA excision in transgenic plants. Nat Biotechnol 19:157–161

Zupan JR, Zambryski P (1995) Transfer of T-DNA from *Agrobacterium* to the plant cell. Plant Physiol 107:1041–1047.

Part II
Transgenic Traits

Chapter 3
Insect Resistance in Corn Through Biotechnology

Graham Head and Dannette Ward

3.1 Introduction

The first corn hybrids with insect resistance traits introduced through biotechnological methods were commercialized in 1996 in the USA (Mendelsohn et al. 2003; James 2006). These products were targeted at lepidopteran pests of corn, particularly stem borers such as the European corn borer *Ostrinia nubilalis* and ear-feeding insects such as the corn earworm *Helicoverpa zea* that are difficult to control using conventional insecticides. Subsequently, a suite of comparable products containing lepidopteran-active insecticidal proteins derived from the bacterium *Bacillus thuringiensis* (Bt) have been introduced. These so-called Bt corn hybrids have since been adopted on tens of millions of hectares (James 2006). In addition, Bt corn hybrids containing coleopteran-active insecticidal proteins that control the larvae of the damaging corn rootworm complex (*Diabrotica* spp.) have been developed. Increasingly corn farmers are purchasing hybrids with combinations of these insect resistance traits (both lepidopteran and coleopteran pest protection), along with herbicide-tolerance traits for improved weed control (James 2006; Brookes and Barfoot 2007). In this chapter, we will describe the nature of current Bt corn products in more detail; the economic and environmental impacts of these technologies on corn-growing in the USA and globally; and the reasons why farmers have adopted these technologies so enthusiastically.

Graham Head
Monsanto Company, 800 North Lindbergh Blvd, St Louis, MO 63167, USA
e-mail: graham.p.head@monsanto.com

Dannette Ward
Monsanto Company, 800 North Lindbergh Blvd, St Louis, MO 63167, USA

A.L. Kriz, B.A. Larkins (eds.), *Molecular Genetic Approaches to Maize Improvement*
Biotechnology in Agriculture and Forestry, Vol. 63

3.2 The Nature of Bt Corn Technologies

Proteins produced by *Bacillus thuringiensis* (Bt) are highly selective against certain species of insects. Transgenic corn expressing crystalline (Cry) insecticidal proteins from Bt were initially developed to control lepidopteran insects. Cry proteins can control the larvae of damaging lepidopteran (moth) species, including the corn borer complex (the European corn borer, southwestern corn borer *Diatraea grandiosella*, and sugarcane borer *Diatraea saccharalis*), corn earworm, fall armyworm *Spodoptera frugiperda* and black cutworm *Agrotis ipsilon*. These insects can cause significant economic damage to corn. As an example, the European corn borer is estimated to cause $1–2 billion worth of damage to corn annually. The Bt proteins produced by the plant impart resistance to insects feeding on or in various parts of the corn plant and are an important differentiator from applied insecticides. For example, once a corn borer enters the stalk, conventional insecticide applications may be greatly reduced in effectiveness because they cannot reach the developing insect. Eventually, enough damage will have occurred that the plant can suffer significant yield losses or become lodged, preventing the corn from maturing and making harvesting difficult.

Several events of transgenic Bt corn have been developed over the past decade and currently there are three registered Bt corn products available to control lepidopteran pests (Table 3.1). The first of these, known as YieldGard Corn Borer (event MON 810), was commercialized in the USA in 1996. In 1997, a second event was commercialized (Bt11) that also targets the corn borer complex. Both of these products use the Cry1Ab protein to protect against corn borers and both are marketed under the YieldGard trade name. More recently, a third event was commercialized and is known as Herculex Corn Borer. This event (TC1507) uses the Cry1F protein to protect the plant from corn pests.

All three of these products provide more effective and consistent control of European corn borer and other lepidopteran insects than insecticides, with less

Table 3.1 Bt corn products that are currently approved for commercial use

Targeted pests	Commercial name	Event name and company	Insecticidal protein/s
Lepidoptera	YieldGard Corn Borer	MON 810, Monsanto	Cry1Ab
	YieldGard	Bt11, Syngenta	Cry1Ab
	Herculex Corn Borer	TC1507, Pioneer & Dow	Cry1F
Coleoptera	YieldGard Rootworm	MON 863, Monsanto	Cry3Bb1
	YieldGard Rootworm VT	MON 88017, Monsanto	Cry3Bb1
	Herculex Rootworm	DAS 59122, Pioneer & Dow	Cry34A/Cry35A
	Agrisure	MIR 604, Syngenta	Cry3Aa
Lepidoptera and Coleoptera	YieldGard Plus	MON 810 × MON 863, Monsanto	Cry1Ab and Cry3Bb1
	Herculex Extra	TC1507 × DAS 59122, Pioneer & Dow	Cry1F and Cry34A/Cry35A

cost than traditional insecticide applications and fewer logistical, health, and environmental concerns. Furthermore, this technology reduces the risk associated with lepidopteran pests like the European corn borer by improving yield stability. Furthermore, other Bt corn events using novel Bt proteins are currently in regulatory review, including a product from Monsanto containing both the Cry1A.105 and Cry2Ab2 proteins and one from Syngenta with the VIP3A and Cry1Ab proteins. The use of multiple Bt proteins in a single product offers the potential for an expanded spectrum of pest control and reduced risk of resistance evolving in the target pests.

In addition, Bt proteins have been incorporated into transgenic corn events to control corn rootworm (CRW) species (Ward et al. 2005). The CRW complex (Western, Northern and Mexican rootworms; *Diabrotica* spp.) is the most destructive set of insect pests of corn in the United States and these pests are therefore the primary target of insecticides used on corn in the USA. When last assessed in 1995, CRW was responsible for the largest expenditure by growers on insect management in corn production systems (Pike et al. 1995). CRW larvae inflict damage to corn plants by feeding on the root tissues, which reduces the ability of the plant to take up water and nutrients from the soil (Reidell 1990). Damaged plants are also prone to lodging, resulting in reduced yield (Spike and Tollefson 1991) and adding significant time to harvesting operations. In addition to the financial costs associated with yield loss, chemical control tactics for CRW also have significant environmental costs. Further, certain CRW species have recently overcome crop rotation control strategies through behavioral adaptations.

The first commercial transgenic maize hybrid designed to control CRW larval feeding was introduced in 2003 in the USA (Table 3.1). The Cry protein contained in the YieldGard Rootworm (event MON 863) product is known as Cry3Bb1. Additional CRW-control Bt products have been developed since that time. A binary protein complex was commercialized in 2006 that utilizes the Cry34Ab1 and Cry35Ab1 proteins. Both proteins are required to elicit control of CRW and this product is known as Herculex Rootworm (event DAS 59122). Monsanto also commercialized a new version of YieldGard Rootworm in 2006 called YieldGard Rootworm VT. This event (MON 88017) also expresses Cry3Bb1 but combines a Roundup herbicide tolerance gene in the same expression cassette to allow faster inbred conversions to be deployed. Finally, a modified Cry3Aa protein was expressed in corn hybrids and was commercialized in 2007. This event (MIR 604) is marketed under the Agrisure brand.

3.3 Adoption of Bt Corn Technologies and Their Impact on Insecticide Use

By 2006, adoption of Bt corn products had grown to 20.1 million ha globally, with about half of this area consisting of the insecticidal traits alone and half in combination were herbicide tolerance (Table 3.2; James 2006). The highest adoption of these

Table 3.2 Global adoption of Bt corn as a single trait and in combination with herbicide tolerance (HT) between 1996 and 2006 (in millions of hectares) (James 2006)

Trait	1996	1997	1998	1999	2000	2001	2002	2003	2004	2005	2006	Total
Bt	0.3	3.0	6.7	7.5	6.8	5.9	7.7	9.1	11.2	11.3	11.1	80.6
Bt and HT	0.0	0.0	0.0	0.0	1.4	1.8	2.2	3.2	3.8	6.5	9.0	27.9

products is in the USA, with significant adoption also in Argentina, Canada, the Philippines, Spain and South Africa, and a small area in Colombia and Honduras. The greatest growth in adoption is occurring in stacked traits, not just of single Bt traits with herbicide tolerance but also with so-called triple stacks of lepidopteran and CRW resistance genes combined with herbicide tolerance now available from several companies (Table 3.1). If current growth trends continue, these triple stacks can be expected to dominate the US marketplace because most farmers are interested in joint solutions to lepidopteran and CRW damage. CRW are primarily a pest in the USA and South America, and more recently Europe, so such products are unlikely to be introduced to Africa and Asia.

The global adoption of Bt corn products has led to substantial reductions in insecticide use on corn, particularly in insecticides that were targeted at CRW in the USA and lepidopteran applications in other countries such as Argentina and the Philippines. For example, in 2005 when adoption of products to control CRW was still relatively low, the estimated reduction in insecticide use in the USA alone from the adoption of Bt corn was 6.67 million lb a.i., of which 4.85 million lb resulted from the use of lepidopteran-control products and 1.82 million lb resulted from use of CRW-control products (Sankula 2006). This corresponds to a 12.4% reduction in the volume of insecticidal active ingredients applied to corn. On a global basis, the estimated reduction in insecticide use over the period from 1996 to 2005 was 15.4 million lb a.i., corresponding to a 4.1% reduction in insecticide volumes (Brookes and Barfoot 2007). In 2005 alone, the global decrease in insecticide volumes was estimated to be 10.8%. Obviously these global numbers will continue to grow dramatically with the rapid adoption of CRW-control and stacked trait products.

These estimates of Bt corn impacts on insecticide use are conservative in another way. Bt corn products are sufficiently effective in their level of control that they may produce some amount of pest population suppression at regional levels. Where the pests in question also feed on crops other than corn (such as the corn earworm that feeds on cotton, soybeans, peanuts and sorghum), additional reductions in insecticide use may occur on these other crops.

The changes in insecticide use associated with the adoption of Bt corn are important for a number of reasons. First, they represent an economic gain for farmers and consequently they are one of the reasons for the rapid adoption of Bt corn products in the USA and globally (Ortman et al. 2001; Pilcher et al. 2002; Sect. 3.4). Second, reductions in the use of broad-spectrum insecticides are beneficial for agroecosystems because they allow greater survival of beneficial non-target insects such as predators and parasitoids that can help to control secondary pests (Sect. 3.6). Third, reductions in insecticide use can also have benefits for the health of farm workers who would otherwise have been exposed to the insecticides. This is particularly true in developing countries such as the Philippines where insecticide applications are made by hand and protective equipment for applicators is very limited (James 2006); comparable reductions in broad-spectrum insecticide applications associated with the adoption of Bt cotton in China have been found to result in fewer insecticide poisonings in farm workers (Huang et al. 2002).

3.4 The Economic Impact of Bt Corn

The rapid adoption of Bt corn products reflects the real and perceived value of these products to farmers. This value comes in several different forms. First, by protecting corn plants from insect damage, Bt corn products increase yields. These yield increases are proportional to the strength of insect infestation and therefore are greatest in countries and regions with consistent, heavy pest pressure, including parts of the US Corn Belt where CRW are a significant problem and tropical countries such as the Philippines where lepidopteran insect pests may complete several generations on a single corn crop. Second, as discussed in Section 3.3, insecticidal input costs are reduced with Bt corn products. Third, Bt corn products are viewed by farmers as simple and convenient to use, saving time that would otherwise be spent in scouting their crop for pests and making insecticide applications (opportunity costs). These savings are harder to quantify but are viewed as important by farmers (Ortman et al. 2001; Pilcher et al. 2002).

Considering only the impacts on yield and insecticide use, the cumulative economic benefit from the use of lepidopteran-control products in the USA alone over the period 1996 to 2005 has been estimated at US $1.92 billion (Table 3.3). Economic gains in other countries conservatively amount to an additional US $400 million (Table 3.3; Brookes and Barfoot 2007; Brookes 2007).

Some preliminary estimates also have been made for the economic benefit from the adoption of CRW products in the USA. For the period 2003 to 2005, the cumulative benefit of these products to farmers has been estimated at US $70.8 million (Table 3.4; Sankula et al. 2005; Sankula 2006). Given that the adoption of these products has increased almost exponentially since 2005, these gains now are much

Table 3.3 Economic returns from Bt corn products targeted at lepidopteran pests (Brookes 2007; Brookes and Barfoot 2007)

Country	Years of assessment	Primary impact	Cumulative impact on farm income
USA	1996–2005	5% yield increase, decreased costs	1.92 billion
Canada	1996–2005	5% yield increase, decreased costs	144 million
Argentina	1998–2005	9% yield increase	159 million
Spain	1998–2005	6.3% yield increase, decreased costs	28 million
South Africa	2000–2005	11% yield increase	59 million
Philippines	2003–2005	24% yield increase	8.5 million

Table 3.4 Economic returns from Bt corn products targeted at the corn rootworm complex (Coleoptera: Chrysomelidae: *Diabrotica* spp.) in the USA (Sankula et al. 2005; Sankula 2006)

Year	Area planted (millions of acres)	Impact on yield (percent per acre)	Impact on insecticide use (million lb a.i.)	Net gain (million US dollars)
2003	0.34	+3	−0.23	2.4
2004	1.32	+3	−0.67	13.4
2005	3.51	+5	−1.82	55

greater, and likely will approach the benefits realized from the use of lepidopteran-control products.

3.5 The Impact of Bt Corn on Grain Quality

Bt corn products have direct and indirect impacts on grain quality, which bring additional economic benefits beyond those discussed in Section 3.4 and have important implications for human and animal health. Lepidopteran insect pests that feed on the corn ear facilitate the invasion of fungal pathogens that produce mycotoxins. In particular, feeding on corn kernels by the European corn borer often leads to infection by fungi in the genus *Fusarium*, including the fumonisin-producing species *F. verticillioides* and *F. proliferatum*. The mycotoxins produced by these fungi can be toxic to livestock and humans, and acceptable mycotoxin thresholds have been established for corn in many countries. For the USA, the cost of corn rejected either for food or for feed has been estimated at US $39 million, and could be as high as US $86 million (Table 3.5; Wu 2006). Of this, most (about $38 million) of the estimated losses are through corn rejected for food. Bt corn products that reduce insect feeding on the corn ear will also reduce levels of fungal infestation and consequently mycotoxin levels. Field studies have shown that Cry1Ab-expressing Bt corn hybrids such as YieldGard Corn Borer that control European corn borer have lower concentrations of fumonisins in kernels compared with their non-transgenic counterparts (Munkvold et al. 1999; Dowd 2001; Clements et al. 2003). For example, Dowd observed average fumonisin concentrations to be 1.8- to 15-fold lower in grain from Bt corn hybrids than comparable non-Bt corn hybrids. Similarly

Table 3.5 Economic impacts of selected mycotoxins in the USA, and the ability of Bt corn to alleviate those impacts (Wu 2006). Values are averages, with 95% confidence intervals in parentheses, in millions of US dollars

	Fumonisin	Deoxynivalenol	Aflatoxin
Market loss	39 (14–86)	52 (17–120)	163 (73–332)
Animal health loss	0.27 (0.05–2)	0	N/A
Total US losses	40 (14–88)	52 (17–120)	163 (73–332)
Benefit from Bt corn	8.8 (2.3–31)	8.1 (2.6–16)	14 (6.2–28)

Munkvold et al. (1999) observed an approximately 8-fold reduction in fumonisin levels. These results demonstrate that Bt hybrids can reduce fumonisin concentrations in grain when European corn borer or other Bt-susceptible pest species infest corn. Wu (2006) estimated the total annual benefit of this mycotoxin reduction by lepidopteran-control Bt corn products in the USA to be US $30 million ($1.50/Bt corn acre).

The benefits from Bt corn adoption for grain quality are likely to increase substantially in the future as triple stacks combining lepidopteran and CRW control predominate because protection from CRW feeding will further increase the ability of plants to withstand fungal invasion. In addition, the introduction of several new Bt corn events with multiple Bt proteins for lepidopteran control can be anticipated over the next 5 years and these events will have broader spectrum lepidopteran control, particularly with respect to ear-feeding pests such as corn earworms. This broader spectrum control should further reduce mycotoxin levels in corn grain, including offering the potential for reducing aflatoxin levels.

Outside the USA, and particularly in Central America, Asia and Africa where relatively more corn is grown for human consumption, reducing mycotoxin levels through the use of Bt corn will have even greater benefits for human health (Gressel et al. 2004). High white corn consumption has been linked to a number of serious presumably mycotoxin-related diseases in these regions. The full extent of these benefits has yet to be realized because of various practical and political hurdles in Africa in particular, but various partnerships between private and public institutions offer great promise.

Another related benefit from lepidopteran-control Bt corn is that commercial hybrids containing Cry1Ab have some protection against several lepidopteran storage pests such as the Indian meal moth *Plodia interpunctella* and the Angoumois grain moth *Sitotroga cerealella* (Sedlacek et al. 2001). These pests can cause substantial damage in grain bins, and these Bt corn products (and future Bt corn events) have the potential to reduce this damage.

3.6 The Environmental Impact of Bt Corn

By facilitating reduced use of broad-spectrum insecticides against pests like CRW and European corn borer, Bt corn products provide a number of indirect environmental benefits in addition to the economic benefits discussed above (in Sects. 3.4 and 3.5). Bt corn products, with their high insecticidal specificity and reduced insecticide use, allow greater survival of a range of beneficial non-target species, including generalist predators that are important for biological control of secondary pests and pollinators. These impacts will be realized both within Bt corn fields and in neighboring fields through reduced insecticide drift.

These environmental benefits have been comprehensively demonstrated through a large number of laboratory and field studies that have looked at the ecological impacts of Bt corn and alternative insect-control practices on non-target invertebrate populations and communities (Head and Dively 2004; Naranjo et al. 2005; Eizaguirre et al. 2006; Brookes and Barfoot 2007; Marvier et al. 2007; Romeis et al. 2006). In many cases, these studies were part of applications for regulatory approval and were conducted prior to the commercialization of products (Mendelsohn et al. 2003). However, for some of the large-scale, multi-year field studies, the work was conducted in commercial fields under typical farmer practices (for example, Naranjo et al. 2005; de la Poza et al. 2005). These studies have involved field corn and sweet corn, and a variety of Bt proteins (including Cry1Ab, Cry1F, Cry3Bb1 and VIP3A) and bio-climatic zones (including US, Spanish and French cropping systems). Collectively, they demonstrate that Bt corn products do not have unexpected toxic effects on non-target species; only the targeted pest species are directly impacted by Bt corn, as would be predicted from knowledge of the mode of action and specificity of Bt proteins (Mendelsohn et al. 2003; Naranjo et al. 2005; Marvier et al. 2007; Romeis et al. 2006). Because of this specificity, Bt corn products effectively preserve local populations of various economically important biological control organisms that can be adversely impacted by broad-spectrum chemical insecticides. For example, Bhatti et al., (2005) demonstrated that a broad array of important non-target species is more abundant on CRW-protected Bt corn than on non-Bt corn treated with a commonly used soil insecticide. As hybrids with both CRW and lepidopteran protection become more common, these environmental benefits will be compounded because insecticide use targeted at both sets of pests will be reduced. The only indirect effects on non-target organisms that have been observed with Bt corn are local reductions in numbers of certain specialist parasitoids whose hosts are the primary targets of Bt crops. Such trophic effects will be associated with any effective pest control technology, whether it be transgenic, chemical, or cultural, as well as with natural fluctuations in host populations.

3.7 Conclusions

Bt corn products already represent a huge success story in agriculture, with unprecedented levels of adoption in the USA and globally, significant documented economic benefits to farmers, and substantial realized environmental benefits. All indicators suggest that current adoption trends will continue, particularly of stacks of Bt corn with herbicide tolerance and triple stacks of CRW and lepidopteran-control traits. Furthermore, new corn events currently in regulatory review promise improved pest control. Beyond these events will be generations of stacked trait products built upon a base of lepidopteran and CRW control but with additional traits related to environmental stress tolerance and enhanced food and feed characteristics.

These prospects are exciting indeed, but they also highlight the need to ensure that Bt corn technologies are sustained in the marketplace. The primary concern in this respect is the potential evolution of Bt resistance in the target insect pests.

Anticipating this concern, biotechnology companies have worked with academic scientists and regulators to design and implement proactive insect resistance management (IRM) programs (EPA 1998). These unique programs include a requirement for farmers growing Bt corn to plant non-Bt corn refuges to support susceptible pest insect populations, as well as resistance monitoring studies to track the success of the programs. It is a testament to these programs that no instances of field resistance to Bt corn have yet been recorded in mainland USA, despite over a decade of high adoption by farmers (Siegfried et al. 2007). In addition, the planned introduction of Bt corn events with multiple Bt proteins to control the same suite of pests will further reduce the risk of target insects evolving resistance (for example, Roush 1998).

One additional outstanding issue is how to ensure that the benefits of these technologies are as broadly available as possible to farmers in the developing world. Even though these products have an obvious technical fit in many countries, the regulatory systems are not always in place to approve such products, and distributional and educational challenges exist when it comes to getting the products in farmers' hands (particularly in Africa). It will be up to both private and public institutions to devise solutions to these problems. Novel partnerships will be needed, along with broad governmental involvement and assistance from international organizations such as the Food and Agricultural Organization of the United Nations (FAO; www.fao.org).

References

Bhatti MA, Duan J, Head GP, Jiang C, McKee MJ, Nickson TE, Pilcher CL, Pilcher CD (2005) Field evaluation of the impact of corn rootworm (Coleoptera: Chrysomelidae) protected Bt corn on ground-dwelling invertebrates. Environ Entomol 34:1325–1335

Brookes G (2007) The benefits of adopting genetically modified insect resistant (Bt) maize in the European Union (EU): first results from 1998–2006 plantings. PG Economics, www.pgeconomics.co.uk

Brookes G, Barfoot P (2007) GM crops: the first ten years – global socio-economic and environmental impacts. ISAAA Brief no.36

Clements MJ, Campbell KW, Maragos CM, Pilcher C, Headrick JM, Pataky JK, White DG (2003) Influence of Cry1Ab protein and hybrid genotype on fumonisin contamination and *Fusarium* ear rot of corn. Crop Sci 43:1283–1293

de la Poza M, Pons X, Farinos GP, Lopez C, Ortego F, Eizaguirre M, Castanera P, Albajes R (2005) Impact of farm-scale Bt maize on abundance of predatory arthropods in Spain. Crop Protect 24:677–684

Dowd PF (2001) Biotic and abiotic factors limiting efficacy of Bt corn in indirectly reducing mycotoxin levels in commercial fields. J Econ Entomol 94:1067–1074

Eizaguirre M, Albajes R, Lopez C, Eras J, Lumbierres B, Pons X (2006) Six years after the commercial introduction of Bt maize in Spain: field evaluation, impact and future prospects. Transgenic Res 15:1–12

EPA (1998) FIFRA Scientific Advisory Panel, Subpanel on *Bacillus thuringiensis* (B.t.) plant-pesticides and resistance management, February 9–10, 1998 (Docket no. OPP 00231)

Gressel J, Hanafi A, Head G, Marasas W, Obilana B, Ochanda J, Souissi T, Tzotzos G (2004) Major heretofore intractable biotic constraints to African food security that may be amenable to novel biotechnological solutions. Crop Protect 23:661–689

Head G, Dively G (2004) Impact of transgenic Bt crops on non-target animal species. In: Koul O, Dhaliwal GS (eds) Transgenic crop protection: concepts and strategies. Science, Enfield, pp 307–324

Huang J, Hu R, Fan C, Pray CE, Rozelle S (2002) Bt cotton benefits, costs, and impacts in China. AgBioForum 5:153–166. http://www.agbioforum.org/v5n4/v5n4a04-huang.pdf

James C (2006) Global status of commercialized biotech/GM crops: 2006. ISAAA Brief no. 35

Marvier M, McCreedy C, Regetz J, Kareiva P (2007) A meta-analysis of effects of Bt cotton and maize on nontarget invertebrates. Science 316:1475–1477

Mendelsohn H, Kough J, Vaituzis Z, Matthews K (2003) Are Bt crops safe? Nat Biotechnol 21:1003–1009

Munkvold GP, Hellmich RL, Rice LG (1999) Comparison of fumonisin concentrations in kernels of transgenic Bt maize hybrids and nontransgenic hybrids. Plant Dis 83:130–138

Naranjo S, Head G, Dively G (2005) Field studies assessing arthropod nontarget effects in Bt transgenic crops: introduction. Environ Entomol 34:1178–1180

Ortman EE, Barry BD, Buschman LL, Calvin DD, Carpenter J, Dively GP, Foster JE, Fuller BW, Hellmich RL, Higgins RA, Hunt TE, Munkvold GP, Ostlie KR, Rice ME, Roush RT, Sears MK, Shelton AM, Siegfried BD, Sloderbeck PE, Steffey KL, Turpin FT, Wedberg JL (2001) Transgenic insecticidal corn: the agronomic and ecological rationale for its use. Bioscience 51:900–903

Pike D (1995) Biologic and economic assessment of pesticide use on corn and soybeans, report 1-CA-95. USDA National Agricultural Pesticide Impact Assessment Program, Washington, DC

Pilcher CD, Rice ME, Higgins RA, Steffey KL, Hellmich RL, Witkowski J, Calvin D, Ostlie KR, Gray M (2002) Biotechnology and the European corn borer: measuring historical farmer perceptions and adoption of transgenic Bt corn as a pest management strategy. J Econ Entomol 95:878–892

Reidell WE (1990) Rootworm and mechanical damage effects on root morphology and water relations in maize. Crop Sci 30:628–631

Romeis J, Meissle M, Bigler F (2006) Transgenic crops expressing *Bacillus thuringiensis* toxins and biological control. Nat Biotechnol 24:63–71

Roush RT (1998) Two toxin strategies for management of insecticidal transgenic crops: can pyramiding succeed where pesticide mixtures have not? Phil Trans R Soc Lond 353: 1777–1786

Sankula S (2006) Quantification of the impacts on US agriculture of biotechnology-derived crops planted in 2005. NCFAP, Washington, DC

Sankula S, Marmon G, Blumenthal E (2005) Biotechnology-derived crops planted in 2004 – impacts on US agriculture. NCFAP, Washington, DC

Sedlacek JD, Komaravalli SR, Hanley AM, Price BD, Davis PM (2001) Life history attributes of Indian meal moth (Lepidoptera: Pyralidae) and Angoumois grain moth (Lepidoptera: Gelechiidae) reared on transgenic corn kernels. J Econ Entomol 94:586–592

Siegfried BD, Spencer T, Crespo AL, Storer NP, Head GP, Owens ED, Guyer D (2007) Ten years of Bt resistance monitoring in the European corn borer: what we know, what we don't know, and what we can do better. Am Entomol 53:208–214

Spike BP, Tollefson JJ (1991) Yield response of corn subjected to western corn rootworm (Coleoptera: Chrysomelidae) infestation and lodging. J Econ Entomol 84:1585–1590

Ward DP, DeGooyer TA, Vaughn TT, Head GP, McKee MJ, Astwood JD, Pershing JC (2005) Genetically enhanced maize as a potential management option for corn rootworm: YieldGard® Rootworm maize case study. In: Vidal S et al. (eds) Western corn rootworm: ecology and management. CAB International, Wallingford, pp 239–262

Wu F (2006) Mycotoxin reduction in Bt corn: potential economic, health and regulatory impacts. Transgenic Res 15:277–289

Chapter 4
Seed Total Phosphate and Phytic Acid

Victor Raboy

4.1 Introduction

Both the total amount and composition of seed phosphorus (P) are important to maize end-use quality. Seed total P represents a major pool in the flux of P through the agricultural ecology. The phosphate stored in seeds of major crops represents a sum equivalent to more than 50% of phosphate fertilizer applied annually worldwide (Lott et al. 2000). Clearly this pool has value as a target in efforts to enhance the management of P in agricultural production. The most abundant form of P in mature crop seeds is phytic acid (*myo*-inositol-1,2,3,4,5,6-P_6 or $InsP_6$). In maize it often represents about 80% of seed total P (Lott et al. 2000; Raboy 2006). Seed-derived dietary phytic acid binds tightly to nutritionally important minerals such as calcium, iron and zinc, and is not efficiently digested by humans and non-ruminants such as poultry, swine and fish (Brinch-Pedersen et al. 2002). As a result most seed-derived phytic acid P consumed by non-ruminants is excreted. Perhaps 40% of maize produced in the USA is used in non-ruminant feeds, and non-ruminant waste P has the potential of contributing to water pollution, an environmental problem requiring additional management (EPA 2002; Sharpley et al. 1994). In the context of human nutrition, excretion of phytic acid can contribute significantly to mineral depletion and deficiency, such as iron and zinc deficiency, in populations that rely on whole grains and legumes as staple foods (Brown and Solomons 1991). It is estimated that more than a billion people suffer iron deficiency, and hundreds of millions suffer zinc and other mineral deficiencies.

There has been substantial progress in the genetics of seed P composition. In the case of forward genetics, a number of *low phytic acid* (*lpa*) mutations have been isolated in maize and other crop species, using mutagenized wild-type germplasm and screens for either reduced seed phytic acid or increased seed inorganic P (screen

Victor Raboy
USDA-ARS, Small Grains and Potato Germplasm Research Unit, 1691 South 2700 West, Aberdeen, ID 83210, USA
e-mail: victor.raboy@ars.usda.gov

A.L. Kriz, B.A. Larkins (eds.), *Molecular Genetic Approaches to Maize Improvement*
Biotechnology in Agriculture and Forestry, Vol. 63

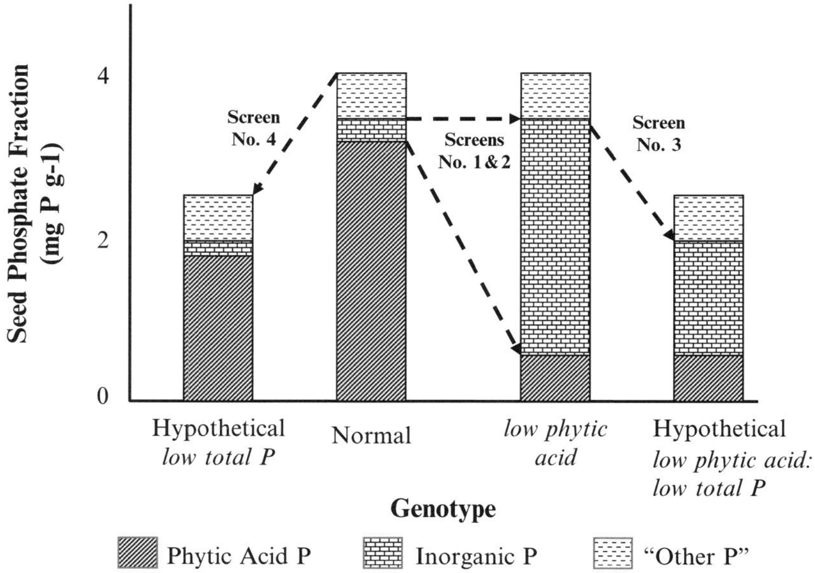

Fig. 4.1 Seed phosphate fractions in four types of genotypes: wild-type (normal), *low phytic acid*, and two hypothetical genotypes that are either *low total P* or *low phytic acid*:*low total P*. The genetics screens used to isolate *lpa* genotypes and those proposed to be used to isolate the hypothetical genotypes are indicated as screens 1–4, respectively

nos 1 and 2, respectively; Fig. 4.1; reviewed in Raboy 2006). These mutations block the synthesis of phytic acid during seed development, but in nearly all cases have little effect on seed total P or the distribution of P in the mature seed. Instead, a far greater proportion of seed total P is found as inorganic P (P_i), resulting in the "high inorganic P" phenotype of *lpa* genotypes (Fig. 4.1). Thus *lpa* alleles alter the chemistry of seed total P, but with one possible exception (barley *lpa*1-1; see below), have not been shown to substantially alter total P concentration. Advances in the molecular biology of phytic acid metabolism and storage have also resulted in reverse genetics approaches to engineering seed P composition (Shi et al. 2007; Stevenson-Paulik et al. 2005).

Numerous animal nutrition studies have shown that the increase in non-phytic acid phosphorus in *lpa* grain, mostly inorganic P, largely represents "available P" to non-ruminants such as poultry, swine and fish (reviewed in Raboy 2006). With proper dietary formulation, non-ruminants can satisfy more of their dietary need for P from *lpa* grain, and excrete concomitantly less P. Studies with human subjects have shown that iron, zinc and calcium consumed in foods prepared with *lpa* maize are about 35–50% more available than they are in foods prepared with normal maize (reviewed in Raboy 2006). Thus the *lpa* trait is desirable from the standpoint of seed utilization in both human and non-ruminant diets.

Ruminants efficiently digest seed total P regardless of its chemistry. Seed phytic acid is not a major issue in this context. However, feeds prepared from grain and legume products might contain ~25% more total P than required by cattle

(Warden and Russell 2004). The excess feed P results in elevated manure P, disposal of which represents an environmental hazard and a management problem (EPA 2002; Volk et al. 2000). Also, developing grain products enhanced for use in ethanol production is a major current objective. A major side-product is "distillers dry grains" (DDGs), use of which in ruminant feeds also results in high manure P levels. Thus reduced seed total P would be desirable for the preparation of "low P" feeds useful in ruminant production, and of value when maize is milled for use in ethanol production.

P is a major plant nutrient important to crop productivity, but available soil P is often limiting to crop growth. Improved plant utilization of soil P is an important goal for international agriculture (Raghothama 1999). Most research in this area has addressed P uptake by roots and in response to P deficiency (Schünmann et al. 2004). Even though seed total P represents a major bottleneck in the flux of P through the agricultural ecosystem, and is very important to major end-uses of grains, there has been relatively little interest and little progress in the genetics of seed total P. Mutations in the barley *lpa*1 gene both block seed phytic acid accumulation and result in a net reduction in seed total P of $\sim 15\%$ (Dorsch et al. 2003; Ockenden et al. 2004), providing a proof-of-principal that it may be possible to engineer reduced seed total P.

Two types of forward genetics screens for "low seed total P" are illustrated in Fig. 4.1. Screen no. 3 (Fig. 4.1, right) is for mutations that reduce the "high inorganic P" phenotype of *lpa* genotypes, one class of which should translate into *low seed total P*, resulting in a *low phytic acid*:*low total P* phenotype. This would be desirable for both ruminant and non-ruminant feeds. Screen no. 4 (Fig. 4.1, left) is a direct screen for reduced seed total P. Such forward genetics screens would probably identify here-to-for unknown genes and functions of interest in the biology of P sensing and transport and of value in engineering desirable levels of seed total P. This chapter will focus on potential targets useful for engineering both the amount of seed total P and its composition.

4.2 Phytic Acid Synthesis, Breakdown and Storage

The pathway to phytic acid during seed development first requires the supply of the substrates phosphate and *myo*-Inositol (Ins) (Fig. 4.2). The supply of phosphate will be discussed below in the context of the genetics of seed total P. Most of the Ins needed for phytic acid synthesis is probably supplied de novo during seed development. The enzyme D-Ins(3)P_1 synthase ("MIPS") is the sole synthetic source of the Ins ring (Loewus and Murthy 2000). Yoshida et al. (1999) demonstrated that MIPS is expressed proximal to the site of phytic acid accumulation during rice seed development. Hitz et al. (2002) demonstrated that the LR33 mutation in a soybean MIPS gene resulted in a block in seed phytic acid accumulation. The product of MIPS activity, Ins(3)P_1, is then hydrolyzed to Ins and phosphate with the activity of Ins monophosphatase ("IMP").

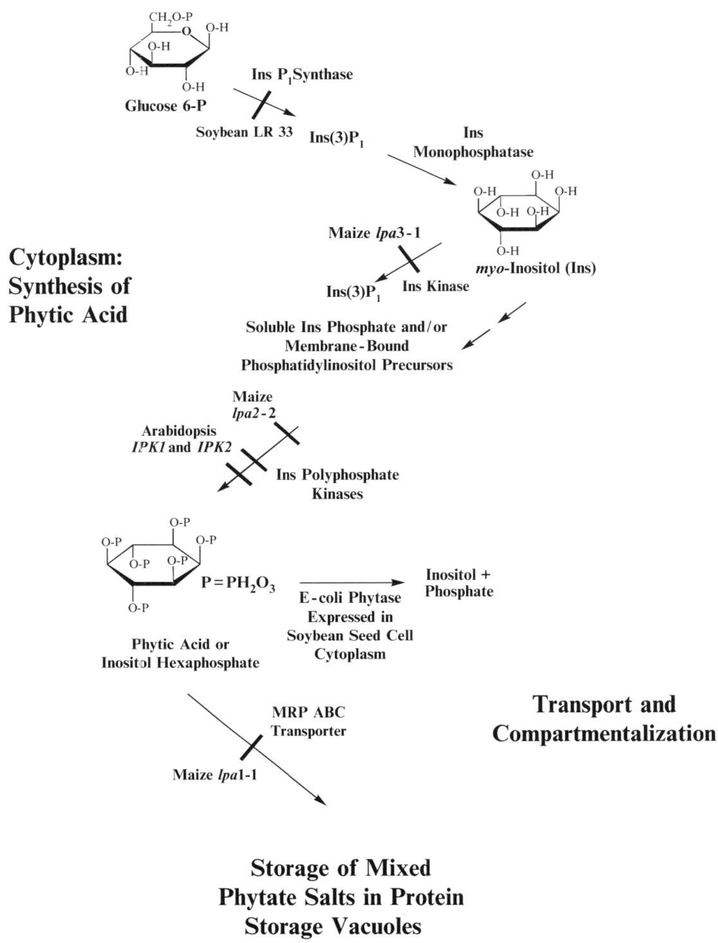

Fig. 4.2 Functions active during seed development useful as targets for engineering the *low phytic acid* trait. Synthesis of seed phytic acid is largely cytoplasmic whereas storage is in a "protein storage vacuole". Higher plant mutations that block phytic acid synthesis or accumulation are indicated by bars across arrows. Soybeans have been engineered to express an *E. coli* phytase in the cytoplasm of developing seeds

Following the synthesis of Ins, Ins *tris* phosphates are produced, either via a pathway involving phosphatidylinositol (PtdIns) lipid intermediates such as PtdIns(4,5)P_2 or via a pathway that uses soluble Ins phosphates intermediates (Fig. 4.2; Raboy 2006; Stephens and Irvine 1990; Stevenson-Paulik et al. 2005; York et al. 1999). The first step in the "soluble Ins phosphate pathway" is the initial phosphorylation of Ins, catalyzed by Ins kinase. These *tris* phosphates are then converted to Ins(1,3,4,5,6)P_5 via the action of two types of Ins polyphosphate kinases, Ins polyphosphate 3-/6- or 5-/6-kinases. Ins(1,3,4,5,6)P_5 is converted to InsP_6 via an Ins polyphosphate 2-kinase. More highly phosphorylated pyrophosphate-containing Ins phosphates may be involved, but the molecular biology of these compounds has not been studied in plant systems yet. Mutations of genes encoding each of the above kinases reduces phytic acid accumulation in seeds. The maize *lpa*3 gene encodes Ins kinase (Shi et al. 2005). The maize *lpa*2 gene encodes an Ins polyphosphate 5-/6-kinase (Shi et al. 2003). The Arabidopsis *IPK2β* gene encodes an Ins polyphosphate 3-/6-kinase, and the Arabidopsis *IPK1* gene encodes an Ins polyphosphate 2-kinase (Stevenson-Paulik et al. 2005).

A variety of alternative names and possibly alternative functions for many of these genes and enzymes exist in the literature. See Shears (2004) and Raboy (2006) for more details. The genetics of phytic acid synthesis during maize seed development is still not completely known. Many of the above functions may be encoded by multi-copy genes, of which only a subset have been studied, and no functions of a purely regulatory nature have been identified yet, so much work remains.

Processes important to phytic acid deposition and storage in seeds also represent a target for engineering seed P chemistry. The bulk of phytic acid that accumulates in seeds is deposited as a mixed salt in inclusions referred to as globoids (Lott et al. 1995). In the cereal grain, phytic acid is deposited primarily as a mixed salt of K and Mg. Globoids are found in one class of storage microvacuoles referred to as protein storage vacuoles (PSVs). In the cereal grains, most phytic acid deposition occurs within the aleurone layer and the germ, consisting of the embryo and scutellum (O'Dell et al. 1972). In maize, approximately 80% of seed total phytic acid is found in the germ, with the remainder in the aleurone layer. In normal cereal grains, the central starchy endosperm contains little or no phosphate or phytic acid at maturity.

A model for the various functions of possible importance to phytic acid deposition is provided in Fig. 4.3 (Bentsink et al. 2003; Shi et al. 2007; Takahashi et al., 2004). A genetics study of *Arabidopsis* lines that differ quantitatively in

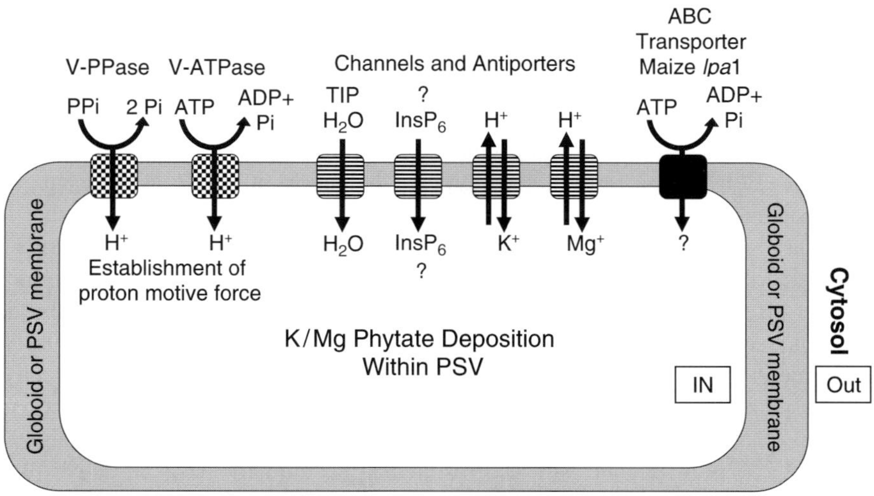

Fig. 4.3 Components and processes of a phytate-accumulating membrane-bound organelle, representing either a protein storage vacuole (*PSV*) or a membrane-bound globoid found within a compound PSV. *V-PPase* Vacuolar inorganic pyrophosphatase; *V-ATPase* vacuolar ATPase; *TIP* tonoplast intrinsic protein. *Questionmarks* indicate purely speculative aspects of the diagram.

the levels of phosphate and Ins P_6 in vegetative and seed tissues identified a quantitative trait locus (QTL) that accounts for a significant amount of the variation observed (Bentsink et al. 2003). Contained within the 99-kb chromosomal segment represented by this QTL were 13 ORFs, one of which encoded a putative vacuolar ATPase (V-ATPase). Bentsink et al., (2003) hypothesized that the variation in phosphate and phytic acid levels observed among the *Arabidopsis* lines in their study was in large part due to variation in phosphate transport caused by heritable differences in the V-ATPase. More recently, Shi et al., (2007) reported that maize *lpa*1 encodes a "multidrug resistance-associated protein" (MRP), a type of "ATP-binding cassette" (ABC) transporter . While the cellular site of the maize *lpa*1 ABC transporter and the specific metabolites involved, precursor or phytic acid, have not been determined yet, it is probably involved in phytic acid transport and storage (Figs. 4.2 and 4.3).

In nearly all cases, homozygosity for an *lpa* allele has some negative impact on seed and plant performance, such as reduced germination, emergence, stress tolerance and seed dry weight accumulation (Bregitzer and Raboy 2006; Meis et al. 2003; Oltmans et al. 2005; Raboy et al. 2000). The net impact is reduced yield (Ertl et al., 1998). This is probably due to the fact that pathways involving phytic acid are part of basic cellular housekeeping and are active in all tissues, vegetative and seed. They are important to a wide array of functions including signal transduction important for stress response, development and phosphate sensing, to DNA repair, RNA editing and mRNA export. In developing seeds the synthesis of phytic acid represents a sink of Ins, itself important to several pathways unique to plant cells, and may also be involved in P homeostasis. It is not surprising therefore that mutations that block the conversion of Ins to Ins P_6 impact Ins levels, which in turn impact sucrose and rafinosaccharide levels, or that these mutations impact starch metabolism via their large effects on seed inorganic P levels (Karner et al. 2004; Raboy et al. 2000). The genes perturbed in *lpa* mutations are therefore important to both seed P chemistry and to processes and functions important to vegetative growth.

One way to alter seed P chemistry but not impact processes important to vegetative tissues is via regulation of gene expression using seed-specific promoters. Shi et al. (2007) engineered a seed-targeted reduction in phytic acid via use of seed-specific promoters and the maize *lpa*1 ABC transporter. They found that in some cases substantial reductions in seed phytic acid were achieved that were accompanied by little effect on seed dry weight. Both this observation and the lack of other effects on seed and seedling function, and subsequent plant growth and function, require more study for confirmation. Another potential activity useful in engineering the low-phytate trait via seed-specific expression is that of phytase enzymes (Brinch-Pedersen et al. 2002). Engineering cytoplasm-specific expression of an *E. coli appA*-encoded phytase in developing soybean seeds achieved large reductions in phytate and also resulted in accumulations of an active phytase (Fig. 4.2; K.D. Bilyeu et al. 2008). Such seeds would be both high available phosphorus/low phytate phosphorus, and have the added advantage of providing an active phytase that upon ingestion would break down phytates from other feed components. Initial studies indicated little effect on seed germination and yield, but confirmation via follow-up studies is required.

4.3 Seed Total P

Seed total P is a function of two processes: P uptake by the parent plant and transport/localization to the developing seed (Fig. 4.4). The developing seed and its immediate environment, the proximal surrounding tissues of the maternal plant, form a microcosm that might represent a parallel to the developing plant and its immediate environment, the soil solution. P uptake at the root–rhizosphere interface and P transport from the maternal plant to the developing seed at the "maternal-filial" interface (Fig. 4.4A) ultimately involve transport from an external apoplast to an internal symplast, and the molecular and regulatory machinery of P uptake and transport might be similar. For brevity a few selected components that are both informative of relevant biology and illustrative of targets for engineering seed total P will be discussed. A recurring theme will be parallels in these two processes.

These transport processes require the maintenance of an electrochemical gradient established by the activity of plasma membrane H^+-ATPases (Sondergaard et al. 2004). Plant genomes contain multiple copies of plasma membrane H^+-ATPases, and studies of *Arabidopsis* identified one copy whose expression is specific to developing seeds (Harper et al. 1994). This may be the vacuolar ATPase

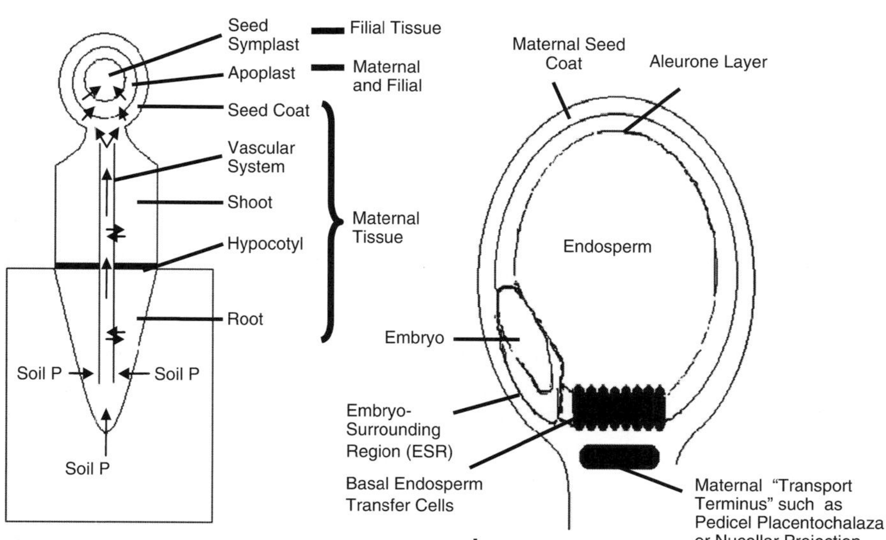

Fig. 4.4 Schematics of phosphate uptake and transport through a plant (**a**) and into the developing cereal caryopsis (**b**). (Modified from Olsen (2004))

identified via QTL analysis of variation in seed P mentioned above (Bentsink et al. 2003), and represents one target for engineering seed total P.

Figure 4.4B illustrates the maternal and filial tissues critical to seed P uptake (Offler et al. 2002; Olsen et al. 2004). These include tissues that represent the maternal "transport terminus", such as the seed coat, the placentochalaza in maize or nucellar projection in barley. Once P arrives at these maternal tissues, its transport into the developing endosperm and aleurone probably occurs at least in part through specially adapted cells in the "basal endosperm transfer layer" (BETL), or the specialized cells of the embryo surrounding region (ESR). Genes have been identified whose expression appears specific to these cell types (Bonello et al. 2000). One candidate that may be involved in P transport is the maize *TCRR-1* gene (Muniz et al. 2006), which is expressed in the BETL and which encodes a "type-A" response regulator involved in the "His-Asp" phosphorelay pathway. This pathway is important to hormonal signal transduction, in particular to the cytokinin signal transduction pathway involved in P sensing (see below).

A major portion of P transport is accomplished via the transporters encoded by the *Pht1* gene family. In most species the *Pht1* gene family is represented by multiple copies which often display different patterns of tissue-specific expression. In *Arabidopsis*, four *Pht1* family members appear to be expressed in the root, but others may be expressed in vegetative tissues and reproductive tissues, indicating possible roles in P transport into seeds (Karthikeyan et al. 2002; Mudge et al., 2002). The two most highly expressed members of this family are *At Pht1;1* and *At Pht1;4*. Loss-of-function alleles of both *Pht1;1* (Shin et al. 2004) and *Pht1;4* (Misson et al. 2004) reduce the uptake of P_i into the shoot, and shoots homozygous for the *Pht;1:Pht;4* double mutant display greater reductions in shoot P_i than does the *Pht;1* single mutant (Shin et al. 2004). However, the impact on seed P of these genetically determined differences in root and plant P has not been reported, reflecting the relative lack of interest in processes that determine seed total P.

The concentration of available inorganic P in soils is relatively low compared with that in root cells (about 10 μM, compared with 10 mM; Rausch 2002). Members of the *Pht1* family that have a high affinity for P are thought important for transport against this concentration gradient, whereas members with a relatively lower affinity for P may be important to P redistribution throughout the plant or into seeds. Of the eight known *Pht1* family members in the barley genome, *Hv Pht1;1* through Hv Pht4*;1* were primarily expressed in the root in response to P starvation and *Pht1;8* was expressed in the root in response to mycorrhizal colonization, whereas *Pht1;6* was primarily expressed in aging leaves and flag leaves (Glassop et al. 2005; Schünmann et al. 2004). The root-specific *HvPht1;1* had a high affinity for P ($K_m \sim 9$ μM), whereas the shoot-specific *HvPht1;6* had a low affinity for P ($K_m \sim 386$ μM; Rae et al. 2003). These data indicate that *HvPht1;6* may be a good target for engineering reduced seed total P. While analyses of patterns of expression of the maize genome's five *Pht1* loci revealed differences in tissue specificity, no candidate for targeted engineering of low seed total P as clear as barley *HvPht1;6* stands out (Nagy et al. 2006).

In addition to the role of P transporters, unloading of nutrients into the apoplast may proceed via electrodiffusional processes that use a poorly selective pore in the seed plasma membrane (van Dongen et al. 2001). If this is the case, movement down a chemical gradient would play a relatively more important role compared with transport that ultimately requires ATP bond energy.

Manipulating plant and seed P composition will require enhanced understanding of the interconnection and coordination involved in the regulation of growth and metabolism in response to a complex array of developmental and environmental signals, varying nutrient needs and varying production and supply of sugars. Three examples relevant to plant and seed P, the Ins phosphate pathways and *lpa* genetics, illustrate this well. Two come out of studies of barley *lpa* mutations and one concerns studies of the *Arabidopsis IPK1*-encoded Ins polyphosphate 2-kinase. These studies also yield clues to targets potentially useful in engineering plant and seed P.

P transport machinery is regulated by mechanisms able to sense plant P needs and is mediated in part by cytokinin signaling (Martin et al. 2000). A major component of the cytokinin-mediated transduction pathway uses a His-Asp phosphorelay mechanism (Sheen 2002). A recent study involving one of the *Arabidopsis* genome's cytokinin pathway receptor histidine kinases (Franco-Zorrilla et al. 2005) demonstrated "crosstalk" between "cytokinin, sugar and Pi-starvation signaling". Similarly, there is coordination and crosstalk between ethylene and sugar signaling pathways (Gazzarrini and McCourt 2003). One approach to understanding the impact of low-phytate mutations on these complex processes is to compare genome-wide gene expression in mutant and wild-type isolines. The first study to do so focused on the barley M955 *lpa* mutation, homozygosity for which blocks phytic acid synthesis throughout seed development, resulting in a net reduction of >90% (Bowen et al. 2007). Using a barley microarray containing 22,000 "probe sets" (genes), 38 probe sets were identified that were consistently differentially expressed during seed development in M955 versus wild-type. Of this small number of functions displaying large changes in expression in the *lpa*M955 tissues, two were part of the His-Asp phosphorelay pathway (a histidine-containing phosphotransfer protein and a histidine kinase response regulator), two were in sugar sensing and transport (barley homologues of *brittle 1* and *sucrose synthase 2*), and one was an ethylene response protein (Bowen et al. 2007).

This indicates interaction between sugar, cytokinin, ethylene and Ins phosphate signaling pathways during seed development, and supports the hypothesis that developing seeds in a microcosm parallels processes important to developing plants. It indicates one way in which *lpa* genotypes might have a negative impact on starch synthesis, via perturbation of sugar sensing and transport. Of greatest relevance here, however, is that in developing M 955 seed, expression of genes encoding two components of the His-Asp phosphorelay signal transduction pathway known to be involved in nutrient-need response in vegetative tissues appear suppressed as compared with wild-type. The barley histidine kinase response regulator suppressed in M955 developing seeds has homology to the maize *TCRR-1* described above. Further studies are required to determine if it is a homologue of the maize gene, if it is expressed in the BETL and if it and the maize gene are directly involved in P sensing important to transport. The most straightforward explanation of this result is that it reflects the inverse relationship between the cytokinin signal transduction and inorganic P concentration; the high inorganic P concentration in developing M 955 seeds suppresses the pathway. However, this suppression of signal transduction has little apparent effect on seed total P concentration, which is similar in M 955 and wild-type. Thus while this finding supports the hypothesis that processes involved in P-sensing and transport in developing seeds parallel those in vegetative tissues, it also raises questions about the importance of given components of these processes. For example, perhaps this finding supports the finding of van Dongen et al. (2001) that P uptake by seeds in part is via poorly selective pores and electrodiffusional processes which would not depend upon seed-specific gene expression in response to seed P concentrations.

Additional evidence of the interrelationship between plant and seed P biology and the Ins phosphate pathways comes from studies of the *Arabidopsis* Ins polyphosphate 2-kinase gene and the barley *lpa*1-1 mutation. The *Arabidopsis IPK1* gene encodes an inositol polyphosphate 2-kinase that is important to both the synthesis of IP_6 in seeds and the ability of the maternal plant to regulate P uptake (Stevenson-Paulik et al. 2005). Mutations in the Arabidopsis *Ipk1* gene not only block seed Ins P_6 synthesis, but also result in plants that take up excess phosphate when supplied with moderate to high nutrient P. There is a parallel in the phenotypes of *Arabidopsis Ipk1* mutants and barley *lpa1-1* in that the latter perturbs seed phytic acid synthesis but also alters P transport, but in this case apparently only in the seed. Homozygosity for barley *lpa*1-1 has no apparent effect on plant P, but results in a reduction of whole-seed total P of about 10–15% (Dorsch et al. 2003). This is the only *lpa* mutation in any species to have this effect. The barley *lpa*1-1 block in seed phytic acid accumulation is aleurone-specific (Ockenden et al. 2004). Barley *lpa*1-1 embryos have $InsP_6$ levels greater than wild-type. There is also a shift in the distribution of total P in barley *lpa*1-1 grain; *lpa*1-1 embryos have about 40% more total P than wild-type, and a combined fraction containing both aleurone and endosperm has about 30% less total P than wild-type. Thus the barley *lpa*1-1 mutation results in an endosperm/aleurone-specific block in phytic acid synthesis and/or accumulation accompanied by a shift in grain total P from endosperm/aleurone to embryo/germ, and these changes somehow are probably also the cause of the net reduction in grain total P.

4.4 Conclusion

The genetics and molecular biology of phytic acid synthesis and storage during seed development have advanced to the point where this field of research is approaching maturation. While much work remains, there is growing knowledge of many if not most of the genes and gene products involved, and this will allow the development of a detailed understanding of this area of biology within the next few years. Studies such as those with the *Arabidopsis IPK1*gene and the barley *lpa*1-1 and M955 mutants clearly illustrate that P sensing, uptake and distribution in the plant and developing seed are interrelated and that the Ins phosphate pathways play a central role in this interrelationship. Studies of these and other functions important to P sensing, uptake and distribution will soon lead to a detailed understanding of those processes important to both the maternal plant and the developing seed. As a result, clearly there are now many tools available for engineering both seed P amount and the chemistry to achieve optimal end-use quality, whether that end use be in foods or feeds for either ruminants or non-ruminants.

References

Bentsink L, Yuan K, Koornneef M, Vreugdenhil D (2003) The genetics of phytate and phosphate accumulation in seeds and leaves of *Arabidopsis thaliana*, using natural variation. Theor Appl Genet 106:1234–1243

Bilyeu KD, Zeng P, Coello P, Zhang ZJ, Krishnan HB, Bailey A, Beuselink PR, Polacco JC (2008) Quantitative conversion of phytate to inorganic phosphorus in soybean seeds expressing a bacterial phytase. Plant Phyisol 146:468–477

Bonello J-F, Opsahl-Ferstad H-G, Perez P, Dumas C, Rogowsky PM (2000) *ESR* genes show different levels of expression in the same region of maize endosperm. Gene 246:219–227

Bowen DE, Souza EJ, Guttieri MJ, Fu J, Raboy V (2007) A low phytic acid barley mutation alters seed gene expression. Crop Sci 47:S149–S157

Bregitzer P, Raboy V (2006) Agronomic effects of four independent low phytate mutations in barley: comparisons of wildtype and low-phytate sib lines. Crop Sci 46:1318–1322

Brinch-Pedersen H, Sørensen LD, Holm PB (2002) Engineering crop plants: getting a handle on phosphate. Trends Plant Sci 7:118–125

Brown KH, Solomons NW (1991) Nutritional problems of developing countries. Infect Dis Clin N Am 5:297–317

Dorsch JA, Cook A, Young KA, Anderson J, Bauman AT, Volkmann CJ, Murthy PPN, Raboy V (2003) Seed phosphorus and inositol phosphate phenotype of barley *low phytic acid* genotypes. Phytochemistry 62:691–706

Environmental Protection Agency (2002) Concentrated Animal Feeding Operations Ruling. http://cfpub.epa.gov/npdes/afo/caforule.cfm. Accessed 24 October 2005

Ertl DS, Young KA, Raboy V (1998) Plant genetic approaches to phosphorus management in agricultural production. J Environ Qual 27:299–304

Franco-Zorrilla JM, Martin AC, Leyva A, Paz-Ares J (2005) Interaction between phosphate-starvation, sugar, and cytokinin signaling in Arabidopsis and the roles of cytokinin receptors CRE1/AHK4 and AHK3. Plant Physiol 138:847–857

Gazzarrini S, McCourt P (2003) Cross-talk in plant hormone signalling: what Arabidopsis mutants are telling us. Ann Bot 91:605–612

Glassop D, Smith SE, Smith FW (2005) Cereal phosphate transporters associated with the mycorrhizal pathway of phosphate uptake into roots. Planta 222:688–698

Harper JF, Manney L, Sussman MR (1994) The plasma membrane H^{+}-ATPase gene family in *Arabidopsis*: genomic sequence of AHA10 which is expressed primarily in developing seeds. Mol Gen Genet 244:572–587

Hitz WD, Carlson TJ, Kerr PS, Sebastian SA (2002) Biochemical and molecular characterization of a mutation that confers a decreased raffinosaccharide and phytic acid phenotype on soybean seeds. Plant Physiol 128:650–660

Karner U, Peterbauer T, Raboy V, Jones DA, Hedley CL, Richter A (2004) *myo*-Inositol and sucrose concentrations affect the accumulation of raffinose family oligosaccharides in seed. J Exp Bot 55:1981–1987

Karthikeyan AS, Varadarajan DK, Mukartira UT, D'Urzo MP, Damsz B, Raghothama KG (2002) Regulated expression of Arabidopsis phosphate transporters. Plant Physiol 130:221–233

Loewus FA, Murthy PPN (2000) *myo*-Inositol metabolism in plants. Plant Sci 150:1–19
Lott JNA, Greenwood JS, Batten GD (1995) Mechanisms and regulation of mineral nutrient storage during seed development. In: Kigel J, Galili G (eds) Seed development and germination. Marcel Dekker, New York, pp 215–235
Lott JNA, Ockenden I, Raboy V, Batten GD (2000) Phytic acid and phosphorus in crop seeds and fruits: a global estimate. Seed Sci Res 10:11–33
Martin AC, del Pozo JC, Iglesias J, Rubio V, Solano R, de la Pena A, Leyva A, Paz-Ares J (2000) Influence of cytokinins on the expression of phosphate responsive genes in *Arabidopsis*. Plant J 24:559–567
Meis SJ, Fehr WR, Schnebly SR (2003) Seed source effect on field emergence of soybean lines with reduced phytate and raffinose saccharides. Crop Sci 43:1336–1339
Misson J, Thibaud M-C, Bechtold N, Raghothama K, Nussaume L (2004) Transcriptional regulation and functional properties of *Arabidopsis* Pht1;4, a high affinity transporter contributing greatly to phosphate uptake in phosphate deprived plants. Plant Mol Biol 55:727–741
Mudge SR, Rae AL, Diatloff E, Smith FW (2002) Expression analysis suggests novel roles for members of the Pht1 family of phosphate transporters in *Arabidopsis*. Plant J 31:341–353
Muniz LM, Royo J, Gomez E, Barrero C, Bergareche D, Hueros G (2006) The maize transfer cell-specific type-A response regulator *ZmTCRR-1* appears to be involved in intercellular signalling. Plant J 48:17–27
Nagy R, Vasconcelos MJV, Zhao S, McElver J, Bruce W, Amrhein N, Raghothama KG, Bucher M (2006) Differential regulation of five Pht1 phosphate transporters from maize (*Zea mays* L.). Plant Biol 8:186–197
Ockenden I, Dorsch JA, Reid MM, Lin L, Grant LK, Raboy V, Lott JNA (2004) Characterization of the storage of phosphorus, inositol phosphate and cations in grain tissues of four barley (*Hordeum vulgare* L.) low phytic acid genotypes. Plant Sci 167:1131–1142
O'Dell BL, de Boland AR, Koirtyohann SR (1972) Distribution of phytate and nutritionally important elements among the morphological components of cereal grains. J Agric Food Chem 20:718–721
Offler CE, McCurdy DW, Patrick JW, Talbot MJ (2002) Transfer cells: cells specialized for a special purpose. Annu Rev Plant Biol 54:431–454
Olsen O-A (2004) Nuclear endosperm development in cereals and *Arabidopsis thaliana*. Plant Cell 16:S214–S227
Oltmans SE, Fehr WR, Welke GA, Raboy V, Peterson KL (2005) Agronomic and seed traits of soybean lines with low-phytate phosphorus. Crop Sci 45:593–598
Raboy V (2006) Seed phosphorus and low-phytate crops. In: Turner BL, Richardson AE, Mullaney EJ (eds) Inositol phosphates: linking agriculture and environment. CAB International, Wallingford, pp 111–132
Raboy V, Gerbasi PF, Young KA, Stoneberg SD, Pickett SG, Bauman AT, Murthy PPN, Sheridan WF, Ertl DS (2000) Origin and seed phenotype of maize *low phytic acid 1-1* and *low phytic acid 2-1*. Plant Physiol 124:355–368
Rae AL, Cybinski DH, Jarmey JM, Smith FW (2003) Characterization of two phosphate transporters from barley; evidence for diverse function and kinetic properties among members of the Pht1 family. Plant Mol Biol 53:27–36
Raghothama KG (1999) Phosphate acquisition. Annu Rev Plant Physiol Plant Mol Biol 50:665–693
Rausch C, Bucher M (2002) Molecular mechanisms of phosphate transport in plants. Planta 216:23–37
Schünmann PHD, Richardson AE, Smith FW, Delhaize E (2004) Characterization of promoter expression patterns derived from the Pht1 phosphate transporter genes of barley (*Hordeum vulgare* L.). J Exp Bot 55:855–865
Sharpley AN, Chapra SC, Wedepohl R, Sims JT, Daniel TC, Reddy KR (1994) Managing agricultural phosphorus for protection of surface waters: issues and options. J Environ Qual 23:437–451
Shears SB (2004) How versatile are inositol phosphate kinases? Biochemical J 377:265–280

Sheen J (2002) Phosphorelay and transcription control in cytokinin signal transduction. Science 296:1650–1652

Shi J, Wang H, Wu Y, Hazebroek J, Meeley RB, Ertl DS (2003) The maize low phytic acid mutant *lpa2* is caused by mutation in an inositol phosphate kinase gene. Plant Physiol 131: 507–515

Shi J, Wang H, Hazebroek J, Ertl DS, Harp T (2005) The maize *low-phytic acid 3* encodes a *myo*-inositol kinase that plays a role in phytic acid biosynthesis in developing seeds Plant J 42:708–719

Shi J, Wang H, Schellin K, Li B, Faller M, Stoop JM, Meeley RB, Ertl DS, Ranch JP, Glassman K (2007) Low-phytic acid and high available phosphorus maize produced by embryo-specific silencing of the *lpa1* gene encoding an MRP ABC transporter. Nat Biotech 25:930–937

Shin H, Shin H-S, Dewbre GR, Harrison MJ (2004) Phosphate transport in *Arabidopsis*: Pht;1 and Pht;4 play a major role in phosphate acquisition from both low- and high-phosphate environments. Plant J 39:629–642

Sondergaard TW, Schulz A, Palmgren MG (2004) Energization of transport processes in plants. Roles of the plasma membrane H^+-ATPase. Plant Physiol 136:2475–2482

Stephens LR, Irvine RF (1990) Stepwise phosphorylation of *myo*-inositol leading to *myo*-inositol hexakisphosphate in *Dictyostelium*. Nature 346:580–583

Stevenson-Paulik J, Bastidas BJ, Chiou S-T, Frye RA, York JD (2005) Generation of phytate-free seeds in *Arabidopsis* through disruption of inositol polyphosphate kinases. Proc Natl Acad Sci 102:12612–12617

Takahashi H, Rai M, Kitagawa T, Morita S, Masumura T, Tanaka K (2004) Differential localization of tonoplast intrinsic proteins on the membrane of protein body type II and aleurone grain in rice seeds. Biosci Biotech Biochem 68:1728–1736

van Dongen JT, Laan RGW, Wouterlood M, Borstlap AC (2001) Electrodiffusional uptake of organic cations by pea seed coats. Further evidence for poorly selective pores in the plasma membrane of seed coat parenchyma cells. Plant Physiol 126:1688–1697

Volk H, Metcalf JA, Withers PJA (2000) Prospects for minimizing phosphorus excretion in ruminants by dietary manipulation. J Environ Qual 29:28–36

Warden BM, Russell WK (2004) Resource allocation in a breeding program for phosphorus concentration in maize grain. Crop Sci 44:753–757

York JD, Odom AR, Murphy R, Ives EB, Wente SR (1999) A phospholipase C-dependent inositol polyphosphate kinase pathway required for efficient messenger RNA export. Science 285:96–100

Yoshida KT, Wada T, Koyama H, Mizobuchi-Fukuoka R, Naito S (1999) Temporal and spatial patterns of accumulation of the transcript of *myo*-inositol-1-phosphate synthase and phytin-containing particles during seed development in rice. Plant Physiol 119:65–72

Chapter 5
Traits and Genes for Plant Drought Tolerance

John Mullet

5.1 Introduction

World population has increased from ∼1 billion to over 6 billion since 1900, and another 4–5 billion people may be added before the human population peaks between 2025–2050 (Khush 1999). At a minimum, crop productivity will need to double on land currently used for agriculture to feed this population (Briggs 1998; Khush 1999). However, the actual increase in demand for agricultural products is likely to be much greater for several reasons. Rapid economic development in China, India and many other parts of the world is leading to higher consumption of animal protein, which requires more grain per calorie consumed compared to grain-based diets. In addition there is growing demand for biofuels and other bioproducts from agriculture. The biomass needed for biofuel production will be derived from grain (starch), sugarcane and sweet sorghum (sugars, bagasse), crop residues (cellulose) and a new generation of dedicated bioenergy crops that produce large amounts of ligno-cellulose. In the USA alone, grain-based ethanol production has increased from ∼1.5 billion gal in 2000 to ∼3.9 billion gal in 2005, consuming approximately ∼14% of the US corn crop (Somerville 2007). The goal is for the USA to produce up to 60 billion gal of biofuels by 2030, which will require significant acreage dedicated to biofuel crops, such as energy cane, bioenergy sorghum, switchgrass, miscanthus and other new crops (Perlack et al. 2005). Minimizing the competition between land used for growing crops for food vs. fuel can only be achieved if crops grown for bioenergy/biofuels are grown on marginal land that is not suitable for food crops and if crop productivity overall is increased.

Meeting the demand for agricultural products for an increasing world population, while preserving wildlife habitat, will require substantial increases in sustainable

---

John Mullet
Department of Biochemistry and Biophysics, Texas A&M University, College Station, TX 77843, USA
e-mail: jmullet@tamu.edu

A.L. Kriz, B.A. Larkins (eds.), *Molecular Genetic Approaches to Maize Improvement*
Biotechnology in Agriculture and Forestry, Vol. 63

productivity on land used currently for agriculture as well as on land that is more marginal due to poorer soil quality and adverse environmental conditions. Potential crop yield is decreased to a significant extent by abiotic constraints and especially by water deficit (Boyer 1982). Unfortunately, there is little opportunity to increase crop productivity through additional irrigation, because water supplies are limited and the demand for water from non-agricultural sectors is increasing (Gleick 2003; Johnson et al. 2001). Therefore, crops grown in environments subject to periodic drought and other abiotic constraints will need to be created through genetic improvement of existing crops that are optimally grown in well-managed systems that also minimize losses due to pests.

5.2 Prospects for Improving Plant Stress Tolerance Through Genetics

There are several reasons to think that crop yield in general, and maize yield in particular, can be increased through genetic improvement of traits influencing adaptation to adverse environments without significant change in the basic biochemistry of photosynthesis and carbon fixation. This thinking was first articulated by Boyer (1982), who analyzed USDA crop yield data in the USA and found that the average of most crops, including corn, was ∼25% or less compared to record yields. He interpreted this to indicate that only a fraction of the genetic potential of most crops was being realized due to abiotic and biotic constraints. Abiotic constraints, and in particular water deficit, has the most significant impact on yield in the USA, where weeds, disease, and insect pressures in general are well managed (Boyer 1982). This observation raised the possibility that crop yield could be significantly increased by ameliorating the negative impact of abiotic stress through a combination of management and genetic improvement of plant adaptation to adverse environments. In fact, for more than 50 years a concerted maize breeding effort has been underway in the USA to increase yield through selection of breeding lines grown in many locations, including regions and years that were subject to abiotic stress. Part of the stress selection came from increased planting density, which also increased productivity per hectare. Remarkably, maize yields in the USA have increased from ∼2 Mg/ha to almost ∼8 Mg/ha since 1940, and it is estimated that about 50% of this increase was due to genetic improvement (Duvick 2005). Moreover, modern maize cultivars were found to be more stress tolerant than earlier genotypes, indicating that genetic improvement for stress adaptation occurred during selection for increased yield across a wide range of environments.

Comparisons of rice, sorghum and maize suggest that the drought tolerance of maize and other cereals can be improved. Rice, a C3 grass largely adapted to high rainfall regions of Asia, is relatively drought sensitive compared to sorghum. Comparisons of sorghum and rice reveal that morphological features, such as greater leaf wax content in sorghum, could be changed in rice to make it more like sorghum in terms of drought adaptation. Upland rice genotypes are better suited to low rainfall climates, indicating potential genetic variation in rice germplasm that can be used to improve drought tolerance. In addition, a growing number of transgenic rice lines have been produced with promising drought tolerance responses that may well lead to significant improvement in rice drought tolerance.

In general, C4 grasses such as sorghum, millet, and corn are better suited to hot dry environments, compared to C3 grasses like rice, due to their ability to concentrate and fix CO_2 in bundle sheath cells, which allows reduced stomatal conductance and water loss during periods of water deficit. Among the cultivated C4 grasses used for grain production, sorghum and pearl millet are generally considered the most drought tolerant and productive in water-limited environments. The underlying drought tolerance of sorghum and millet is consistent with their centers of origin in Africa, where these species evolved under selection in water-limiting environments for millions of years. Within sorghum germplasm there is considerable genetic variation for adaptation and response to water limitation and in the mechanisms and traits that contribute to drought tolerance (Borrell et al. 2006). The type and duration of drought stress (early season, intermittent, or terminal), soil type and other factors characteristic of the various agro-ecological regions in Africa have apparently selected sorghum genotypes for a range of drought tolerance traits. This indicates that even in sorghum there is significant potential for drought tolerance improvement by pyramiding beneficial alleles for tolerance traits from different genotypes. In corn, there has been significant yield improvement in water-limited environments. The high sensitivity of maize to water limitation during anthesis was ameliorated to some degree by selecting genotypes that have shorter anthesis-silking intervals (Ribaut and Ragot 2007). There are several promising maize transgenic events that are likely to enhance drought tolerance in this crop (i.e., Nelson et al. 2007). In summary, there is significant potential for further improvement of maize productivity in water-limited environments through breeding and by the transfer of genes/alleles/traits from sorghum, millet, and other plants to maize.

The base of knowledge about specific genes and pathways that contribute to tolerance and adaptation to abiotic stress in general, and drought tolerance in particular, is expanding rapidly. This holds promise that additional genes and mechanisms for stress tolerance can be identified, properly engineered, and deployed to increase yield still further in drought-prone regions. Additional reading on this topic can be found in many excellent reviews and books that provide an in-depth analysis of plant adaptation and response to adverse environments at the whole plant to genome levels (e.g., Seki et al. 2007; Tuberosa and Salvi 2006; articles in Ribaut 2006; Zhang et al. 2004; Himmelbach et al. 2003; Shinozaki et al. 2003; Bray 2002; Blum 1996). Following a brief description of the physiological and developmental framework for evaluating drought tolerance traits, this chapter will examine the approaches being used to identify traits and genes for drought tolerance and the methods required to optimize their utilization to increase the productivity of maize in drought-prone areas.

5.3 Physiological/Developmental Framework for Assessing the Role and Potential Utility of Genes and Traits for Drought Tolerance

Plant drought tolerance mechanisms for the most part can be categorized into those that allow escape, tolerance, or avoidance of water deficit. Escape mechanisms allow plants to complete their life cycle when water is plentiful. Genetic variation in phenology and time to flowering can be optimized so that the plant's life cycle matches the available water supply in a given ecological region. This is especially important for grain species, because they are most susceptible to water deficit during the reproductive phase. Early flowering genotypes with short seasons reduce the probability of yield loss due to terminal drought; however, these genotypes sacrifice yield in years when water is available and could support a longer growing season. Our knowledge of the genes and pathways that control flowering has advanced to the point where, by using transgenes that are activated by an external chemical treatment, it is possible to engineer crops that could be induced to flower. This would be a very useful management tool allowing control of flowering time depending on water supply.

Tolerance mechanisms allow plant tissues to withstand dehydration. This type of tissue "hardening" occurs through the accumulation of proteins such as the dehydrins (hydrophilins) and heat shock proteins, and a wide range of compatible solutes (e.g., polyols, glycine betaine, proline, inositol). Plants also increase the level and activity of enzymes and pathways that protect tissues from the generation of potentially damaging reactive oxygen species (ROS) that are generated during periods of water limitation and stomatal closure (i.e., ROS protective systems, GABA shunt, photorespiration). Dehydration tolerance traits are especially important in turf grasses, forages, and biofuel crops, where retention of leaf function, regrowth following periods of relatively severe water deficit, and biomass accumulation, per se, are the main determinants of yield.

Drought avoidance mechanisms constitute the third and probably the most important category of drought tolerance traits that impact grain yield, because they impact water supply and utilization. Root system traits and responses are particularly important, because variation in root development, morphology, and function affect water supply (i.e., root depth/spatial organization, growth response to water deficit, aluminum tolerance, water extraction capability, phenology of root growth/suberization/turnover). There is also a suite of traits that modulate water utilization (i.e., C3/C4 or CAM photosynthesis, variation of stomatal distribution and conductance, leaf cuticle properties (wax, hairs, boundary layers), hydraulic conductivity, leaf architecture (thickness, size, area, rate of appearance, leaf rolling, erectness), and canopy architecture). Drought avoidance traits have a significant impact on yield, because they help plants maintain good water status, allowing continued photosynthesis, growth, and development.

The importance of various plant responses and adaptations to water-limiting environments and the genes that control them depend in part on the stage of plant development they impact. For example, dehydration tolerance is important during seedling establishment if tolerance helps improve stand establishment and minimizes the need for replanting. Fortunately, most grasses recover rapidly from transient periods of water deficit. Increased tolerance of leaf tissues to ROS helps minimize damage to the photosynthetic apparatus during water deficit and this should improve overall biomass accumulation by maintaining photosynthetically active green leaf area. However, during the vegetative phase, older leaves normally senesce when shaded and their constituents are mobilized and re-utilized for the production of new leaves at the top of the canopy. Therefore, plants can recover from a transient water deficit that partially damages one set of leaves through the production of additional leaves during the normal course of development. Because tolerance mechanisms are usually induced under conditions of plant water deficit that close stomata and inhibit photosynthesis, these mechanisms may protect preformed structures and accelerate recovery, but their impact on overall biomass accumulation and yield can be relatively small if water limitation is transient. In fact, constitutive expression of some tolerance mechanisms may actually decrease productivity and increase the risk of plant death by reducing the plant's ability to avoid more severe water loss (i.e., by preventing leaf senescence and leaf loss, which is a key mechanism for slowing water utilization in some plants). In contrast, traits that help plants avoid water deficit, such as the establishment of a deep rooting system, are likely to have a greater impact on yield, assuming in the case of deep roots that water is available in the soil profile and soil water content is recharged annually (Jordan et al. 1983; Sinclair and Muchow 2001). Similarly, the drought avoidance 'stay-green' trait in sorghum has a significant impact on yield in water-limited environments, because this response improves plant water status, photosynthetic activity, and nitrogen uptake in water-limited environments during the reproductive phase (Borrell et al. 2001).

5.4 Identification and Testing Gene/Trait Leads for Drought Tolerance

Gene/trait leads that might improve drought tolerance and increase yield in crop plants grown in water-limited environments have been discovered in several ways: (1) by identifying genes that are induced/repressed in response to water deficit; (2) by analyzing mutants that show modified response to water limitation; (3) by ectopic expression of transcription factors and other genes, followed by screens for improved plant performance under water-limiting conditions; (4) by dissecting physiological traits followed by QTL mapping and cloning genes/alleles found in germplasm collections that modify the traits; and (5) by comparative analysis of drought-sensitive and drought-tolerant species. Genes that show modified expression in response to water deficit are the easiest to identify. However, determining the importance of a specific inducible gene with regard to yield in water-limiting environments is challenging. QTL analysis of alleles in germplasm that modulate traits known to impact plant responses to water limitation allows early assessment of a gene's impact on yield through breeding. However, the trait-QTL to gene approach is also challenging, because phenotyping a plant's response to water limitation is difficult in non-uniform genetic backgrounds and map-based cloning is labor intensive. Industrial-scale testing of thousands of plant genes in transgenic plants is now feasible. The critical questions surrounding this approach involve how best to express transgenes so that an impact on yield in water-limited environments can be assessed, and what screens to use to identify promising gene leads (Verslues et al. 2006).

Genes and pathways that are modulated through changes in gene expression in response to water deficit have been characterized in numerous plant species. Genome-wide analysis using microarrays or digital expression analysis of plants exposed to water deficit have identified thousands of genes representing a wide range of biochemical functions that are modulated by water deficit (Buchanan et al. 2005). These studies have identified large suites of genes that are modulated by several different signaling pathways (regulons). One of these pathways starts with perception of water deficit through reduction of cell turgor, and this leads to accumulation of the plant hormone abscisic acid (ABA). ABA in turn activates a signaling pathway that reduces stomatal aperture, contributes to differential root/shoot growth (Sharp et al. 2004), and modulates gene expression. Genes that are activated in response to increasing levels of ABA as a consequence of water deficit include those that encode the dehydrins/LEAs (hydrophilins), oleosins and heat shock proteins, enzymes involved in protection from ROS and pests, and proteins that contribute to the accumulation of compatible solutes (proline, glycine betaine, polyols, trehalose, and other sugars, etc.). A large number of genes with altered expression are involved in a range of metabolic and transport functions that are altered when transpiration, photosynthesis, and growth is reduced in response to water deficit.

Identification of the signaling pathways that mediate changes in gene expression in response to water deficit and the suites of genes connected to these pathways provide the starting point for attempts to improve drought tolerance through transgenics. Attempts have been made to increase tolerance by activating entire response pathways by over-expressing genes that modulate hormone levels (ethylene/ACC synthase; Young et al. 2004), transcription factors (Hu et al. 2006; Cabello et al. 2007; Xu et al. 2007; Karaba et al. 2007; Qin et al. 2007), or genes that encode steps in signal transduction pathways (Shou et al. 2004; Catala et al. 2007). Other groups have constructed transgenic plants that have high expression of downstream genes, including the dehydrins (LEAs) (Xiao et al. 2007), heat shock proteins (Sato and Yokoya 2007), aquaporins (Lian et al. 2004), and genes that encode enzymes that increase the level of compatible solutes (Vendruscolo et al. 2007; Liu et al. 2007). Many of these studies report that ectopic expression of the target pathways and genes increases stress tolerance, as measured by the ability of plants to withstand short periods of relatively severe water deficit (survival). More recently, field experiments have been carried out in an attempt to determine if expression of specific genes improves yield when water deficit occurs during the reproductive phase, due to the sensitivity of this stage to water deficit. Following these preliminary screens, promising leads require full-scale testing in elite backgrounds over multiple environments and years to determine the impact of transgenic modifications on yield. In recent years, large numbers of transcription factors and other plant genes have been expressed in Arabidopsis, and the resulting plants screened for a wide range of phenotypes, including response to water deficit (Nelson et al. 2007). This type of unbiased screen is capable of identifying a wide spectrum of genes, including those that modulate traits that are not regulated in response to water deficit. This approach led to the isolation of the transcription factor, NF-Y, which increases maize drought tolerance (Nelson et al. 2007).

The dehydrins and LEA proteins were an early target for ectopic gene expression and drought tolerance engineering, and greater accumulation of these proteins has been correlated with increased tissue dehydration tolerance. However, ectopic expression of these proteins using strong constitutive promoters can inhibit growth and produce morphological defects. Similar attempts to increase compatible solutes by constitutive expression of genes that encode rate-limiting steps in some biochemical pathways also showed deleterious symptoms (protection but growth inhibition). This is not surprising and is consistent with the fact that plants normally restrict high expression of these proteins and pathways to conditions of water deficit. When these same genes were subsequently placed under the control of promoters that are activated in water-deficient plants, growth defects largely disappeared, while protection conferred by the genes was retained. Overall, this approach appears very promising, although it will require large-scale multi-location and environment testing over many years to fully optimize gene expression by yield optimization.

The identification and pyramiding of naturally occurring alleles that confer improved plant productivity in water-limiting environments represents a second promising approach to improve yield in adverse environments. This approach requires large-scale screening of germplasm to detect allelic variation in drought tolerance traits, followed by population development, QTL mapping, and map-based gene cloning. Marker-assisted selection can be carried out in parallel with map-based cloning to provide an early assessment of the impact of alleles and traits. Recent success with marker-assisted selection of maize for a shorter anthesis-silking interval (Ribaut and Ragot 2007) and QTL-based analysis of root traits demonstrated the potential of this approach (Landi et al. 2007). The sorghum 'stay green' trait provides a good example of the QTL to marker/gene approach for enhancing drought tolerance. The sorghum stay-green trait is correlated with improved grain yield and lodging resistance under terminal drought conditions (Rosenow and Clark 1995). Borrell et al. (2001) found that the 'stay green' trait in the B35 inbred increased yield of hybrids $\sim 47\%$ in terminal drought conditions in Australia without a negative impact on yield in well-watered environments. The stay-green trait results in retention of green leaf area and photosynthetic activity for a longer time during grain filling under terminal drought stress conditions. However, it is important to emphasize that the stay-green trait found in B35 does not block normal leaf senescence; it only delays its onset and/or slows its rate under water-limiting conditions. Therefore, the carbon and nitrogen in the leaf are mobilized and used for grain production. Studies show that the stay-green QTL alters the plant's physiology prior to and after anthesis, resulting in higher leaf nitrogen content and larger stem mass prior to stress imposition. As a consequence, the stay-green trait increases lodging resistance, which is especially important for high grain yield. Four major QTLs for the stay-green trait have been identified and mapped in sorghum populations derived from crosses with B35 (BTx642), a primary source of stay-green alleles. Moreover, in most combinations the B35 alleles act with varied levels of dominance, making them especially useful for hybrid production. Fine mapping experiments are in progress to isolate the genes that modulate the sorghum stay-green trait for utilization in maize and other cereals (Harris et al. 2007).

5.5 Deployment of Drought Tolerance Genes and Genotype

Early attempts to enhance plant drought tolerance by increasing the expression of a gene or pathway using strong constitutive promoters were not very successful, possibly because high expression interferes with normal metabolism and development. Expression of putative stress tolerance genes using promoters that are activated in response to water deficit (or ABA) has been more successful in enhancing tolerance without secondary effects. Further improvement may require differential expression in roots vs. shoots, tissue- or cell-specific expression, or expression specific to a developmental stage. For example, like most plants sorghum closes its stomata in response to water deficit during vegetative growth in order to avoid more severe dehydration and possible death. However, sorghum stomata become nearly insensitive to water deficit following anthesis. This change in sensitivity allows continued CO_2 fixation and grain filling even under water-limiting conditions that might otherwise close stomata, inhibit photosynthesis, reduce sugar levels, and cause complete loss of reproductive structures. Therefore, once traits and genes for drought tolerance are identified, utilization of these genes needs to be optimized in terms of expression level, tissue/cell-specific expression, and expression during plant development.

The optimal deployment of genotypes with suites of drought tolerance traits into the various agro-ecological regions of production represents one of the most important opportunities for increasing yield. Plant simulation models have advanced to the point where the impact of traits can be predicted as a function of environment (Hammer et al. 2004; Sinclair and Muchow 2001). Information from weather stations, soil maps, remote sensing technology, and yield monitoring can be integrated into simulation models and used to predict and measure the impact of trait/gene modifications on yield. This approach provides a way to simulate and then test plant responses to different types and severity of water limitation and the impact of a wide range of traits on yield within these regimes. The deployment and testing of genotypes that vary in the suite of traits used in the models in geo-referenced locations will provide important insight into the connections between genes, traits, and their impact on yield. Overall, the identification of genes that modulate biochemical responses and traits that improve plant productivity in water-limited environments has advanced rapidly in the last 10 years. In the future, increasing effort will focus on optimal expression of key genes/traits and the deployment of the resulting crops into environments where the expressed traits will have maximum impact on yield.

References

Blum A (1996) Crop responses to drought and the interpretation of adaptation. Plant Growth Reg 20:135–148

Borrell A, Hammer G, Oosterom EV (2001) Stay-green: a consequence of the balance between supply and demand for nitrogen during grain filling? Ann Appl Biol 138:91–95

Borrell A, Jordan D, Mullet J, Henzell B, Hammer G (2006) Drought adaptation in sorghum. In: Ribaut JM (ed) Drought adaptation in cereals. Haworth Press, Binghamton, New York, pp 335–400

Boyer JS (1982) Plant productivity and environment. Science 218:443–448

Bray EA (2002) Classification of genes differentially expressed during water-deficit stress in *Arabidopsis thaliana*: an analysis using microarray and differential expression data. Ann Bot (Lond) 89:803–811

Briggs SP (1998) Plant genomics: more than food for thought. Proc Natl Acad Sci USA 95(5):1986–1988

Buchanan CD, Lim S, et al. (2005) Sorghum bicolor's transcriptome response to dehydration, high salinity and ABA. Plant Mol Biol 58(5):699–720

Cabello JV, Dezar CA, et al. (2007) The intron of the *Arabidopsis thaliana* COX5c gene is able to improve the drought tolerance conferred by the sunflower Hahb-4 transcription factor. Planta 226(5):1143–1154

Catala R, Ouyang J, et al. (2007) The Arabidopsis E3 SUMO ligase SIZ1 regulates plant growth and drought responses. Plant Cell 19(9):2952–2966

Duvick D (2005) The contribution of breeding to yield advances in maize (*Zea mays* L.). Adv Agron 86:83–145

Gleick PH (2003) Global freshwater resources: soft-path solutions for the 21st century. Science 302(5650):1524–1528

Hammer GL, Sinclair TR, et al. (2004) On systems thinking, systems biology, and the in silico plant. Plant Physiol 134(3):909–911

Harris K, Subudhi PK, et al. (2007) Sorghum stay-green QTL individually reduce post-flowering drought-induced leaf senescence. J Exp Bot 58(2):327–338

Himmelbach A, Yang Y, et al. (2003) Relay and control of abscisic acid signaling. Curr Opin Plant Biol 6(5):470–479

Hu H, Dai M, et al. (2006) Overexpressing a NAM, ATAF, and CUC (NAC) transcription factor enhances drought resistance and salt tolerance in rice. Proc Natl Acad Sci USA 103(35):12987–12992

Johnson N, Revenga C, et al. (2001) Ecology. Managing water for people and nature. Science 292(5519):1071–1072

Jordan W, Dugas WA, Shouse PJ (1983) Strategies for crop improvement for drought-prone regions. Agric Water Manage 7:281–299

Karaba A, Dixit S, et al. (2007) Improvement of water use efficiency in rice by expression of HARDY, an Arabidopsis drought and salt tolerance gene. Proc Natl Acad Sci USA 104(39):15270–15275

Khush GS (1999) Green revolution: preparing for the 21st century. Genome 42(4):646–655

Landi P, Sanguineti MC, et al. (2007) Root-ABA1 QTL affects root lodging, grain yield, and other agronomic traits in maize grown under well-watered and water-stressed conditions. J Exp Bot 58(2):319–326

LeNoble ME, Spollen WG, et al. (2004) Maintenance of shoot growth by endogenous ABA: genetic assessment of the involvement of ethylene suppression. J Exp Bot 55(395):237–245

Lian HL, Yu X, et al. (2004) The role of aquaporin RWC3 in drought avoidance in rice. Plant Cell Physiol 45(4):481–489

Liu HL, Dai XY, et al. (2007) Over-expression of OsUGE-1 altered raffinose level and tolerance to abiotic stress but not morphology in Arabidopsis. J Plant Physiol 164(10):1384–1390

Nelson DE, Repetti PP, et al. (2007) Plant nuclear factor Y (NF-Y) B subunits confer drought tolerance and lead to improved corn yields on water-limited acres. Proc Natl Acad Sci USA 104(42):16450–16455

Perlack RD, Wright LL, Turhollow A, Graham R, Stokes B, Erbach D (2005) Biomass as feedstock for a bioenergy and bioproducts industry: the technical feasibility of a billion-ton annual supply. US Department of Energy and Oak Ridge National Laboratory, Tech Rep ORNL/TM-2006/66, www1.eere.energy.gov/biomass/pdfs/final_billionton_vision_report2.pdf

Qin F, Kakimoto M, et al. (2007) Regulation and functional analysis of ZmDREB2A in response to drought and heat stresses in *Zea mays* L. Plant J 50(1):54–69

Ribaut JM (ed) (2006) Drought adaptation in cereals. Haworth Press, Binghamton, New York

Ribaut JM, Ragot M (2007) Marker-assisted selection to improve drought adaptation in maize: the backcross approach, perspectives, limitations, and alternatives. J Exp Bot 58(2):351–360

Rosenow DT, Clark LE (1995) Drought and lodging resistance for a quality sorghum crop. In: Proceedings of the 50th Annual Corn and Sorghum Industry Research Conference 1995, pp 82–97

Sato Y, Yokoya S (2007) Enhanced tolerance to drought stress in transgenic rice plants overexpressing a small heat-shock protein, sHSP17.7. Plant Cell Rep 27(2):329–334

Seki M, Umezawa T, et al. (2007) Regulatory metabolic networks in drought stress responses. Curr Opin Plant Biol 10(3):296–302

Shinozaki K, Yamaguchi-Shinozaki K, et al. (2003) Regulatory network of gene expression in the drought and cold stress responses. Curr Opin Plant Biol 6(5):410–417

Shou H, Bordallo P, et al. (2004) Expression of the Nicotiana protein kinase (NPK1) enhanced drought tolerance in transgenic maize. J Exp Bot 55(399):1013–1019

Sinclair TR, Muchow RC (2001) System analysis of plant traits to increase grain yield on limited water supplies. Agron J 93:263–270

Somerville C (2007) Biofuels. Curr Biol 17(4):R115–R119

Tuberosa R, Salvi S (2006) Genomics-based approaches to improve drought tolerance of crops. Trends Plant Sci 11(8):405–412

Vendruscolo EC, Schuster I, et al. (2007) Stress-induced synthesis of proline confers tolerance to water deficit in transgenic wheat. J Plant Physiol 164(10):1367–1376

Verslues PE, Agarwal M, et al. (2006) Methods and concepts in quantifying resistance to drought, salt and freezing, abiotic stresses that affect plant water status. Plant J 45(4):523–539

Xiao B, Huang Y, et al. (2007) Over-expression of a LEA gene in rice improves drought resistance under the field conditions. Theor Appl Genet 115(1):35–46

Xu ZS, Xia LQ, et al. (2007) Isolation and molecular characterization of the *Triticum aestivum* L. ethylene-responsive factor 1 (TaERF1) that increases multiple stress tolerance. Plant Mol Biol 65(6):719–732

Young TE, Meeley RB, et al. (2004) ACC synthase expression regulates leaf performance and drought tolerance in maize. Plant J 40(5):813–825

Zhang JZ, Creelman RA, et al. (2004) From laboratory to field. Using information from Arabidopsis to engineer salt, cold, and drought tolerance in crops. Plant Physiol 135(2):615–621

Chapter 6
Biotechnology Approaches to Improving Maize Nitrogen Use Efficiency

Stephen Moose and Fred E. Below

6.1 The Importance of Improving Nitrogen Use Efficiency and a Biotechnology Approach

Nitrogen (N) is an essential and often limiting nutrient to plant growth. Maize grain yields are highly responsive to supplemental N, leading to annual application of an estimated 10 million metric tons of N fertilizer to the maize crop worldwide (FAO 2004). Nearly all cultivated maize in developed countries receives some form of N fertilizer and N use is increasing in developing countries, where its impacts on raising grain yields from nutrient-poor soils are greatest. The extensive use of N fertilizer not only increases crop input costs, but also can negatively impact soil, water and air quality at both local and ecosystem scales (Tilman et al. 2002). The manufacture of N fertilizer is an energy-intensive process that is becoming increasingly costly, due to the use of natural gas as both a reactant and heat source for the conversion of atmospheric N_2 to anhydrous ammonia (NH_3). For these reasons, reducing the amount of supplemental N used in maize production will have significant positive economic and environmental benefits to world agriculture.

Nitrogen use efficiency (NUE) can be defined in a variety of ways that emphasize different components of the soil and plant system (reviewed in Good et al. 2004) or economic returns to fertilizer use. In cereal crops like maize, agronomic NUE is most simply expressed as the ratio of grain yield to N fertilizer supplied. Comparisons of maize grain yields and N fertilizer usage on a global basis lead to estimates of maize NUE ranging from 25–50% (Raun and Johnson 1999; Tilman et al. 2002),

Stephen Moose
Department of Crop Sciences and Institute for Genomic Biology, University of Illinois at Urbana-Champaign, 389 Edward R. Madigan Laboratory, 1201 W. Gregory Drive, Urbana, IL 61801, USA
e-mail: smoose@uiuc.edu

Fred E. Below
Department of Crop Sciences, University of Illinois at Urbana-Champaign, 389 Edward R. Madigan Laboratory, 1201 W. Gregory Drive, Urbana, IL 61801, USA

A.L. Kriz, B.A. Larkins (eds.), *Molecular Genetic Approaches to Maize Improvement*
Biotechnology in Agriculture and Forestry, Vol. 63

indicating that more than half the fertilizer N applied in maize crop production is lost to the environment. Thus, there is considerable opportunity for enhancing maize NUE.

Despite the importance of NUE, genetic improvement of this trait has primarily been an indirect outcome of breeding for higher grain yields. Past progress in selecting genotypes with improved NUE has been hampered by the complexity of the genetic network regulating whole plant N metabolism, environmental interactions, and the labor-intensive nature of NUE evaluations. Demand by farmers for the highest yields and the desire of breeding programs to reduce variability due to N stress have also favored the selection of maize hybrids adapted to high N input environments (Castleberry et al. 1984; Purcino et al. 1998), except in the case of germplasm adapted to low-fertility soils in tropical environments (e.g., Worku et al. 2007). Direct evaluation of NUE in maize breeding populations and inbred lines has also been problematic, due to the weak correlation between maize inbreds and hybrids for grain yield and its response to N (Bertran et al. 2003).

One of the most widely publicized claims of early agricultural biotechnology was the development of "N-fixing" maize, which would be capable of assimilating atmospheric N within plant cells, as do nodulating legumes that support symbiotic associations with N-fixing microbes. However, the development of such "N-fixing maize" is unlikely, though biotechnology has and will continue to play an important role in improving NUE. Biotechnology can be applied to enhance the discovery and validation of genes controlling NUE and its component traits, to develop molecular markers for accelerating breeding progress independent of growth environment, and to introduce transgenes that modify key physiological processes contributing to NUE. The tools of biotechnology thus can help overcome some of the previous challenges to improving NUE.

6.2 The Biology of Maize NUE

NUE in maize has been intensively studied. Much of this research has focused on productivity and physiological responses of maize hybrids to different N management practices, as observed in numerous agronomic evaluations conducted in a wide range of environments (Giller et al. 2004). Other studies have documented genetic variation for N-responsive traits among diverse sources of maize germplasm (reviewed in Gallais and Coque 2005). NUE is governed by interactions between soil N levels, N availability due to microbial activity in the rhizosphere, and the ability of the maize plant to assimilate and use acquired N for plant growth. Thus, measuring the genetic component of NUE requires the characterization of biological responses to N under carefully controlled environmental conditions.

Although agronomic evaluations are critical to demonstrating commercial improvement, such studies often suffer from spatial and temporal variability in N availability to the plant and yield little information about biological mechanisms. Alternative experimental approaches have thus been developed to characterize physiological and molecular responses to N. These include soil fertigation, greenhouse hydroponics, and in vitro kernel culture systems that permit more precise

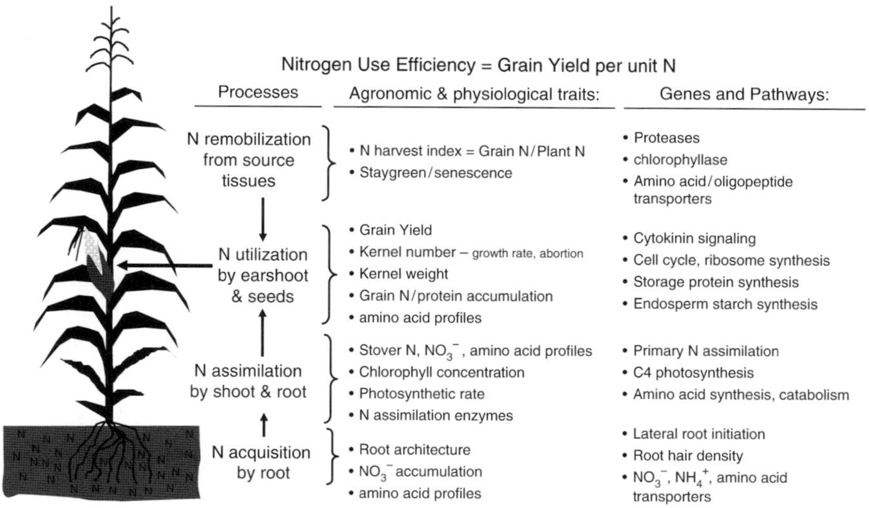

Fig. 6.1 Major physiological processes, observed phenotypes, and genetic pathways associated with maize nitrogen use efficiency

manipulation of N supply to developing plants. Activity assays of enzymes associated with N metabolism and measurement of N-containing compounds such as chlorophyll, nitrate or amino acids are dynamic metabolic indicators of plant N status. When coupled with ^{15}N isotopic labeling, estimates of N fluxes can also be obtained. Recent advances in molecular biology, genetics, and functional genomics have also been applied to better understand maize N metabolism and identify genes whose expression or activities might be modified to improve NUE (reviewed in Good et al. 2004).

NUE is influenced by the complex interplay between many physiological processes. These include N uptake from soil, assimilation into amino acids that serve as N carriers throughout the plant, and N transport from source to sink tissues throughout plant development, especially during the grain-filling period. Signaling and regulatory pathways that integrate plant N status and plant growth also play important roles. Figure 6.1 illustrates important processes, traits, pathways, and classes of genes that are either demonstrated or expected contributors to maize NUE. We draw upon a number of excellent reviews (Below 2002; Good et al. 2004; Hirel et al. 2005a, b, 2007) to briefly summarize our current knowledge on N uptake, metabolism, and transport in maize, which provides a context for how biotechnology might modify these processes for enhanced NUE.

6.2.1 N Uptake and Assimilation

N primarily enters the maize plant through the roots, whose size and branching patterns define the volume of soil solution supplying N. Membrane-associated transporters mediate uptake of nitrate, and to a lesser extent ammonia, with evidence for both constitutive low-affinity and induced high-affinity systems. N is rapidly assimilated into amino acids either within the root or after transport to shoot tissues. In maize, the majority of N assimilation occurs in leaves, powered by the high rates of C-fixation resulting from C4 photosynthesis, which produces different pools of carbon acceptors for N and a reduction in potential volatilization of N compared to plants that perform C3 photosynthesis. The pathways and genes encoding key enzymes for primary N assimilation in maize are similar to those that have been extensively studied in model systems such as *Arabidopsis* and tobacco. Nitrate is rapidly reduced to ammonia which is then combined with glutamate to form glutamine, a metabolically active amino acid that is an amino-donor to many other reactions of N metabolism. Glutamine, glutamate, alanine, aspartate, and the relatively inert storage amino acid asparagine form the bulk of amino acids found in source tissues. Key enzymes involved in primary N assimilation and interconversion among the major amino acid carriers thus include nitrate reductase, glutamine synthetase, alanine aminotransferase, aspartate aminotransferase, and asparagine synthetase.

Maize exhibits determinate shoot development and, unlike many plant species, maize vegetative biomass is not highly responsive to N. Instead, leaves accumulate chlorophyll and abundant photosynthetic or N assimilation enzymes such Rubisco, PEP carboxylase, and glutamine synthetase (GS) as temporary storage forms of N. Under conditions of high N availability, leaf vacuolar concentrations of nitrate and amino acids also rise. Rapid measures of N uptake efficiency that are often used in agronomic evaluations include combustion analysis of total N concentration in leaf and stem (stover) tissues and hand-held chlorophyll meters that detect far-red light reflectance.

6.2.2 N Transport

Labeling experiments with ^{15}N show that N metabolism is highly dynamic. However, N uptake and recycling proceed through four distinct developmental phases. Early seedling growth is primarily supported by N released from the breakdown of seed reserves, which must be accounted for in studies of N metabolism in maize seedlings. The majority of plant N is accumulated during subsequent vegetative development. N uptake from soil continues for a brief period immediately after flowering and then declines, with newly acquired N being directed to developing seeds as a physiological sink. Though the proportion varies among maize genotypes and with environmental conditions, the majority of N that supports seed development is remobilized from leaves, stalk, and earshoot tissues. Phenotypes associated with continued N accumulation during the grain filling period include enhanced grain protein concentration and delayed plant senescence characterized by the "staygreen" leaf phenotype. N remobilization is a three-stage process, where free amino acids, then amino acids from protein breakdown, followed by recycled chlorophyll are transported to the developing seeds. The flow of N compounds from vegetative source to reproductive sink tissues is not continuous, being exported from the phloem into the pedicel and placental-chalazal region for subsequent uptake through the basal endosperm transfer cells.

Though a well-studied physiological process, the molecular events associated with N remobilization are poorly understood. Initial events likely involve proteases (and aminotransferases), as well as GS and glutamate dehydrogenase (GDH) to rapidly recycle released ammonia, each of which funnel N into preferred transport amino acids. In addition, the chlorophyll degradation pathway and amino acid/oligopeptide transporters are expected to play important roles in moving N from source to sink tissues.

6.2.3 N Utilization by Kernels

N utilization efficiency is the ratio of grain yield to plant N and indicates the response of reproductive sink capacity and growth to acquired N. The component trait that is most sensitive to and highly correlated with maize grain yield response to N is kernel number. Final kernel number is the product of the number of potential ovules on the ear, the proportion of ovules that are successfully fertilized, and those ovules that complete kernel development. Figure 6.2 illustrates the impact of moderate N stress on each of these kernel number component traits for a typical maize hybrid. Ear length and the total number of ovules are similar, but may exhibit differences under severe N stress. Both ears also formed a small number of ovules at their tip that were not pollinated, due the slower growth of silks relative to pollen availability. The delay between pollen shed and silk emergence, also know as anthesis-silking interval (ASI), often becomes more pronounced under severe N stress. The region

Fig. 6.2 Ears at physiological maturity from the B73 × Mo17 hybrid grown with sufficient N (*left*) or under N-stress conditions (*right*). The regions of the ear that illustrate different physiological effects of N stress are indicated

of approximately 10 kernel rows between this unpollinated zone and where kernels have fully developed illustrates the effects of kernel abortion due to N stress. These ovules were pollinated and initiated, but did not complete kernel development. Though much is known about how N maintains basic cellular processes such as the cell cycle, ribosome production and nucleic acid synthesis in *E. coli* and yeast, this information is generally lacking in higher plants.

Within the maize seed, N is used initially to support continued growth and high rates of starch synthesis in the endosperm. N is also deposited in storage proteins, particularly the zeins in the endosperm and globulin in the embryo, which provide a source of N during seed germination. Grain protein concentration increases in response to N supply over a greater range than grain yield and is controlled by maternal genotype, indicating that storage protein synthesis is highly sensitive to the N status of source tissues. Many studies in cereal crops have also found a strong negative relationship between grain protein and starch concentrations, as well as grain protein concentration and yield (e.g. Uribelarrea et al. 2004), suggesting that C and N seed storage pathways compete directly for assimilate and energy supply.

6.2.4 Regulation of N-Associated Processes

The regulation and coordination of N metabolism, transport, and partitioning is poorly characterized in plants, and even less so in maize. Unlike drought or temperature stress, N-deficiency does not directly elicit strong localized or systemic stress responses. Instead, N assimilation is regulated by metabolic indicators of carbon/nitrogen balance, such as activation by simple sugars and feedback inhibition by the amino acid products of primary N assimilation. N metabolism is also sensitive to light and hence energy status, being more active during the day than night. Because of their central role as N carriers, both the amounts and profiles of amino acids appear to play important roles in N signaling, with the ratio of metabolically active glutamine relative to the inert transport form asparagine potentially serving as a signal of plant N status (Seebauer et al. 2004).

Growth regulators, particularly N-containing cytokinins and polyamines, also serve as secondary signals to coordinate developmental responses to N (Sakakibara et al. 2006). The downstream effectors of these signaling pathways are not clearly defined, but some N- and/or C-responsive regulatory proteins have been identified from studies in *Arabidopsis* (see Good et al. 2004). These include the GATA transcription factor GNC, a putative glutamate receptor, and a protein related to the central cyanobacterial N sensor PII protein. Building on work from yeast, *Arabidopsis* components have been identified for the TOR complex that regulates ribosome synthesis in response to nutrient status (Deprost et al. 2007). Additional genes that may contribute to nutrient-responsive changes in root architecture include the MADS-box factor ANR1 and high-affinity nitrate transporters that appear to also act as nitrate sensors.

6.3 Candidate Genes for Enhancing Maize NUE

The large number of genes and interacting pathways associated with maize NUE phenotypes suggests that directed changes in one component of the N network may not be sufficient to effect significant changes in overall NUE. Thus, functional genomics approaches that provide genome-scale perspectives on N-responsive gene networks are likely to identify key control points for modification via biotechnology. Efforts to integrate quantitative trait mapping, RNA expression and metabolite profiling, and transgene testing are beginning to discover candidate genes for improving maize NUE (Hirel et al. 2007).

6.3.1 Quantitative Trait Loci

Like grain yield, NUE is a complex trait for which considerable genetic variation exists within maize germplasm. Identifying quantitative trait loci (QTL) for maize N response traits is important for at least four reasons. First, the genetic architecture of NUE and its component traits can be determined. Second, greater success in map-based cloning makes QTL mapping feasible as a discovery approach for genes that have a proven impact on NUE-associated traits. Third, if molecular markers can be identified that are predictive of N response phenotypes, they may be used to characterize genetic performance independently from N supply, thus decreasing phenotyping costs. Finally, knowledge of NUE QTL will have value in combining transgenes with genetic backgrounds that maximize trait expression and stability.

Previous studies have identified QTL controlling NUE and some of their component traits (Agrama et al. 1999; Bertin and Gallais 2001; Hirel et al. 2001; Gallais and Hirel 2004), and additional experiments with higher-resolution mapping populations are in progress (S. Moose and F. Below, unpublished). In addition to agronomic parameters, QTL have been associated with N metabolites and enzyme activities. Heritabilities and genetic correlations for NUE traits are generally high (>0.5) when measured on plants grown with different levels of N supply, though some QTL can only be identified in either high or low N environments. Specific QTL have been associated with multiple NUE component traits, indicating that the underlying genes may act at key nodes of the N response network. Several QTL co-localize with candidate genes encoding enzymes for primary N metabolism, for example cytosolic glutamine synthetase (Bertin and Gallais 2001).

Another potential germplasm source for the discovery of genes associated with NUE is the Illinois long-term selection experiment for grain protein concentration (Moose et al. 2004). More than a century of selection has produced the known phenotypic extremes for grain protein, which are also associated with changes in whole plant N metabolism and NUE (Uribelarrea et al. 2007). The Illinois High Protein (IHP) genotype exhibits dramatically increased N uptake, whereas the Illinois Low Protein (ILP) genotype has enhanced N utilization by developing kernels. QTL affecting grain protein concentration, and presumably some aspects of NUE, have already been identified in a population derived from a cross between IHP and ILP (Dudley et al. 2007).

6.3.2 RNA Expression Profiling

Genome-scale surveys of N-responsive gene expression have been conducted in model microbial as well as plant systems, including *Arabidopsis* (Scheible et al. 2004; Wang et al. 2003) and rice (Lian et al. 2006). Hundreds of genes have been identified as N-responsive in seedling roots and shoots, with many of these genes also showing interactive effects of C and N (Palenchar et al. 2004). Though some of these genes are known to be involved in plant N metabolism, many have yet to be functionally defined. Extending RNA profiling experiments to maize reproductive tissues during the key period when kernel number is determined will likely yield additional candidate genes that are important for NUE.

N-responsive genes identified via RNA expression profiling experiments also offer sources for expression regulatory elements that are both N-responsive and active in key target tissues or cell types, which may be used in subsequent transgenic experiments. Furthermore, the finding that microRNAs regulate phosphate and sulfur uptake and homeostasis in *Arabidopsis* (Chiou 2007) raises the possibility that small RNAs could also be key regulators of N metabolism.

6.3.3 Transgenes for Improving Maize NUE

Any transgene that improves grain yield can be considered to indirectly impact NUE. The recent introduction of maize hybrids with transgenic resistance to root feeding by corn rootworm (*Diabrotica* spp.) larvae is one such example. The larger and healthier root system that results from reduced damage due to insect feeding may possibly lead to greater N uptake. Similarly, the projected release of transgenic maize hybrids with enhanced drought tolerance could indirectly increase both N uptake and N utilization.

Two transgenic approaches have been reported that directly modify expression of the genes involved in N metabolism in maize. The promoter of the maize senescence-enhanced protease gene driving the expression of *Agrobacterium* isopentenyl transferase presumably increased cytokinin levels and conferred a "stay-green" phenotype, delaying senescence, loss of photosynthetic activity, and days to flowering (Robson et al. 2004). Under low N conditions, N remobilization from lower leaves appeared to be reduced and led to leaf yellowing in upper leaves. In another study, constitutive overexpression of a cytosolic GS isoform increased kernel number and grain yield approximately 30% relative to non-transgenic sibling plants (Martin et al. 2006). However, these effects have only been observed for early generation transgenic lines in greenhouses and need to be confirmed for maize hybrids grown under low N conditions in the field.

Due to the greater ease of transformation, a large number of genes have been tested for their impacts on N metabolism in *Arabidopsis*, tobacco, and more recently rice. Genes for ammonia and nitrate transporters, nitrate reductase, and many of the key enzymes for primary N assimilation have been overexpressed and in some cases downregulated by gene silencing. These modifications often lead to changes in the amount and form of accumulated N, particularly under low N conditions, and in some cases have also increased plant growth. Three noteworthy examples are an increase in grain weight among rice plants that overexpress NADH-GOGAT (Tabuchi et al. 2007), higher seed N concentration in *Arabidopsis* with enhanced asparagine synthetase (Lam et al. 2003), and greater yields under low N conditions for canola and rice that overexpress alanine aminotransferase (Good et al. 2007). Another rice study demonstrated an increase in grains per panicle when cytokinin oxidase gene expression was reduced in the inflorescence (Ashikari et al. 2005). A similar approach may increase kernel number in maize.

As illustrated in Fig. 6.1 and by the examples described above, there are many opportunities for modifying gene expression to enhance component traits of NUE. In addition to available plant and microbial genes, other genetic engineering tools such as gene shuffling and RNAi offer a wide variety of methods to improve NUE. It is evident from initial attempts that the complexity of NUE will likely require the introduction of multiple genes that coordinately target N uptake, transport, and utilization to achieve significant gains. Strategies to improve N uptake and primary N assimilation have advanced to the proof-of-concept phase in a number of plant species and should be readily applied to maize in the near future. However, methods for optimizing N remobilization and N utilization by developing kernels have yet to be defined, but once discovered will have significant impact on increasing maize NUE.

6.4 Commercialization of Maize Hybrids with Improved NUE

Due to the high cost of research and development of biotechnology traits, sufficient commercial potential must be present to warrant significant investment. Like resistance to herbicides and insect pests that have proven their commercial success, improved NUE offers both economic and environmental benefits on a global scale and in nearly all agricultural systems. Capturing value from improved NUE may occur in two ways. Trait value may be defined by the savings achieved from lowering the amount of N required (N_{req}) to obtain a target grain yield, and by increasing the response of grain yield (GY_{NR}) to a given amount of applied N. From a trait deployment perspective, these strategies are not mutually exclusive, but the biological modifications used to achieve them may differ.

These concepts are illustrated by the responses of both grain yield and plant N accumulation to increasing rates of N fertilizer for two maize hybrids, one that was widely grown in the US Corn Belt in 1980 and the second representing a more recent leading hybrid from 2005. Both of these hybrids were grown on the same field site in east central Illinois at their optimal population density during the 2004–2006 growing seasons. Since 1980, N applications have stabilized at approximately

150 kg N/ha, yet average US grain yields have increased approximately 20%, reflecting potential improvement in NUE. Transgenic hybrids with enhanced NUE would be expected to minimally provide similar increases in grain yield per unit N.

The N response curves for these two hybrids are essentially parallel, with the newer hybrid producing higher grain yields at all N rates and a lower N_{req} to achieve maximum grain yield. Importantly, the newer hybrid produces higher grain yield even in the absence of applied fertilizer, which is indicative of greater tolerance to N deficiency stress and would be desirable in low-input production systems. Thus, the 2005 hybrid would be advantageous for NUE in a broad range of scenarios compared to the 1980 hybrid. Interestingly, the higher yields of the 2005 hybrid were obtained without any significant change in the total amount of N accumulated per plant, even without supplemental fertilizer. This observation suggests that enhanced N utilization is the primary physiological basis for NUE improvement in this comparison.

Future improvements in maize NUE via biotechnology approaches must accelerate the incremental breeding gains evident from Fig. 6.3. Three features of the N response curves suggest specific agronomic targets for biotechnology efforts aimed at improving NUE. First, it is apparent that raising GY_0 leads to proportional increases in grain yield at higher N rates, which indicates that transgenes that elevate GY when N is most limiting will be effective for NUE over a broad range of N supply. Second, N uptake per plant has remained constant, which implies that

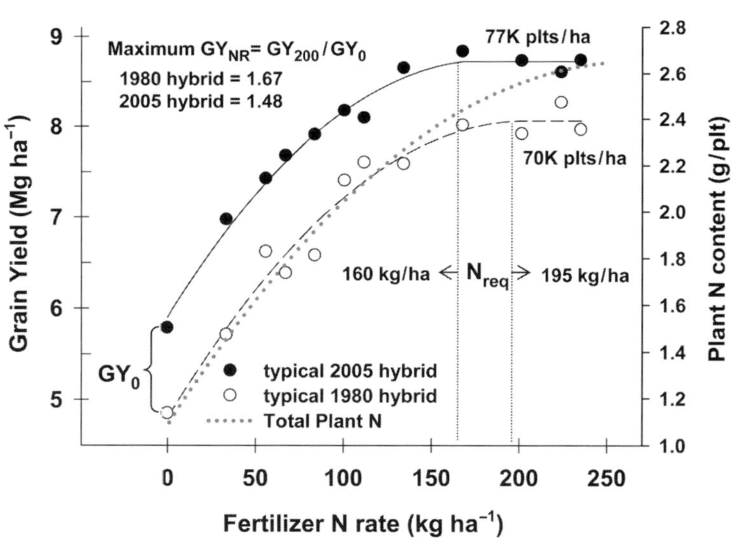

Fig. 6.3 Agronomic parameters that distinguish nitrogen use efficiency in maize hybrids. Mean response curves for grain yield (*GY*) and plant N accumulation are shown for two maize hybrids grown in replicated field trials at their recommended optimum population densities with different amounts of supplemental N, in each of the 2002–2005 seasons. GY_0 Grain yield without supplemental N; GY_{NR} response of grain yield to N; N_{req} amount of supplemental N to achieve maximum GY

improvement of N uptake via breeding approaches may have reached an upper limit, requiring biotechnology strategies to promote greater N uptake efficiency. Finally, the largest increases in NUE will be obtained when a higher GY_0 is coupled with a stronger GY_{NR}, yet GY_{NR} typically decreases as GY_0 increases, as observed in Fig. 6.3. Thus, a significant challenge to any strategy for improving maize NUE must be how to both increase GY_0 and at least maintain and hopefully increase GY_{NR}.

In addition to scientific factors and strategies for capturing value from maize hybrids with improved NUE, commercialization will also be impacted by both the intellectual property landscape surrounding relevant biotechnology innovations and issues that may arise in gaining government regulatory approvals. Recent inspection of issued US patents finds relatively few that both claim and reduce in practice the use of genes for improved NUE-associated traits in plants, with most examples describing enhanced N accumulation in model plant species such as *Arabidopsis* and tobacco. However, the number of pending applications has greatly increased during the past year, suggesting intensifying activity among major agricultural biotechnology companies in this area.

Transgenic approaches to enhanced NUE in maize are not expected to pose any significant new food safety or environmental risks that would require additional regulatory review beyond demonstrating the safety of the genetic elements employed. Many of the current strategies to enhance N uptake are expected to change the amounts and profiles of amino acids accumulated in vegetative and possibly seed tissues. Furthermore, many pathways of secondary metabolism derive from amino acid precursors. As a result, compositional and metabolite profiling studies will need to be performed that compare the magnitude and potential impact of transgene-mediated changes relative to existing natural variation. Germplasm such as Illinois High Protein and its greatly enhanced N accumulation may be valuable in this regard.

In summary, the prospects are bright for the development and commercialization of maize hybrids with transgenic improvements in NUE. Many of the physiological processes associated with NUE are well characterized and genes have been identified that favorably modify target pathways, particularly for N uptake and assimilation. Efforts to identify genes for optimizing N remobilization and utilization for grain yield are also in progress. A greater focus on an integrated approach to understanding NUE is evident in both public- and private-sector research that combines the latest advances in genomics with traditional agronomic and physiological evaluations. Each of the major agricultural biotechnology companies, along with smaller firms such as Arcadia Biosciences, Ceres, and Evogene, have indicated NUE as a high-priority target trait for their research and development programs. Both Monsanto and DuPont/Pioneer indicate promising trait leads that demonstrate proof-of-concept, which could lead to commercialization of enhanced NUE hybrids within the next decade. Though perhaps not as fascinating as the early promises of "N-fixing" maize, the use of biotechnology to improve NUE of maize and other cereal crops will offer dramatic economic and environmental benefits to world agriculture.

References

Agrama HAS, Zakaria AG, Said FB, Tuinstra M (1999) Identification of quantitative trait loci for nitrogen use efficiency in maize. Mol Breed 5:187–195

Ashikari M, Sakakibara H, Lin S, Yamamoto T, Takashi T, Nishimura A, Angeles ER, Qian Q, Kitano H, Matsuoka M (2005) Cytokinin oxidase regulates rice grain production. Science 309:741–745

Below FE (2002) Nitrogen metabolism and crop productivity. In: Pessarakli M (ed) Handbook of plant and crop physiology. Marcel Dekker, New York, pp 385–406

Bertin P, Gallais A (2001) Genetic variation for nitrogen use efficiency in a set of recombinant inbred lines. II. QTL detection and coincidences. Maydica 46:53–68

Bertran FJ, Beck D, Banziger M, Edmeades GO (2003) Genetic analysis of inbred and hybrid grain yield under stress and non-stress environments in tropical maize. Crop Sci 43:807–817

Castleberry RM, Crum CW, Krull CF (1984) Genetic yield improvement of U.S. maize cultivars under varying fertility and climatic environments. Crop Sci 24:33–36

Chiou TJ (2007) The role of microRNAs in sensing nutrient stress. Plant Cell Environ 30: 323–332

Deprost D, Yao L, Sormani R, Moreau M, Leterreux G, Nicolai M, Bedu M, Robaglia C, Meyer C (2007) The *Arabidopsis* TOR kinase links plant growth, yield, stress resistance and mRNA translation. EMBO Reps 8:864–870

Dudley JW, Clark D, Rocheford TR, LeDeaux JR (2007) Genetic analysis of corn kernel chemical composition in the random mated 7 generation of the cross of generations 70 of IHP × ILP. Crop Sci 47:45–57

FAO (2004) FAOSTAT. http://faostat.fao.org/

Gallais A, Coque M (2005) Genetic variation and selection for nitrogen use efficiency in maize: a synthesis. Maydica 50:531–537

Gallais A, Hirel B (2004) An approach to the genetics of nitrogen use efficiency in maize. J Exp Bot 55:295–306

Giller KE, Chalk P, Dobermann A, Hammond L, Heffer P, Ladha JK, Nyamudeza P, Maene L, Ssali H, Freney J (2004) Emerging technologies to increase the efficiency of use of fertilizer nitrogen. In: Mosier AR, Syers KJ, Freney JR (eds) Agriculture and the nitrogen cycle: assessing the impacts of fertilizer use on food production and the environment. Island Press, Washington, pp 35–51

Good A, Depauw M, Kridl J, Theodoris G, Shrawat AK (2007) Nitrogen-efficient monocot plants. US Patent Applic 20070162995

Good AG, Shrawat AK, Muench DG (2004) Can less yield more? Is reducing nutrient input in to the environment compatible with maintaining crop production? Trends Plant Sci 9:597–605

Hirel B, Bertin P, Quillere I, Bourdoncle W, Attagnant C, Dellay C, Gouy A, Cadiou S, Retailliau C, Falque M, Gallais A (2001) Towards a better understanding of the genetic and physiological basis for nitrogen use efficiency in maize. Plant Physiol 125:1258–1270

Hirel B, Andrieu B, Valadier M, Renard S, Quillere I, Chelle M, Pommel B, Fournier C, Drouet J (2005a) Physiology of maize II: identification of physiological markers representative of the nitrogen status of maize (*Zea mays*) leaves during grain filling. Physiol Plant 124: 178–188

Hirel B, Martin A, Terce-Laforgue T, Gonzalez-Moro M, Estavillo J (2005b) Physiology of maize I: a comprehensive and integrated view of nitrogen metabolism in a C4 plant. Physiol Plant 124:167–177

Hirel B, Le Gouis J, Ney B, Gallais A (2007) The challenge of improving nitrogen use efficiency in crop plants: towards a more central role for genetic variability and quantitative genetics within integrated approaches. J Exp Bot 58:2368–2387

Lam HM, Wong P, Chan HK, Yam KM, Che L, Chow CM, Coruzzi GM (2003) Overexpression of the ASN1 gene enhances nitrogen status in seeds of Arabidopsis. Plant Physiol 132: 926–935

Lian X, Wang S, Zhang J, Feng Q, Zhang L, Fan D, Li X, Yuan D, Han B, Zhang Q (2006) Expression profiles of 10,422 genes at early stage of low nitrogen stress in rice assayed using a cDNA microarray. Plant Mol Biol 60:617–631

Martin A, Lee J, Kichey T, Gerentes D, Zivy M, Tatout C, Dubois F, Balliau T, Valot B, Davanture M, Tercé-Laforgue T, Quilleré I, Coque M, Gallais A, Gonzalez-Moro M-B, Bethencourt L, Habash DZ, Lea PJ, Charcosset A, Perez P, Murigneux A, Sakakibara H, Edwards KL, Hirel B (2006) Two cytosolic glutamine synthetase isoforms of maize are specifically involved in the control of grain production. Plant Cell 18:3252–3274

Moose SP, Dudley JW, Rocheford TR (2004) Maize selection passes the century mark: a unique resource for 21st century genomics. Trends Plant Sci 9:358–364

Palenchar PM, Kouranov A, Lejay LV, Coruzzi GM (2004) Genome-wide patterns of carbon and nitrogen regulation of gene expression validate the combined carbon and nitrogen (CN)-signaling hypothesis in plants. Genome Biol 5:R91

Purcino AAC, Arellano C, Athwal GS, Huber SC (1998) Nitrate effect on carbon and nitrogen assimilating enzymes of maize hybrids representing seven eras of maize breeding. Maydica 43:83–94

Raun WR, Johnson GV (1999) Improving nitrogen use for cereal production. Agron J 91:357–363

Robson PR, Donnison IS, Wang K, Frame B, Pegg SE, Thomas A, Thomas H (2004) Leaf senescence is delayed in maize expressing the *Agrobacterium IPT* gene under the control of a novel maize senescence-enhanced promoter. Plant Biotech J 2:101–112

Sakakibara H, Takei K, Hirose N (2006) Interactions between nitrogen and cytokinin in the regulation of metabolism and development. Trends Plant Sci 11:440–448

Scheible WR, Morcuende R, Czechowski T, Fritz C, Osuna D, Palacios-Rojas N, Schidelasch D, Thimm O, Udvardi MK, Stitt M (2004) Genome-wide reprogramming of primary and secondary metabolism, protein synthesis, cellular growth processes, and the regulatory infrastructure of Arabidopsis in response to nitrogen. Plant Physiol 136:2483–2499

Seebauer JR, Moose SP, Fabbri BJ, Crossland LD, Below FE (2004) Amino acid metabolism in maize earshoots: implications for assimilate preconditioning and nitrogen signaling. Plant Physiol 136:4326–4334

Tabuchi M, Abiko T, Yamaya T (2007) Assimilation of ammonium ions and reutilization of nitrogen in rice (*Oryza sativa* L.). J Exp Bot 58:2319–2327

Tilman D, Cassman KG, Matson PA, Naylor R, Polasky S (2002) Agricultural sustainability and intensive production practices. Nature 418:671–677

Uribelarrea M, Below FE, Moose SP (2004) Grain composition and productivity of maize hybrids derived from the Illinois Protein Strains in response to variable nitrogen supply. Crop Sci 44:1593–1600

Uribelarrea M, Moose SP, Below FE (2007) Divergent selection for grain protein affects nitrogen use in maize hybrids. Field Crops Res 100:82–90

Wang R, Okamoto M, Xing X, Crawford NM (2003) Microarray analysis of the nitrate response in Arabidopsis roots and shoots reveals over 1,000 rapidly responding genes and new linkages to glucose, trehalose-6-phosphate, iron and sulfate metabolism. Plant Physiol 132: 556–567

Worku M, Banziger M, Schulte auf'm Erley, G., Friesen D, Diallo AO, Horst WJ (2007) Nitrogen uptake and utilization in contrasting nitrogen efficient tropical maize hybrids. Crop Sci 47: 519–528

Chapter 7
Enhancement of Amino Acid Availability in Corn Grain

Alan L. Kriz

7.1 Introduction

As modern corn hybrids were bred for higher yields, the composition of the grain has inadvertently trended to higher starch content at the expense of protein (Scott et al. 2006). Moreover since corn grain protein is deficient in certain nutritionally essential amino acids, this reduction in grain protein level has further reduced the nutritional quality of the grain. One approach to address this problem is to increase the nutritional quality of corn grain protein, particularly by enhancing the content of essential amino acids, such as lysine and tryptophan.

The most limiting amino acid in corn grain, with respect to the dietary needs of monogastric animals, is lysine. Therefore, enhancement of lysine content is a primary target for improving grain quality. The poor nutritional quality of corn protein is mostly caused by the amino acid composition of endosperm proteins. Corn protein has a lysine content of 2.7%, which is well below the recommendation by FAO (FAO/WHO/UNU 1985) for human nutrition. Although the germ protein has an adequate lysine content (5.4%) in whole grain, this is diluted by the much more abundant endosperm proteins, which have an average lysine content of only about 1.9%. This is because 60–70% of endosperm protein consists of zeins, which contain few or no lysine residues (Coleman and Larkins 1999). Similarly, the absence of tryptophan residues in zein proteins is the reason for the low tryptophan content of corn protein. Therefore, modification of the grain protein profile through approaches such as zein reduction and expression of lysine-rich proteins could significantly improve the balance of amino acids. Alternatively, the lysine content of the grain could be increased by elevating the level of free lysine in the kernel.

Alan L. Kriz
BASF Plant Science, LLC, 26 Davis Drive, Research Triangle Park, NC 27709, USA
e-mail: alkriz@comcast.net

A.L. Kriz, B.A. Larkins (eds.), *Molecular Genetic Approaches to Maize Improvement* 79
Biotechnology in Agriculture and Forestry, Vol. 63

7.2 Increased Lysine Accumulation Through Deregulation of Metabolic Pathways

Enhancement of seed lysine content through deregulation of lysine anabolic and catabolic pathways has been extensively reviewed (e.g., Galili 2004; Falco et al. 1995; Stepansky et al. 2006; Azevedo et al. 2006) and will not be discussed in detail here. Two of the key enzymes involved in lysine accumulation are dihydrodipicolinate synthase (DHDPS) and the bifunctional enzyme lysine-ketoglutarate reductase/saccharopine dehydrogenase (LKR/SDH). DHDPS catalyzes the condensation of β-aspartic semialdehyde with pyruvate, which is the first committed step in lysine biosynthesis. In plants, DHDPS activity is regulated by lysine levels through the mechanism of feedback inhibition. At the other end of the lysine pathway, LKR/SDH degrades lysine in a two-step reaction: the first is reduction of lysine and α-ketoglutarate to saccharopine (the LKR reaction), and the second step is conversion of saccharopine to glutamate and α-aminoadipic semialdehyde (the SDH reaction) (Kemper et al. 1998; Stepansky et al. 2006).

Increases in corn grain lysine content have been achieved through engineering of both lysine anabolism and lysine catabolism. Perhaps not surprisingly, the anabolic approaches have met with the greatest success. Huang et al. (2005) described deregulation of lysine biosynthesis by expression of a lysine-insensitive version of DHDPS from *Corynebacterium glutamicum* – referred to as CordapA – under control of the corn germ-preferred *Globulin1* promoter (Belanger and Kriz 1991). Hemizygous grain generated from a transgenic corn line expressing this recombinant gene construct exhibited a 40-fold increase in free lysine content, from 43 to 1838 ppm; this translated to a total grain lysine content of 0.43%, in comparison to the control grain which contained 0.26% lysine (Huang et al. 2005). A construct similar to that employed in the above study was used to generate the high-lysine transgenic corn event LY038 (Lucas et al. 2004), which received a determination of nonregulated status by the Animal and Plant Health Inspection Service, United States Department of Agriculture (Federal Register 71:5801–5802, 2006).

Manipulation of lysine catabolism also results in an increase in free lysine in corn grain. Houmard et al. (2007) used an RNA interference (RNAi) strategy designed to suppress expression of the LKR/SDH gene in corn endosperm. Grain from plants homozygous for the LKR/SDH RNAi transgene exhibited up to a 20-fold increase in free lysine content, relative to transgene-negative controls. While these increases in free lysine content are lower than those observed for germ-targeted expression of CordapA, they provide important insights into mechanisms of lysine accumulation in plant seeds. The finding that suppression of lysine catabolism in the endosperm is sufficient to allow for accumulation of significant levels of lysine in the grain is consistent with earlier work that pointed to the importance of lysine degradation in the endosperm with respect to regulation of lysine levels in cereal grains (Arruda et al. 2000; Azevedo et al. 2004). Furthermore, in contrast to the situation observed for CordapA in the corn germ, expression of a lysine-insensitive DHDPS in corn endosperm does not appear to result in lysine accumulation in the grain (Mazur et al. 1999). Taken together, these results indicate that the primary mechanism by which lysine levels are regulated in the corn endosperm is catabolic, while in the germ such regulation likely occurs by anabolic mechanisms.

7.3 Modification of Corn Grain Protein Profiles

7.3.1 Distribution of Proteins in the Corn Grain

To provide an understanding of the nature of the protein profile of corn grain, it is appropriate to first discuss the general distribution of various types of proteins in the seed. The basis of seed protein classification dates to the work of T.B. Osborne, who fractionated proteins from a variety of seeds on the basis of differential solubility (Osborne 1908). The basic scheme involved sequential extraction of seed meal with the following: water, yielding albumins; 10% NaCl, yielding globulins; 70% ethanol, yielding prolamins; and dilute acid or alkali, yielding glutelins. Fractionation of whole-grain corn meal by the Osborne solubility scheme shows that the protein is comprised of 40% prolamins (zeins), 30% glutelins, and 20% albumins plu globulins (Wall and Paulis 1978).

Zeins are by far the best characterized of the corn grain protein groups, and they can be further fractionated into the distinct classes of α (19 and 22 kD), β (15 kD), γ (16, 27 and 50 kD), and δ (10 and 18 kD) (reviewed by Holding and Larkins, this volume). All zein classes are localized in the endosperm of the grain. The globulins are the next best characterized group of corn grain proteins, and this fraction is primarily comprised of polypeptides of 60–70 kD (encoded by the *Glb1* gene) and of 45 and 27 kD (encoded by the *Glb2* gene) (Kriz 1989; Kriz and Wallace 1991). Globulins are primarily localized in the germ portion of the grain.

7.3.2 Zein Reduction

Since the first report of the effect of the *opaque2* mutation on corn grain lysine content (Mertz et al. 1964), many studies have indicated that a reduction in zein content results in a redistribution of nitrogen in the endosperm to other protein fractions, as well as the accumulation of free amino acids. This is generally attributed to the importance of α-zein proteins as a nitrogen sink in the grain – if that sink is not available to the kernel (as is the case in zein-reduction mutants), available nitrogen will be assimilated into other proteins in the grain. As described below, evidence exists for an increase of non-zein proteins in both the endosperm and the germ as a response to zein reduction. Since these alternative N-sink proteins contain at least some amount of lysine, the effective lysine content of the grain is elevated. Commercialization of *opaque2* corn hybrids has met with limited success, primarily due to reduced yield and grain quality issues associated with the softer kernel phenotype (Holding and Larkins, this volume). Targeted genetic engineering strategies could provide an effective means to modify the protein profile of corn grain as a means of enhancing its nutritional quality in a manner that is amenable to US agricultural practices.

7.3.2.1 Effect of *opaque2* Homozygosity on Protein Fractions in Corn Grain

A good deal of work has focused on total protein content in endosperm of both normal (N) and *opaque2* kernels, and most of these show little difference in protein content in the two kernel types. However, almost all of these studies reported data only on endosperm, with no data provided on whole grain protein content, and these data were almost exclusively obtained from analysis of inbred lines. Some examples include Mertz et al. (1964), who reported 8.69% endosperm protein in both normal (N) and *opaque2* kernels; Tsai et al. (1978), who reported 9.8% endosperm protein in normal, 10.9% in *opaque2*; and 8.4% total grain protein in normal, 8.9% in *opaque2*. Some studies did show a reduction in total endosperm protein content: 11.8% in N, 10.1% in *opaque2*, in the inbred line Oh43; 8.5% in N, 7.3% in the W22 *opaque2* inbred line (Misra et al. 1972); and 12.7% in N, 11.1% in *opaque2* (Nelson et al. 1965). The largest reduction in endosperm protein content in *opaque2* kernels relative to normal that has been reported is 19% (Gentinetta et al. 1975). This variation may be due to an effect of environment on total grain protein content or the genotypic differences of the inbred lines.

Gupta et al. (1975) surveyed both inbreds and hybrids made with *opaque2* conversions for total grain protein content (endosperm plus germ). Some materials showed an increase in grain protein in *opaque2* versions, some showed a decrease, and some no change; mean values showed that there was no significant difference in protein content due to *opaque2* homozygosity in hybrids (in inbreds, mean protein content was 11.72% grain protein for N, 12.66% in *opaque2*; in hybrids the means were 10.85% for N, 9.94% for *opaque2*). Hadjinov et al. (1972) surveyed six *opaque2* conversions (BC4), and found that on average the *opaque2* versions contained 95% of total grain protein in comparison to normal versions of those inbreds.

Clearly, these studies imply there is a redistribution of nitrogen in the grain from zeins to different protein fractions in zein-reduction mutants, such as *opaque2*. Levels of globulins and albumins are elevated in *opaque2* kernels (Dierks-Ventling 1981, 1983; Sodek and Wilson 1971; Misra et al. 1972; Landry et al. 2002; Landry and Moreaux 1982; Nelson 1969). Both Dierks-Ventling (1981) and Puckett and Kriz (1991) demonstrated by SDS-PAGE the elevation of Glb1-encoded proteins in *opaque2* kernels over N kernels.

It should also be noted that in the initial description of modification of *opaque2* kernels to a vitreous phenotype (albeit a rather crude experiment that involved the splitting of kernels into opaque and vitreous sectors), the authors concluded that "...the differences observed <in lysine content> on the whole kernel basis were due to germ tissue that was almost entirely a part of the opaque fraction" (Paez et al. 1969). Although it has since been demonstrated that differences in endosperm protein and endosperm free amino acids (e.g. Wang and Larkins 2001; Sodek and Wilson 1971) certainly contribute to the overall redistribution of nitrogen content in *opaque2* kernels, such redistribution is also accounted for by an increase in the levels of germ proteins (Puckett and Kriz 1991).

7.3.2.2 Zein Reduction by Transgenic Means

Huang et al. (2004) expressed a 19-kD α-zein construct in both sense and antisense orientation in transgenic corn plants. These plants exhibited an opaque kernel phenotype very similar to that of *opaque2*. Biochemical analysis of the grain proteins by SDS-PAGE and mass spectrometry (MALDI-TOF) indicated that the opaque kernels exhibited a reduction in 19-kD α-zeins. No difference was observed in total grain protein content in bulked kernels from transgenic and wild-type hybrids. Amino acid analysis demonstrated that bulked kernels from transgenic hybrids contained significantly ($p < 0.001$) higher levels of total lysine, methionine, tryptophan, and aspartate, and lower levels of proline and leucine, in comparison to wild-type hybrids.

In a similar study, Segal et al. (2003) used a 22-kD α- zein RNAi construct to generate transgenic maize plants. They observed an opaque kernel phenotype, a reduction in 22-kD α-zein proteins relative to the untransformed control, an increase in grain lysine, glycine, and aspartate plus asparagine content, and a decrease in leucine and alanine content. No data were provided with respect to total grain protein content.

In an extension of the studies described above, Huang et al. (2006) used chimeric double-stranded RNA constructs to target both the 19-kD and 22-kD α-zein sequences. Ears were generated from homozygous transgenic plants expressing these constructs. Kernels from these ears exhibited an opaque phenotype, a reduction in both 19-kD and 22-kD α-zeins, an increase in total grain lysine, tryptophan, arginine, histidine, and glycine content, and a reduction in total grain leucine content. Although variation in total grain protein content was observed in the transgenic ears, no significant reduction in grain protein was observed in kernels from homozygous transgenic ears in comparison to kernels from corresponding wild-type ears. Moreover, the zein-reduced kernels exhibited a lysine/protein ratio of 7.23%, compared to a 2.83% lysine/protein ratio in wild-type kernels. The authors concluded that the lysine increase in zein-reduced kernels results from accumulation of lysine-containing non-zein proteins. Since the zein-reduced kernels also contained higher levels of free amino acids than the wild-type kernels, the authors suggest that inhibition of zein synthesis during kernel development resulted in an increase in amino acid pools which subsequently drives the synthesis of non-zein proteins, so that other proteins serve as a nitrogen sink in the absence of the "normal" α-zein sink.

As described above, two approaches have been used to modify the overall amino acid composition of corn grain: zein-reduction, which appears to shift protein accumulation from α-zeins to other protein fractions, and deregulation of amino acid biosynthesis as a means to over-produce a specific amino acid. To determine whether there were any synergies between these two different approaches, Huang et al. (2005) used the original 19-kD α-zein reduction events (ZR) from Huang et al. (2004) and generated F1 seed by crossing ZR plants with transgenic plants that exhibited deregulation of the lysine biosynthetic pathway through expression of a *Corynebacterium* DHDPS (CordapA) transgene in the germ. In this experiment, progeny kernels were uniformly hemizygous for both the ZR and the CordapA transgenes. When comparing kernels from ZR + CordapA plants with those from CordapA plants alone, they observed an increase in CordapA expression in the presence of the ZR construct over that observed in plants containing CordapA alone. This is likely due to the response of the Glb1 promoter, which was used to drive CordapA expression, to the reduced zein phenotype, since upregulation of the native Glb1 gene was observed during development of *opaque2* kernels (Puckett and Kriz 1991).

7.3.2.3 Improved Protein Digestibility and Amino Acid Uptake by Zein Reduction

The biological utilization of a protein depends not only on its absolute amino acid composition, but also on its digestibility. The structure of the protein can influence the availability of the amino acids it contains. Zeins, which are hydrophobic in nature and form water-insoluble protein bodies, have been shown to be the major pepsin- and trypsin/chymotrysin-indigestible proteins of uncooked and cooked maize flour (Hamaker et al. 1987). Therefore, it is postulated that corn grain with a reduced zein content and an increase in non-zein proteins, such as occurs with the *opaque2* mutant, has improved protein digestibility.

Certain cereal proteins are also characterized by excessive concentrations of some amino acids, which can affect protein utilization through a condition known as amino acid antagonism (Harper et al. 1970). In the case of corn where an excess of leucine is found, the utilization of isoleucine and valine is depressed (Harper et al. 1955). This is because the high leucine concentration stimulates the degradation of branched-chain amino acids, including isoleucine and valine, which makes them unavailable for protein synthesis (May et al. 1991). In addition to being low in lysine and tryptophan, α-zeins contain up to 19.5% (mole percent) of leucine (Coleman and Larkins 1999). Therefore, it is not surprising that in α-zein-reduced kernels, besides the increase in lysine and tryptophan, a significant reduction in leucine is also observed (Huang et al. 2006).

In 1992, the World Food Organization of the United Nations (FAO) published a summary of collective studies on maize for human nutrition (FAO 1992). It also included several studies on *opaque2* maize and QPM (*opaque2*-derived maize or Quality Protein Maize; see Sect. 7.3.2.4). The results suggest their potential contribution towards improving the diet of maize-eating populations. For example, when the nutritional value of maize protein and other cereal proteins was compared, common maize had a relatively low protein quality (32.1% casein), similar to other cereals, except rice. However, both *opaque2* maize and QPM had a protein quality (96.8% and 82.1% casein, respectively) that is far superior to common maize and other cereal grains. In another example, the protein quality of common maize, *opaque2* maize, and QPM was evaluated for protein digestibility, net nitrogen utilization, biological value, and nitrogen source retention for children fed with these different types of maize. The casein values of *opaque2* maize and QPM were slightly lower, but significantly higher than the values for common maize.

In developed countries, corn gain is mainly used as animal feed, and improving its protein quality could have considerable commercial impact with regard to monogastric livestock and poultry (Johnson et al. 2001). Swine feeding studies have shown that QPM diets resulted in better performance than regular corn diets in terms of weight gain (Sullivan et al. 1989) and reduction in the requirement of soybean meal supplementation (Burgoon et al. 1992).

7.3.2.4 Zein Reduction and Kernel Hardness

Whether originating from natural mutations or by transgenic means, there is little doubt that corn grain with reduced zein content has a better nutritional value, but this leads to an opaque kernel phenotype. In contrast to normal vitreous kernels, opaque kernels are soft and starchy, which makes them prone to damage during grain handling and storage. To overcome these shortcomings, corn breeders have successfully restored normal endosperm texture to *opaque2* mutants, making them hard and translucent. The resulting QPM varieties are comparable with wild-type, but have a higher lysine content (Prasanna et al. 2001). However, creating QPM lines is of limited use in large-scale commercial agricultural practices in developed countries. This is mainly due to the relatively large number of *opaque2* modifier genes required to achieve a hard endosperm phenotype, and, concomitantly, to the difficulty of integrating these modifiers along with the recessive *opaque2* mutant allele into commercial breeding programs. Analysis of modified *opaque2* lines has, nonetheless, yielded important information for potential modification of opaque endosperm through genetic engineering.

The level of 27-kD γ-zein has been implicated in the opaque endosperm modification. Its accumulation was found to correlate directly with modifier gene dosage (Geetha et al. 1991; Lopes and Larkins 1991), and genetic mapping of *opaque2* modifiers revealed a linkage between the gene encoding the 27-kD γ-zein and an *opaque2* modifier QTL (Lopes et al. 1995). In addition to the increased level of 27-kD γ-zein, altered starch structure in *mo2* (modified *opaque2*, derived from QPM) was also observed (Gibbon et al. 2003). Through proteomic and sequence analyses between *opaque2* and *mo2*, granule-bound starch synthase I and starch synthase IIb were identified as other possible factors affected by *opaque2* modifiers (Gibbon and Larkins 2005).

7.4 Expression of Lysine-rich Proteins in Corn Grain

Seed-specific expression of genetically engineered proteins with a high lysine concentration has been explored to increase the lysine content of corn grain. In principle, the higher the lysine concentration in the transgenic protein, the less required to achieve a certain percentage of lysine increase.

A number of genes encoding naturally occurring proteins have been used directly to increase the lysine content in corn kernels. Milk proteins are attractive choices because of their balanced amino acid profiles and excellent digestibility. Expression of a codon-optimized α-lactalbumin fused with a zein signal sequence or a zein signal sequence and an endoplasmic reticulum (ER) retention motif achieved lysine levels of 0.01–0.05% of kernel weight using the maize *Ubi-1* promoter (Yang et al. 2002). Similarly, α-lactalbumin driven by the maize 27-kD γ-zein promoter was found to accumulate to 0.003–0.095% of kernel weight and increased kernel lysine content by 29–47% (Bicar et al. 2008). It is unclear whether the increased lysine was directly contributed by α-lactalbumin, since the lysine increase outweighed the amount of α-lactalbumin accumulated. A high lysine pollen-specific protein, sb401, from potato has also been expressed in corn endosperm using a 19-kD α-zein promoter (Yu et al. 2004). This resulted in an increase in lysine (16.1–54.8%) and total protein content (11.6–39%) in R1 transgenic seeds. It is likely that sb401 did not account for most of the increased protein content. A 1% accumulation of sb401, which contains 19% lysine (w/w), in corn kernels (10% total protein), would add 1900 ppm of lysine to the grain. This is a significant lysine increase, considering that the average lysine content in inbred kernels is only 2500 ppm.

To significantly impact corn lysine content, the quantity of transgenic protein accumulated is as important as its quality. High levels of endosperm expression by zein promoters, along with modifications of the transgene using maize-preferred codons or fusing with a zein signal peptide and ER-retention sequences, have been employed to maximize the synthesis and accumulation of transgenic proteins in corn endosperm, as described in the previous examples. The addition of a prolamin mRNA 3' untranslated region (UTR) could also enhance the accumulation of transgenic proteins by directing the mRNA to protein body ER (Hamada et al. 2003). Alternatively, transgenic proteins could be anchored to starch granules by fusing with the starch-binding domain from a *Bacillus circulans* protein (Ji et al. 2003).

Expression of modified, lysine-rich zeins has several advantages when combined with zein reduction. First, it provides a nitrogen sink for excess amino acids in zein-reduced kernels. Second, in theory the modified zeins could be targeted to endosperm protein bodies for high levels of accumulation. Finally, restoration of protein bodies could improve the hardness of zein-reduced kernels. Two classes of zeins have been investigated for the possibility of high-lysine modifications. α-Zeins are known to be the most significantly reduced in *opaque2* kernels (Coleman and Larkins, 1999), and therefore are candidates for lysine enhancement. A genetically engineered lysine-containing 19-kD α-zein was successfully synthesized and assembled into structures similar to maize protein bodies in *Xenopus* oocytes (Wallace et al. 1988), and specific domains within a 22-kD α-zein that bound preferentially to other zeins were identified in the yeast two-hybrid system (Kim et al. 2002). These observations suggest the possibility of engineering high-lysine α-zeins by domain swapping. Similarly, other studies have shown that the 27-kD γ-zein can have lysine-rich sequences inserted and can accumulate in protein bodies of transiently transformed maize endosperms (Torrent et al. 1997). As mentioned above, the 27-kD γ-zein is an *opaque2* modifier candidate.

Recently, a novel approach to enrich the lysine content of corn grain by endosperm-specific expression of an *Arabidopsis* lysyl tRNA synthetase (KRS) was reported (Wu et al. 2007). Excess KRS in transgenic endosperm promoted incorrect acylation of tRNAs with lysine, and caused translational incorporation of lysine into zeins at nonsense codons. Endosperm proteins of the transgenic kernels had up to a 26% increase in lysine content. However, the practical applications of this approach are a cause of greater regulatory concern. It randomly alters endosperm proteins by replacing some of their amino acids with lysine, resulting in novel proteins that could have unpredictable consequences.

References

Arruda P, Kemper EL, Papes F, Leite A (2000) Regulation of lysine catabolism in higher plants. Trends Plant Sci 5:324–330

Azevedo RA, Damerval C, Lea PJ, Landry J, Bellato CM, Meinhardt LW, Le Guilluox M, Delhaye S, Toro AA, Gaziola SA, Varisi VA, Gratao PL (2004) Endosperm protein synthesis and lysine metabolism in distinct opaque maize seed mutants. Funct Plant Biol 31:339–348

Azevedo RA, Lancien M, Lea PJ (2006) The aspartic acid metabolic pathway, an exciting and essential pathway in plants. Amino Acids 30:143–162

Belanger FC, Kriz AL (1991) Molecular basis for allelic polymorphism of the maize Globulin-1 gene. Genetics 129:863–872

Bicar EH, Woodman-Clikeman W, Sangtong V, Peterson JM, Yang SS, Lee M, Scott MP (2008) Transgenic maize endosperm containing a milk protein has improved amino acid balance. Transgenic Res 17:59–71

Burgoon KG, Hansen JA, Knabe DA, Bockholt AJ (1992) Nutritional value of quality protein maize for starter and finisher swine. J Anim Sci 70:811–817

Coleman CE, Larkins BA (1999) The prolamins of maize. In: Shewry PR, Casey R (eds) Seed proteins. Kluwer, Dordrecht, pp 109–139

Dierks-Ventling C (1981) Storage proteins in *Zea mays* (L.): interrelationship of albumins, globulins and zeins in the *opaque-2* mutation. Eur J Biochem 120:177–182

Dierks-Ventling C (1983) Lysine biosynthesis and utilization during seed development of normal and *opaque-2 Zea mays* L. Planta 157:233–238

Falco SC, Guida T, Locke M, Mauvais J, Sanders C, Ward RT, Webber P (1995) Transgenic canola and soybean seeds with increased lysine. Biotechnology 13:577–582

FAO (1992) Maize in human nutrition. FAO, Rome

FAO/WHO/UNU (1985) Energy and protein requirements. Report of a joint FAO/WHO/UNU expert consultation. WHO, Geneva

Galili G (2004) New insights into the regulation and functional significance of lysine metabolism in plants. Ann Rev Plant Physiol Plant Mol Biol 53:27–43

Geetha KB, Lending CR, Lopes MA, Wallace JC, Larkins BA (1991) *Opaque-2* modifiers increase γ-zein synthesis and alter its spatial distribution in maize endosperm. Plant Cell 3: 1207–1219

Gentinetta E, Maggiore T, Salamini F, Lorenzoni C, Piole F, Soave C (1975) Protein studies in 46 *opaque-2* strains with modified endosperm texture. Maydica 20:145–164

Gibbon BC, Larkins BA (2005) Molecular genetic approaches to developing quality protein maize. Trends Genet 21:227–233

Gibbon BC, Wang X, Larkins BA (2003) Altered starch structure is associated with endosperm modification in Quality Protein Maize. Proc Natl Acad Sci USA 100:15329–15334

Gupta D, Kovacs I, Gaspar L (1975) Protein quality traits and their relationship with yield and yield components of *opaque-2* and analogous normal maize hybrids and inbred lines. Theor Appl Genet 45:341–348

Hadjinov MI, Zima KI, Normov AA (1972) Changes in weight, protein and lysine content in *opaque-2* kernels of corn during backcrossing. Maize Genet Coop Newsl 46:101–104

Hamada S, Ishiyama K, Sakulsingharoj C, Choi SB Wu Y, Wang C, Singh S, Kawai N, Messing J, Okita TW (2003) Dual regulated RNA transport pathways to the cortical region in developing rice endosperm. Plant Cell 15:2265–2272

Hamaker BR, Kirleis AW, Butler LG, Axtell JD, Mertz ET (1987) Improving the in vitro digestibility of sorghum with reducing agents. Proc Natl Acad Sci USA 84:626–628

Harper AE, Benton DA, Elvehjem CA (1955) L-leucine, an isoleucine antagonist in the rat. Arch Biochem Biophys 57:1–12

Harper AE, Benevenga NJ, Wohlhueter RM (1970) Effects of ingestion of disproportionate amounts of amino acids. Physiol Rev 50:428–558

Houmard NM, Mainville, JL, Bonin CP, Huang S, Luethy MH, Malvar TM (2007) High lysine corn generated by endosperm specific suppression of lysine catabolism using RNAi. Plant Biotechnol J 5(5):605–614

Huang S, Adams WR, Zhou Q, Malloy KP, Voyles DA, Anthony J, Kriz AL, Luethy MH (2004) Improving nutritional quality of maize proteins by expressing sense and antisense zein genes. J Agric Food Chem 52(7):1958–1964

Huang S, Kruger DE, Frizzi A, D'Ordine RL, Florida CA, Adams WR, Brown WE, Luethy MH (2005) High-lysine corn produced by the combination of enhanced lysine biosynthesis and reduced zein accumulation. Plant Biotechnol J 3:555–569

Huang S, Frizzi A, Florida CA, Kruger DE, Luethy MH (2006) High lysine and high tryptophan transgenic maize resulting from the reduction of both 19- and 22-kD α-zeins. Plant Mol Biol 61:525–535

Ji Q, Vincken JP, Luc CJM, Suurs LCJM, Visser RGF (2003) Microbial starch-binding domains as a tool for targeting proteins to granules during starch biosynthesis. Plant Mol Biol 51:789–801

Johnson LA, Hardy CL, Baumel CP, Yu T-H, Sell JL (2001) Identifying valuable corn quality traits for livestock feed. Iowa State University, Ames

Kemper EL, Cord-Neto G, Capella AN, Goncalves-Butruile M, Azevedo RA, Arruda P (1998) Structure and regulation of the bifunctional enzyme lysine-oxoglutarate reductase-saccharopine dehydrogenase in maize. Eur J Biochem 253:720–729

Kim CS, Woo Y, Clore AM, Burnett RJ, Carneior NP, Larkins BA (2002) Zein protein interactions, rather than the asymmetric distribution of zein mRNAs on endoplasmic reticulum membranes, influence protein body formation in maize endosperm. Plant Cell 14:655–672

Kriz AL (1989) Characterization of embryo proteins encoded by the maize Glb genes. Biochem Genet 27:239–251

Kriz AL, Wallace NH (1991) Characterization of the maize *Globulin-2* gene and analysis of two null alleles. Biochem Genet 29:241–254

Landry J, Moreaux T (1982) Distribution and amino acid composition of protein fractions in *opaque-2* maize grains. Biochemistry 21:1865–1869

Landry J, Delhaye S, Damerval C (2002) Effect of the *opaque-2* gene on accumulation of protein fractions in maize endosperm. Maydica 47:59–66

Lopes MA, Larkins BA (1991) Gamma-zein content is related to endosperm modification in quality protein maize. Crop Sci 31:1655–1662

Lopes MA, Takasaki K, Bostwick DE, Helentjaris T, Larkins BA (1995) Identification of two *opaque2* modifier loci in Quality Protein Maize. Mol Gen Genet 247:603–613

Lucas DM, Glenn KC, Bu J-Y (2004) Petition for determination of nonregulated status for lysine maize LY038. http://www.aphis.usda.gov/brs/aphisdocs/04 22901p.pdf

May RC, Piepenbrock N, Kelly RA, Mitch WE (1991) Leucine-induced amino acid antagonism in rats: muscle valine metabolism and growth impairment. J Nutr 121:293–301

Mazur B, Krebbers E, Tingey S (1999) Gene discovery and product development for grain quality traits. Science 285:372–375

Mertz ET, Bates LS, Nelson OE (1964) Mutant gene that changes protein composition and increases lysine content of maize endosperm. Science 145:279–280

Misra PS, Jambunathan R, Mertz ET, Glover DV, Barbosa HM, McWhirter KS (1972) Endosperm synthesis in maize mutants with increased lysine content. Science 176:1425–1427

Nelson OE (1969) Genetic modification of protein quality in plants. Adv Agron 21:171–174

Nelson OE, Mertz ET, Bates LS (1965) Second mutant gene affecting the amino acid pattern of maize endosperm proteins. Science 150:1469–1470

Osborne TB (1908) Our present knowledge of plant proteins. Science 28:417–427

Paez AV, Helm JL, Zuber MS (1969) Lysine content of *opaque-2* maize kernels having different phenotypes. Crop Sci 9:251–252

Prasanna BM, Vasal SK, Kassahun B, Singh NN (2001) Quality protein maize. Curr Sci 81:1308–1319

Puckett JL, Kriz AL (1991) Globulin gene expression in *opaque-2* and *floury-2* mutant maize embryos. Maydica 36:161–167

Scott MP, Edwards JW, Bell CP, Schussler JR, Smith JS (2006) Grain composition and amino acid content in maize cultivars representing 80 years of commercial maize varieties. Maydica 51:417–423

Segal G, Song R, Messing J (2003) A new opaque variant of maize by a single dominant RNA-interference-inducing transgene. Genetics 165:387–397

Sodek L, Wilson CM (1971) Amino acid composition of proteins isolated from normal, *opaque-2*, and *floury-2* corn endosperms by a modified Osborne procedure. J Agric Food Chem 19:1144–1150

Stepansky A, Less H, Angelovici R, Aharon R, Zhu X, Galili G (2006) Lysine catabolism, an effective versatile regulator of lysine level in plants. Amino Acids 30:121–125

Sullivan JS, Knabe DA, Bockholt AJ, Gregg EJ (1989) Nutritional value of quality protein maize and food corn for starter and growth pigs. J Anim Sci 67:1285–1292

Torrent M, Alvarez I, Geli MI, Dalcol I, Ludevid D (1997) Lysine-rich modified γ-zeins accumulate in protein bodies of transiently transformed maize endosperms. Plant Mol Biol 34:139–149

Tsai CY, Larkins BA, Glover DV (1978) Interaction of the *opaque-2* gene with starch-forming mutant genes on the synthesis of zein in maize endosperm. Biochem Genet 16:883–896

Wall JS, Paulis JW (1978) Corn and sorghum grain proteins. Adv Cereal Sci Technol 2:135–219

Wallace JC, Galili G, Kawata EE, Cuellar RE, Shotwell MA, Larkins BA (1988) Aggregation of lysine-containing zeins into protein bodies in *Xenopus* oocytes. Science 240:662–664

Wang X, Larkins BA (2001) Genetic analysis of amino acid accumulation in *opaque-2* maize endosperm. Plant Physiol 125:1766–1777

Wu XR, Kenzior A, Willmot D, Scanlon S, Chen Z, Topin A, He SH, Acevedo A, Folk WR (2007) Altered expression of plant lysyl tRNA synthetase promotes tRNA misacylation and translational recoding of lysine. Plant J 50(4):627–636. doi: 10.111/j.1365-313X.2007.03076.x

Yang SH, Moran DL, Jia HW, Bicar EH, Lee M, Scott MP (2002) Expression of a synthetic porcine α-lactalbumin gene in the kernels of transgenic maize. Transgenic Res 11:11–20

Yu J, Peng P, Zhang X, Zhao Q, Zhy D, Sun X, Liu J, Ao G (2004) Seed-specific expression of a lysine rich protein *sb401* gene significantly increases both lysine and total protein content in maize seeds. Mol Breed 14:1–7

Chapter 8
Over-expression of Novel Proteins in Maize

Elizabeth E. Hood and John A. Howard

8.1 Introduction

8.1.1 Why Over-produce Proteins?

One of the newest applications for over-expressed proteins in plants, including corn, is as a bio-factory for the production of vaccines, pharmaceuticals and industrial enzymes (Howard and Hood 2005). For these applications, many of the basic molecular and cellular techniques that are useful to study input traits can be applied to boost protein accumulation for the most cost-effective production model. This chapter focuses on this bio-factory application of proteins in maize.

8.1.2 What Do We Want in a Host for Over-production?

There is no one "ideal" host for the over-production of proteins for pharmaceutical, vaccine or industrial applications. Instead there are a number of characteristics of each recombinant host that may have different advantages depending on the protein and application. Some of the characteristics in a recombinant plant host include molecular, environmental and production characteristics that allow high protein accumulation, safety of the product, low production cost, and the ability to ensure that the product line would not enter the food or feed supply. These have been reviewed previously (Howard and Hood 2005) and maize is an excellent choice for fulfilling

Elizabeth E. Hood
Arkansas State University, P O Box 2760, State University, AR 72467, USA
e-mail: ehood@astate.edu

John A. Howard
Applied Biotechnology Institute, Building 36, Cal Poly State University, San Luis Obispo, CA 93407, USA

A.L. Kriz, B.A. Larkins (eds.), *Molecular Genetic Approaches to Maize Improvement*
Biotechnology in Agriculture and Forestry, Vol. 63

many of these desired qualities, which is why it has been pursued by several groups for this purpose.

8.2 Expression Technology

8.2.1 What Is the Protein Being Expressed and How Much is Accumulated?

Many factors bear on the basic questions of how much of the recombinant protein can be expressed and whether the protein has been modified in any way from what was intended. These various factors are interrelated, but for the purposes of this discussion we will address them separately as the inherent properties of the protein itself and where the protein is expressed in the plant. The tools currently available to achieve the desired results are also discussed.

8.2.2 Protein Characteristics

The inherent properties of the protein itself are among the most critical factors enabling accumulation of proteins in any host. These may include susceptibility to proteases, thermal stability, tertiary structure, and potential effects on cellular metabolism.

One common problem with over-expression of recombinant proteins in bacterial systems is proper folding of the protein and the formation of inclusion bodies (Buchner 2002). While protein folding problems may also occur in maize, some significant differences from bacteria make this less common. In maize, many different tissues and organelles where protein expression can be directed provide multiple environments. For example, while a protein may not fold properly in one organelle or tissue type, another target site can be selected that provides a better environment which permits high accumulation. Furthermore, no reference of proteins in inclusion bodies in plants has been seen, perhaps because in whole plants the foreign protein is not secreted into a culture medium but rather sequestered into organelles or apoplastic space between cells.

One potential problem for any protein production system is the susceptibility of the target protein to protease attack (Doran 2006). Protease cleavage sites can be exposed on the foreign protein and quickly attacked by endogenous proteases, leading to degradation. Foreign proteins often can be modified to eliminate cleavage sites unless the specific amino acid is critical for structure or function, or because the protein is destined for pharmaceutical use, leading to a different protein sequence that would require preparation of a different, separate regulatory package. Maize may provide an advantage for certain proteins in that the seed tissue is rich in protease inhibitors (Halim et al. 1973), potentially providing protection.

Thermally stable proteins tend to accumulate more readily in recombinant hosts including maize. However, because most of the proteins selected for accumulation in maize to date were chosen because they do not express well in other systems, this concept has not been extensively tested.

Most proteins possess some type of biological activity, which is why they provoke interest. The question arises, however, how likely is this activity to interfere with other metabolic processes in the host cell? At low concentrations many of these activities are inconsequential, but when expressed at high concentrations they can limit the expression of the protein or lead to various physiological limitations, perhaps even causing cell death. Examples include proteases, glycosidases, phosphatases, and redox enzymes. Trypsin is a protease that is difficult to express at high levels in microbes and in maize (Woodard et al. 2003; Király et al. 2006), presumably interfering with cellular metabolism. Laccase creates free radicals that interfere with many different metabolites, leading to altered growth characteristics (Hood et al. 2003). Avidin binds biotin, an important vitamin, and when expressed at high concentrations it can cause male sterility or cell death (Hood et al. 1997). On the other hand, several pharmaceutical antibodies have been shown to accumulate without great difficulty in maize (Stoger et al. 2004), particularly if they are specific to animal cell receptors, having no function in the plant and therefore acting as inert proteins.

8.2.3 Molecular and Cellular Characteristics

8.2.3.1 Where Foreign Proteins Are Expressed

Foreign proteins in maize can be targeted to specific organelles within specific tissues to provide a multitude of options. Each of these locations can affect accumulation and post-translational modification of the protein. One intracellular location may work in one tissue but not in another. The following examples illustrate these principles.

8.2.3.2 Protein Integrity

Changes made to the protein are due primarily to the post-translational modifications that occur in intracellular locations, but tissue type can greatly influence this effect. The modifications can be desired, inconsequential or lethal, and can include glycosylation, amino acid modification and/or degradation of the protein.

Primary amino acid sequence is important, but, in order to target the protein to certain intracellular locations, an additional amino acid targeting sequence must be attached to the primary protein sequence (Bednarek and Raikhel 1992). For proteins secreted to the cell wall, or targeted to the endoplasmic reticulum or the vacuole, these targeting sequences must be later cleaved at the N terminus of the protein. Rules predict where these sequences will be cleaved (Watson 1984), but there is still some uncertainty in precise cleavage. The signal sequence from barley alpha amylase has been used successfully for many proteins in maize (Zhong et al. 1999; Hood et al. 2003; Woodard et al. 2003), but in the case of brazzein (Lamphear et al. 2005) the cleavage was incorrect by one amino acid. This did not appear to have any adverse effect on the function, but any change to the primary sequence could be problematic for regulatory requirements for food and drug products. Vacuolar targeting sequences have been used with proteins expressed in maize (Caimi et al. 1996), but these may not be fully cleaved, adding additional amino acids to the mature protein as well.

Glycosylation is usually not critical for enzymatic function but may play a major role in proper folding and protection of the protein, particularly in pharmaceutical applications in blood. Maize behaves as other plants in that it does not possess the machinery to add sialic acid residues to proteins but adds a xylose and a different linkage for fucose compared with animal proteins and these sequences have fewer branched sugars off the backbone structure (Rayon et al. 1998). The carbohydrate sequence for maize-produced proteins (Samyn-Petit et al. 2001, 2003) has been studied in detail and its structure appears to be as expected for that of plant-derived glycosylation. These structures are similar to the native sequence on the foreign protein (with the above noted exceptions). An antibody produced in maize was also shown to have no apparent functional difference with these modified sugars (Ludwig et al. 2004). Some bacterial or cytoplasmic proteins have been engineered into maize with secretory signals, enabling them to be glycosylated. This can lead to loss of activity as in the case of GUS (unpublished results), or a reduction in the amount of cross reactivity with a vaccine antigen candidate (unpublished results). O-linked glycosylation is rarely reported, but in at least one case (Woodard et al. 2003) maize performed O-glycosylation where none was observed in the native animal sequence. However, no functional or enzymatic difference could be observed in the protein.

8.2.3.3 How Much Protein Is Accumulated?

Foreign protein synthesis involves factors such as the specific rates of transcription and translation and general features of the tissue such as how much available energy is diverted to protein synthesis. Recombinant proteins represent a new demand on the cell's energy, requiring a net increase to allow for new protein synthesis; otherwise the recombinant protein is made at the expense of native proteins. When recombinant proteins were expressed at very high levels in soybeans, they competed for transcription and translation machinery and available amino acids, thereby changing the composition of native proteins (Beach et al. 1998). Therefore steps were taken to increase the amino acid pools in the cell (Beach and Tarczynski 2000). Although not documented to the same extent in maize, no evidence exists for a net increase in protein content when a recombinant protein is present, and thus the situation is likely similar to that in soybeans.

The cost of recovery of proteins is inversely proportional to the concentration of the protein in the biomass, and the amount of energy allotted for protein synthesis can vary dramatically between crops (Howard and Hood 2005). Since maize seed has a higher protein content than leaves, this makes the seed a better target for keeping the cost low.

Within the seed, foreign proteins can be specifically targeted to the embryo, endosperm or pericarp. The endosperm is the largest portion of the seed and contains most of the seed protein, making this a logical target. Endosperm protein is largely water-insoluble storage protein, most notably zeins, making these easy to separate from the foreign protein. Maize seeds have low levels of phenolics, and thus do not interfere with downstream purification, making this a practical target for highly purified proteins. The approach has been used successfully for the recovery of several proteins including lipase (Gu and Glatz 2007).

The embryo is the other logical target tissue for foreign protein in the seed. It occupies a much smaller portion of the seed (~10%). Embryo tissue has the advantage of higher protein content on a weight basis than the endosperm, which can lower the cost of extraction, assuming separation of the endosperm from the embryo (see Sect. 8.3). The embryo contains oil that can be extracted from the embryo (germ) without denaturing the recombinant protein, using conventional extraction techniques but at lower temperatures than in routine processing plants (Kusnadi et al. 1998a). Several proteins have been treated in this manner, including trypsin (Woodard et al. 2003) and aprotinin (Zhong et al. 1999).

Subcellular location can have a significant effect on expression of protein within the seed. For laccase, several intracellular locations were tested (Hood et al. 2003). The conclusions were that cell wall targeting was a preferred location while cytoplasm and nuclear targets were not. Cell wall targeting is a recurring theme and has correlated with high expression for a number of proteins, including cellulase (Hood et al. 2007), trypsin (Woodard et al. 2003), aprotinin (Zhong et al. 1999; Delaney et al. 2003), vaccine antigens (Lamphear et al. 2002; Streatfield et al. 2003), and brazzein (Lamphear et al. 2005). In some cases, such as cellulase, the endoplasmic reticulum and vacuole targets also show good expression (Hood et al. 2007). Cytoplasmic expression usually provides moderate to low expression in these cases, but if a non-glycosylated protein is required without the option of modifying its protein sequence, then cytoplasmic targeting is a viable option.

While seed represents the greatest effort to date for expressing foreign proteins, biomass tissue is a possible target with some unique advantages. First, a large amount of biomass is present and unused since the plant is harvested for its grain; thus, it is possible to use biomass to produce protein extremely inexpensively. The high mass content, however, results in a higher cost of extraction when required. Also, proteins can degrade if left in the crop during dry down or may require colder temperatures during extraction to prevent active proteases from degrading the protein. Omnipresent phenolics also can interfere with purification, creating disincentives to produce highly purified proteins such as pharmaceuticals. One of the main disadvantages, however, is that the tissue is metabolically active before dry-down, and therefore the foreign protein has a greater potential to interfere with the normal function of the cell. One example is trypsin; when expressed with a constitutive promoter inducing expression in leaf tissue, plants could not survive, presumably because of proteolytic activity (Woodard et al. 2003).

8.2.4 General Tools to Effect Accumulation

Maize has the advantage of being one of the most studied higher plants, and consequently has some of the best characterized regulatory sequences and genetic variants available. A reliable promoter is critical. First, we can consider having the protein expressed in all of the tissues in the plant. One of the most used promoters in plant biology has been the constitutive Cauliflower Mosaic Virus 35S promoter (Benfey and Chua 1990), which also works well for maize. However, the level of seed expression from this promoter is very low in maize and it is more suitable for leaf expression. Therefore, the constitutive promoter of choice for much maize work to date has been that from the ubiquitin gene (Streatfield et al. 2004). This promoter exhibits not only good expression in the leaf but also excellent expression in seed tissue, including the first commercial products expressed in maize (Hood et al. 1997).

In practice, several factors generally make constitutive expression not the first choice. The protein can have detrimental effects on a tissue, which limits expression in other tissues. Examples include the unexpected effect of male sterility with avidin (Hood et al. 1997) or the detrimental effect on plant health of trypsin (Woodard et al. 2003) when these two genes are over-expressed with a constitutive promoter. Therefore a variety of seed preferred promoters have been tested – globulin being the most commonly used for embryo expression (Stoger et al. 2002; Streatfield et al. 2007a, b), although others show excellent expression as well (Streatfield et al. 2006). Zein promoters have been successful for endosperm expression (Russell and Fromm 1997) and used to drive expression of the commercial candidate version of lipase (Gu and Glatz 2007).

Inducible promoters offer the advantage of only expressing the gene of interest at a specific time that may be controlled externally. In this way, the protein will only be accumulated on demand and has the potential to alleviate any prolonged toxicity effects on the plant. For example, a protein that may be toxic to the plant could be engineered into the plant by using a regulated promoter, and expression could then be triggered and the plant harvested shortly afterward to avoid any long-term effects on plant growth. Triggers include chemical (Jepson et al. 1999), wound (Cordero et al. 1994), light (Matsuoka et al. 1993), and heat (Marrs et al. 1993) induction.

Occasionally, a gene encoding the zymogen form of an enzyme, e.g., trypsinogen (Woodard et al. 2003), is necessary to control enzymatic activity until the pro sequence is removed. This strategy was used to accumulate trypsin in maize and may have very useful applications in the expression and recovery of other foreign proteins.

Other DNA sequences are critical for expression, such as leader sequences, terminators, and enhancers. Several elements of translation including codon usage also can greatly increase the level of expression. These have been discussed in other reviews and apply to novel proteins as well (Streatfield 2007).

Transformation events in maize exhibit a wide variation in the level of expression, partially due to the location of insertion of the gene into the chromosome (Peach and Velten 1991). Maize also shows a high degree of variation in expression within a single insertion site (Hood et al. 2003). The mechanism for this is unknown, but for practical applications it is useful to know that the level of expression can increase at least 10-fold from what is observed in the first generation of seed collected after transformation (see Sect. 8.3). Gene dosage can also increase the level of expression through vectors with multiple transcription units (Schöffl et al. 1993), crossing two different alleles to each other, or when selfed, homozygous plants are produced (unpublished results). When carried to the extreme, however, too many copies may cause gene silencing (Matzke and Matzke 1995).

The use of germplasm to increase expression of the foreign gene has received little attention. It is well understood for microbial organisms that selected hosts of the same species can have a great influence on the expression of recombinant protein. In the most obvious approach, this can be applied in maize by using the same rules for creating two compatible inbred parents, and then making hybrid seed. Introgression of a transgene into maize variants, such as high oil or reduced-zein (e.g., opaque-2) lines, also increases recombinant protein expression (Hood et al. 2002a). In addition, specific germplasm may be used to complement the foreign gene. For example, high oil lines were used to help overcome the detrimental effect that laccase had on germination (Hood et al. 2003).

Agronomic practices can also influence the expression of foreign genes. When metal ions are sprayed on plants containing laccase, much more active laccase could be obtained during development. However, if it is desirable to keep the activity of the protein, such as manganese peroxidase (Clough et al. 2006), silent in the cell during growth, one may want to limit exposure of the transgenic plant to soil types containing manganese, and only add the mineral to the plant material after harvest, possibly eliminating enzyme activity within the cell during development.

8.3 Production

Maize can be grown in many environments and geographic locations but is usually limited by seasonal variations. Choices for growing in alternative locations rather than just the Corn Belt of North America are based on the value of the recombinant protein versus the cost of growing the crop to gain an additional season. This is often more of an issue during line development for hybrids because the number of generations is important in order to obtain isogenic lines for hybrid crosses as quickly as possible. However, once the inbreds are developed into parental lines, growing the crop in non-local environments is a cost issue of the recombinant protein versus the volumes required for annual production targets. For smaller volume pharmaceutical crops, confinement requirements will drive the decision of where to grow the plants. However, some industrial crops may be deregulated with respect to both food and feed use, and thus identity preservation will be utilized to protect the crop from mixing with commodities, thereby offering lower cost alternatives for growing the crop.

One requirement for regulated transgenic crops is that the equipment be dedicated and not used for non-transgenic crops. This is cost effective if the equipment can be used year after year for recombinant crops in the same location (Howard and Hood 2007). Once the crop has been harvested, it must be transported to its use destination, whether for processing into a final product or application to an industrial process. Transportation of maize grain is unlikely to affect the quality of the recombinant proteins, which have been shown to be stable for years in mature dry corn seed (Hood and Woodard 2005). The major concern of storing corn seed for future processing of recombinant protein is whether post-harvest diseases or insects will lower the quality of the grain.

Milling corn seed can be a convenient way of increasing the concentration of the protein in the target tissue. Wet or dry milling can separate the germ from the endosperm, allowing the recovery of starch, oil, or germ protein as co-products to recover revenue (Watson 1988). The separation can effectively increase the concentration of the recombinant protein in a specific tissue, particularly if the germ is the target tissue, because germ represents approximately 10% of the total seed weight. If all the recombinant protein is in the germ, a 5- to 10-fold enrichment can be achieved through milling (Howard and Hood 2005). A new milling technology is being marketed through Cereal Process Technology (www.cerealprocess.com) that improves separation of bran, germ, and starch. Although it was developed to improve ethanol yields from starch, the process can yield cleaner germ, bran, and endosperm for protein recovery from any of the tissues. Process engineering experimentation should be performed to ensure that the target protein will not be denatured by the process (Kusnadi et al. 1998a, b).

The protein product may be in the stalks and leaves rather than the seed for such products as cellulases in biomass targeted for conversion to sugars for ethanol (Sticklen 2006; Torney et al. 2007; http://www.agrivida.com/). Although this approach will likely have application in the future, short-term application may be slowed by the need for regulated growth of large tracts of transgenic biomass with the target enzymes.

The seed may be used as a direct delivery vehicle for the protein product, thereby eliminating the extraction and purification process traditionally involved with protein production. Because maize grain has GRAS (*Generally Recognized As Safe*) status, it provides a convenient way to introduce foreign proteins into the diet of animals and humans without having to purify the protein. One example of this is oral vaccine delivery (Streatfield and Howard 2003), but this concept applies to any orally delivered food or feed product. In addition, since maize is a major feedstock the concept may also apply to industrial enzymes where the grain is used (Howard and Hood 2007). For proteins that require extraction and purification, several other features favor seed production, including: high content of protease inhibitors and richness in carbohydrates, factors that aid the stabilization of proteins; dormancy, so that the protein does not compete for active metabolic reactions as is the case with green tissue; and the protein can be stored in the seed for years without loss of activity, making transport and storage possible at ambient temperatures (Kusnadi et al. 1998a, b).

Extraction and purification will be necessary when direct delivery of the product in corn meal is not possible. The buffer and conditions for extraction and purification will be product specific and depend on the amount of biomass and the stability of the protein product. Previously described processes will have an impact on this step – concentration of protein in the tissue and milling for tissue enrichment – so that the volume and steps can be minimized. Purification is dependent on the protein, of course, but can be affected by interfering agents that co-extract with the product or whether cGMP conditions must be met for a pharmaceutical product (Howard and Hood 2005).

The current rules for confinement of regulated transgenic maize are stringent and can be found at http://www.access.gpo.gov/nara/cfr/waisidx_05/7cfr340_05.html. Production of plants under appropriate regulation will obviously be more expensive than if the transgenic event is deregulated. Moreover, the cost of deregulation may be more than the product is worth, particularly if the acreage for the product is small, as is true for some pharmaceutical products. However, the Specialty Crops Regulatory Initiative is developing a paradigm for deregulation assistance for small market biotechnology crops (http://www.csrees.usda.gov/nea/biotech/in_focus/biotechnology_if_initiative.html). In this way, even transgenic corn with specialty applications can be assisted in deregulation for application to new, specialized markets.

8.3.1 Confinement

Much of the controversy surrounding genetically modified plants for protein overproduction has to do with having non-food proteins in food or feed. Certainly this issue is important and should be addressed, particularly for pharmaceutical compounds. The perception of risk must be met with data that address the risk factors and show that these products pose no or low risk, or can be contained adequately if they do pose a risk. Various techniques can be employed to assess risk level (Howard and Donnelly 2004) and to confine a crop at the appropriate level to prevent any inadvertent exposure in cases where risk is present. Confinement can be achieved by a number of techniques such as a creating a buffer zone around the transgenic crop, using male sterile lines, delaying planting to prevent nicking with surrounding maize crops, and/or mechanical detasseling. The cost of each of these methods must be weighed against recovery in product sales versus the perceived risk.

Recently, Howard and Hood (2007) published a model production method for pharmaceutical and industrial products that not only complies with confinement regulations, but also allows for preservation of product identity. This method, in combination with other steps, such as marking transgenic plants to visually identify the product-containing seeds, provides for segregation and identification of the specialty crop, making it much less likely to be inadvertently introduced into the food supply. Eventually, corn may be seen as a food organism, similar to yeast or eggs, that is used to produce pharmaceuticals but that can be kept distinct from the food supply.

8.4 Examples of Products

Products from transgenic plant systems have been reviewed often and details will not be presented here (Daniell et al. 2001; Fischer et al. 2003, 2004; Twyman et al. 2003, 2005; Horn et al. 2004; Stoger et al. 2004; Howard and Hood 2005; Ma et al. 2005). Some advantages of maize-derived products include freedom from human pathogens, low cost production, and the ability to directly deliver a safe product. However, utilizing plants as a production system is a developing industry that could be competitive in the next few years (Hood and Woodard 2005; Hood 2004). A summary of product examples specifically from maize is presented in Table 8.1.

Table 8.1 Product examples expressed in maize tissues. *TGEV* Transmissible gastroenteritis virus; *TSP* total soluble protein

Protein use	Maize tissue	Comments	Reference
Pharmaceutical			
Aprotinin	Seed/embryo	10% TSP	Delaney et al. 2003
Monoclonal antibody	Most tissues	0.3% TSP	Hood et al. 2002b
Monoclonal antibody	Most tissues		Whaley and Zeitlin 2000
Blood proteins	Seed/endosperm		Samyn-Petit et al. 2001
Lipase	Seed/endosperm		Gu and Glatz 2007
Vaccine			
TGEV	Most tissues	50 mg/kg	Streatfield et al. 2001
Lt-B	Most tissues Seed/embryo		Chikwamba et al. 2002; Streatfield et al. 2002
Hepatitis B	Seed/embryo	10% TSP	Streatfield 2005
HIV-gp120	Seed/embryo		Horn 2002
Industrial			
Trypsin	Seed/embryo	3.3% TSP	Woodard et al. 2003
Laccase	Seed/embryo	0.8% TSP	Hood et al. 2003
Manganese peroxidase	Seed/embryo	15% TSP	Clough et al. 2006
E1 cellulase	Seed/embryo	16% TSP	Hood et al. 2007
Cellobiohydrolase	Seed/embryo	16% TSP	Hood et al. 2007
Cellobiohydrolase	Seed/endosperm		Miles et al. 2007
E1 cellulase	Stalks and leaves		Biswas et al. 2006
Reagent			
Avidin	Most tissues	25% TSP	Hood et al. 1997
Beta-glucuronidase	Most tissues	0.7% TSP	Witcher et al. 1998
Polymers			
Polylactic acid	Seed	Synthetic enzymes	http://www.natureworksllc.com/

8.5 Future Prospects

The future of the industry of non-food products depends on stewardship of the early products from a regulatory standpoint as well as ensuring all data are openly available for scrutiny. The industry's future also depends on regulatory guidelines based on scientific data to determine appropriate levels of risk and benefit. While attention to date has focused on minimizing risk, more attention in the future will be placed on benefits. Unlike the current crop improvement products from maize, non-food products will provide the public with direct benefits they can easily recognize, such as orally delivered health products, environmentally friendly solutions to caustic chemicals, and an alternative supply of ethanol.

Some of these new products can be moved into the marketplace by using existing technologies and need only to pass the rigors of product development and regulatory approval. As the ability to accumulate foreign proteins in maize increases, many more products become plausible and existing product candidates become more cost effective. Currently maize has shown cost effectiveness for products at protein accumulation levels of 0.1–1% of seed weight. Compared with other recombinant hosts, however, this is still very low. If expression were to achieve levels of only 10% of the seed weight, it would dramatically change the cost of these products and provide much more incentive for the development of new products. Furthermore, it seems theoretically possible that a 10% diversion of energy to the seed for a novel protein should not cause problems with seed development, assuming the protein itself has no metabolic impact. Therefore, we see a need for continued effort in increasing expression in order to broaden this new industry.

References

Beach L, Tarczynski MC (2000) Methods of increasing accumulation of essential amino acids in seeds. US Patent 6,127,600, 3 October 2000

Beach L, Orman B, Townsend J, Thomas L (1998) Reduction of endogenous seed protein levels in plants. US Patent 5,850,024, 15 December 1998

Bednarek SY, Raikhel NV (1992) Intracellular trafficking of secretory proteins. Plant Mol Biol 20:133–150

Benfey PN, Chua N-H (1990) The cauliflower mosaic virus 35S promoter: combinatorial regulation of transcription in plants. Science 16:959–966

Biswas GCG, Ransoma C, Sticklen M (2006) Expression of biologically active *Acidothermus cellulolyticus* endoglucanase in transgenic maize plants. Plant Sci 171:617–623

Buchner J (2002) Introduction: the cellular protein folding machinery. Cell Mol Life Sci 59:1587–1588

Caimi PG, McCole LM, Klein TM, Kerr PS (1996) Fructan accumulation and sucrose metabolism in transgenic maize endosperm expressing a *Bacillus amyloliquefaciens* SacB gene. Plant Physiol 110:355–363

Chikwamba R, Cunnick J, Hathaway D, McMurray J, Mason H, Wang K (2002) A functional antigen in a practical crop: LT-B producing maize protects mice against *Escherichia coli* heat labile enterotoxin (LT) and cholera toxin (CT). Transgenic Res 11:479–493

Clough RC, Pappu K, Thompson K, Beifuss K, Lane J, Delaney DE, Harkey R, Drees C, Howard JA, Hood EE (2006) Manganese peroxidase from the white-rot fungus *Phanerochaete chrysosporium* is enzymatically active and accumulates to high levels in transgenic maize seed. Plant Biotechnol J 4:53–62

Cordero MJ, Raventós D, San Segundo B (1994) Expression of a maize proteinase inhibitor gene is induced in response to wounding and fungal infection: systemic wound-response of a monocot gene. Plant J 6:141–150

Daniell H, Streatfield SJ, Wycoff K (2001) Medical molecular farming: production of antibodies, biopharmaceuticals and edible vaccines in plants. Trends Plant Sci 6:219–226

Delaney D, Jilka J, Barker D, Irwin P, Poage M, Woodard S, Horn M, Vinas A, Beifuss K, Barker M, Wiggins B, Drees C, Harkey R, Nikolov Z, Hood E, Howard J (2003) Production of aprotinin in transgenic maize seeds for the pharmaceutical and cell culture markets. In: Vasil IK (ed) Proc 10th IAPTC&B Congress. Kluwer, Dordrecht, pp 393–394

Doran PM (2006) Foreign protein degradation and instability in plants and plant tissue cultures. Trends Biotechnol 24:426–432

Fischer R, Twyman RM, Schillberg S (2003) Production of antibodies in plants and their use for global health. Vaccine 21:820–825

Fischer R, Stoger E, Schillberg S, Christou P, Twyman RM (2004) Plant-based production of biopharmaceuticals. Curr Opin Plant Biol 7:152–158

Gu Z, Glatz CE (2007) Recovery of recombinant dog gastric lipase from corn endosperm extract separation. Sci Technol 42:1195–1213

Halim AH, Mitchell HL, Wassom CE (1973) Trypsin inhibitors of corn (*Zea mays*). Trans Kans Acad Sci 76:289–293

Hood E (2004) Bioindustrial and biopharmaceutical products from plants: In Proceedings of the 4th international crop science congress: new directions for a diverse planet, Brisbane. www.cropscience.org.au

Hood EE, Woodard S (2002) Industrial proteins produced from plants. In: Hood EE and Howard JA (eds) Plants as factories for protein production. Kluwer, Dordrecht, pp 119–135

Hood EE, Woodard SL (2005) Commercialization of a protein product from transgenic maize. Natl Agric Biotechnol Council 17:147–157

Hood EE, Witcher DR, Maddock S, Meyer T, Baszczynski C, Bailey MR, Flynn P, Register J, Marshall L, Bond D, Kulisek E, Kusnadi A, Evangelista R, Nikolov Z, Wooge C, Mehigh RJ, Hernan R, Kappel WK, Ritland D, Li C-P, Howard JA (1997) Commercial production of avidin from transgenic maize: characterization of transformant, production, processing, extraction and purification. Mol Breed 3(4):291–306

Hood EE, Howard JA, Delaney DE (2002a) Method of increasing heterologous protein expression in plants. US Patent application 10/252732

Hood EE, Woodard SL, Horn ME (2002b) Monoclonal antibody manufacturing in transgenic plants – myths and realities. Curr Opin Biotechnol 13:630–635

Hood EE, Bailey MR, Beifuss K, Magallanes-Lundback M, Horn ME, Callaway E, Drees C, Delaney DE, Clough R, Howard JA (2003) Criteria for high-level expression of a fungal laccase gene in transgenic maize. Plant Biotechnol J 1:129–140

Hood EE, Love R, Lane J, Bray J, Clough R, Pappu K, Drees C, Hood KR, Yoon S, Ahmad A, Howard JA (2007) Subcellular targeting is a key condition for high-level accumulation of cellulase protein in transgenic maize seed. Plant Biotechnol J 5(6):709–719

Horn ME (2002) Expression of an SIV protein in transgenic maize for use as an edible vaccine and reagent supply. In: McElrath MJ, et al. (eds) Proc Keystone Symp, HIV-1 protection and control by vaccination (X8), 5–11 April, Colorado

Horn ME, Woodard SL, Howard JA (2004) Plant molecular farming: systems and products. Plant Cell Rep 22:711–720

Howard JA, Donnelly KC (2004) A quantitative safety assessment model for transgenic protein products produced in agricultural crops. J Agric Environ Ethics 17:545–558

Howard JA, Hood EE (2005) Bioindustrial and biopharmaceutical products produced in plants. In: Sparks D (ed) Advances in agronomy, vol 85. Academic, San Diego, pp 91–124

Howard JA, Hood EE (2007) Methods for growing nonfood products in transgenic plants. Crop Sci 47:1255–1262

Jepson I, Martinez A, Sweetman JP (1999) Chemical-inducible gene expression systems for plants – a review. Pesticide Sci 54:360 – 367

Király O, Guan L, Szepessy E, Tóth M, Kukor Z, Sahin-Tóth M (2006) Expression of human cationic trypsinogen with an authentic N terminus using intein-mediated splicing in aminopeptidase P (pepP) deficient *Escherichia coli*. Protein Expr Purif 48:104–111

Kusnadi AR, Evangelista RL, Hood EE, Howard JA, Nikolov ZL (1998a) Processing of transgenic corn seed and its effect on the recovery of recombinant beta-glucuronidase. Biotechnol Bioeng 60:44–52

Kusnadi AR, Hood EE, Witcher DR, Howard JA, Nikolov ZL (1998b) Production and purification of two recombinant proteins from transgenic corn. Biotechnol Prog 14:149–155

Lamphear BJ, Streatfield SJ, Jilka JM, Brooks CA, Barker DK, Turner DD, Delaney DE, Garcia M, Wiggins B, Woodard SL, Hood EE, Tizard IR, Lawhorn B, Howard JA (2002) Delivery of subunit vaccines in maize seed. J Control Release 85:169–180

Lamphear BJ, Barker DK, Brooks CA, Delaney DE, Lane JR, Beifuss K, Love R, Thompson K, Mayor J, Clough R, Harkey R, Poage M, Drees C, Horn ME, Streatfield SJ, Nikolov Z, Woodard SL, Hood EE, Jilka JM, Howard JA (2005) Expression of the sweet protein brazzein in maize for production of a new commercial sweetener. Plant Biotechnol J 3:103–114

Ludwig DL, Witte L, Hicklin DJ, Prewett M, Bassi R, Burtrum D, Pereira DS, Jimenez X, Fox F, Saxena B, Zhou Q, Ma Y, Kang X, Patel D, Barry M, Kussie P, Zhu Z, Russell DA, Petersen WL, Jury TP, Gaitan-Gaitan F, Moran DL, Delannay X, Storrs BS, Tou J, Zupec ME, Gustafson KS, McIntyre J, Tarnowski SJ, Bohlen P (2004) Conservation of receptor antagonist anti-tumor activity by epidermal growth factor receptor antibody expressed in transgenic corn seed. Human Antibodies 13:81–90

Ma JK, Barros E, Bock R, Christou P, Dale PJ, Dix PJ, Fischer R, Irwin J, Mahoney R, Pezzotti M, Schillberg S, Sparrow P, Stoger E, Twyman RM (2005) Molecular farming for new drugs and vaccines. Current perspectives on the production of pharmaceuticals in transgenic plants. EMBO Rep 6:593–599

Marrs KA, Casey ES, Capitant SA, Bouchard RA, Dietrich PS, Mettler IJ, Sinibaldi RM (1993) Characterization of two maize HSP90 heat shock protein genes: expression during heat shock, embryogenesis, and pollen development. Dev Genet 14:27–41

Matsuoka M, Tada Y, Fujimura T, Kano-Murakami Y (1993) Tissue-specific light-regulated expression directed by the promoter of a C4 gene, maize pyruvate, orthophosphate dikinase, in a C3 plant, rice. Proc Natl Acad Sci USA 90:9586–9590

Matzke MA, Matzke A (1995) How and why do plants inactivate homologous (trans)genes? Plant Physiol 107:679–685

Miles S, Arellano S, Krebs H, Nelson A, Batie C, Betts S (2007) Requirement of an ER-directed signal peptide for accumulation of active *Trichoderma reesei* cellobiohydrolase I in transgenic corn seed. In: Proc Plant Molecular Biology Retreat, 28–30 Sept. ASPB, Chicago

Peach C, Velten J (1991) Transgene expression variability (position effect) of CAT and GUS reporter genes driven by linked divergent T-DNA promoters. Plant Mol Biol 17:49–60

Rayon C, Lerouge P, Faye L (1998) The protein N-glycosylation in plants. J Exp Bot 49:1463–1472

Russell DA, Fromm ME (1997) Tissue-specific expression in transgenic maize of four endosperm promoters from maize and rice. Transgenic Res 6:157–168

Samyn-Petit B, Gruber V, Flahaut C, Wajda-Dubos JP, Farrer S, Pons A, Desmaizieres G, Slomianny MC, Theisen M, Delannoy P (2001) N-glycosylation potential of maize: the human lactoferrin used as a model. Glycoconj J 18:519–527

Samyn-Petit B, Wajda Dubos JP, Chirat F, Coddeville B, Demaizieres G, Farrer S, Slomianny MC, Theisen M, Delannoy P (2003) Comparative analysis of the site-specific N-glycosylation of human lactoferrin produced in maize and tobacco plants. Eur J Biochem 270:3235–3242

Schöffl F, Schröder G, Kliem M, Rieping M (1993) An SAR sequence containing 395 bp DNA fragment mediates enhanced, gene-dosage-correlated expression of a chimaeric heat shock gene in transgenic tobacco plants. Transgenic Res 2:93–100

Sticklen M (2006) Plant genetic engineering to improve biomass characteristics for biofuels. Curr Opin Biotechnol 17:315–319

Stoger E, Sack M, Fischer R, Christou P (2002) Plantibodies: applications, advantages and bottlenecks. Curr Opin Biotechnol 13:161–166

Stoger E, Schillberg S, Twyman RM, Fischer R, Christou P (2004) Antibody production in transgenic plants. Methods Mol Biol 248:301–318

Streatfield SJ (2005) Oral hepatitis B vaccine candidates produced and delivered in plant material. Immunol Cell Biol 83:257–262

Streatfield SJ (2007) Approaches to achieve high-level heterologous protein production in plants. Plant Biotechnol J 5:2–15

Streatfield SJ, Howard JA (2003) Plant-based vaccines. Int J Parasitol 33:479–493

Streatfield SJ, Jilka JM, Hood EE, Turner DD, Bailey MR, Mayor JM, Woodard SL, Beifuss KK, Horn ME, Delaney DE, Tizard IR, Howard JA (2001) Plant-based vaccines: unique advantages. Vaccine 19:2742–2748

Streatfield SJ, Mayor JM, Barker DK, Brooks C, Lamphear BJ, Woodard SL, Beifuss KK, Vicuna D, Massey L-A, Horn ME, Delaney DE, Nikolov ZL, Hood EE, Jilka JM, Howard JA (2002) Development of an edible subunit vaccine in corn against enterotoxigenic strains of *Escherichia coli*. In Vitro Cell Dev Biol Plant 38:11–17

Streatfield SJ, Lane JR, Brooks CA, Barker DK, Poage ML, Mayor JM, Lamphear BJ, Drees CF, Jilka JM, Hood EE, Howard JA (2003) Corn as a production system for human and animal vaccines. Vaccine 21:812–815

Streatfield SJ, Magallanes-Lundback ME, Beifuss KK, Brooks CA, Harkey RL, Love RT, Bray J, Howard JA, Jilka JM, Hood EE (2004) Analysis of the maize polyubiquitin-1 promoter heat shock elements and generation of promoter variants with modified expression characteristics. Transgenic Res 13:299–312

Streatfield SJ, Love RT, Bray J (2006) Globulin 2 regulatory region and method of using same. US Patent 7,112,723, 26 September 2006

Streatfield SJ, Love RT, Bray J (2007a) Embryo preferred promoter and method of using same. US Patent 7,183,109, 27 February 2007

Streatfield SJ, Love RT, Bray J (2007b) Globulin-1 regulatory region and method of using same. US Patent 7,169,967, 30 January 2007

Torney F, Moeller L, Scarpa A, Wang K (2007) Genetic engineering approaches to improve bioethanol production from maize. Curr Opin Biotechnol 18:193–199

Twyman RM, Stoger E, Schillberg S, Christou P, Fischer R (2003) Molecular farming in plants: host systems and expression technology. Trends Biotechnol 21:570–578

Twyman RM, Schillberg S, Fischer R (2005) Transgenic plants in the biopharmaceutical market. Expert Opin Emerg Drugs 10:185–218

Watson ME (1984) Compilation of published signal sequences. Nucleic Acids Res 12:5145–5164

Watson SA (1988) Corn marketing, processing and utilization. In: Sprague GF, Dudley JW (eds) Corn and corn improvement, 3rd edn, vol 18. American Society of Agronomy, Crop Science Society of America, Soil Science Society of America, Madison, pp 885–942

Whaley KJ, Zeitlin L (2000) Transgenic commensals as mucosal protectants. Nat Biotechnol 18:1038–1039

Witcher DR, Hood EE, Peterson D, Bailey M, Bond D, Kusnadi A, Evangelista R, Nikolov Z, Wooge C, Mehigh R, Kappel W, Register I, James C, Howard JA (1998) Commercial production of β-glucuronidase (GUS): a model system for the production of proteins in plants. Mol Breed 4:301–312

Woodard SL, Mayor JM, Bailey MR, Barker DK, Love RT, Lane JR, Delaney DE, McComas-Wagner JM, Mallubhotla HD, Hood EE, Dangott LJ, Tichy SE, Howard JA (2003) Maize (*Zea mays*)-derived bovine trypsin: characterization of the first large-scale, commercial protein product from transgenic plants. Biotechnol Appl Biochem 38:123–130

Zhong GY, Peterson D, Delaney DE, Bailey MR, Witcher DR, Register III JC, Bond D, Li C-P, Marshall L, Kulisek E, Ritland D, Meyer T, Hood EE, Howard JA (1999) Commercial production of aprotinin in transgenic maize seeds. Mol Breed 5:345–356

Chapter 9
Global Regulation of Transgenic Crops

Bruce M. Chassy

9.1 Regulatory Oversight of Transgenic Maize

Globally, transgenic maize comprised about 25% of the 102 million hectares of transgenic cropland planted in 2006 by more than 10 million farmers in 21 countries (James 2007). These transgenic maize plants contain inserted gene(s) expressing a variety of Cry proteins that confer resistance to stem borers and rootworms. Approximately 45% of the transgenic maize planted also contains inserted gene(s) that mediate herbicide tolerance (James 2007). Transgenic crop varieties must be granted pre-market approval by regulatory authorities, and more than 43 transgenic maize varieties have been approved to date (Agbios GM Database, http://www.agbios.com/dbase.php, accessed 14 January 2008). It must be shown that novel transgenic crops are safe for agriculture and the environment prior to their commercialization and planting. Since maize is widely used as animal feed and in human food, the pre-market regulatory approval process also evaluates the safety of transgenic maize as food and feed. This chapter describes the evolution of the regulatory paradigm and regulations applied to transgenic crops around the world with emphasis on maize, briefly highlights differences in approaches between nations, details the scientific considerations of the regulatory review process, focusing primarily on food safety issues that have been of concern to consumers around the globe, and concludes with an assessment of the consequences and impact of the stringent global regulation of transgenic crops. The chapter will also review claims regarding potential adverse effects of transgenic plants that received attention in the media and which have shaped negative public perceptions about transgenic crops.

Bruce M. Chassy
40 NSRC, 1101 West Peabody, University of Illinois, Urbana, IL 61801, USA
e-mail: bchassy@uiuc.edu

A.L. Kriz, B.A. Larkins (eds.), *Molecular Genetic Approaches to Maize Improvement*
Biotechnology in Agriculture and Forestry, Vol. 63

9.1.1 Development of a Regulatory Paradigm and Rationale

Uncertainty about the potential consequences of the newly developed techniques of in vitro genetic engineering prompted the scientific and regulatory community to take a precautionary approach to recombinant DNA (rDNA) technology and genetic engineering. This concern resulted in the creation of National Institutes of Health (NIH) Guidelines and a Recombinant DNA Advisory Committee (McHughen and Smythe 2008; Chassy 2007). There were also non-technical concerns about the consequences of genetic engineering (Chassy 2007). Perhaps as a consequence, when it became clear during the 1980s that transgenic plants intended for use in agriculture were being developed in the USA and elsewhere, the National Academy of Sciences (NAS) evaluated potential risks associated with DNA technology and recommended a science-based rationale for safety assessment (NAS 1987). The NAS concluded that the use of rDNA technology posed "risks of the same kind" as those encountered in conventional plant breeding. The conclusion was in large part based on the long history of plant genetic modification that took place over millennia and resulted in the development of modern crop plants. Indeed, most modern crops no longer resemble their wild ancestors and would not exist had there not occurred extensive and repeated human intervention (Parrott 2005) – domestic crop plants are in a very real sense "human-made." Most can no longer survive in the wild and require cultivation by humans.

The Coordinated Framework for the Regulation of Biotechnology was announced in the USA in 1986. Under this framework three regulatory agencies (the USDA, EPA, and FDA) were asked to base regulatory review upon existing legislative authority provided under the Plant Protection Act (USDA), the Federal Insecticide, Fungicide, and Rodenticide Act (EPA), and the Federal Food, Drug, and Cosmetic Act (FDA). In practice, new varieties must always gain approval of USDA since the agency has authority for both plant protection and potential environmental impacts, while EPA has authority only where the transgenic plant contains insecticidal traits, and FDA is concerned only that new crops pose a "reasonable certainty of no harm" as a food or feed.

It is noteworthy that FDA explicitly stated the intent to regulate the safety of products rather than the technology used to develop it (FDA 1993) – sometimes stated as to "regulate the product not the process used to create it." FDA also determined that mandatory labels cannot be required on a transgenic food or ingredient unless the product poses a safety concern or has some other material difference from its conventional counterpart, about which consumers have a right to be informed. To date, no transgenic products have been required to be labeled by the FDA. This can be attributed to the fact that FDA would not approve a product about which there remained residual *science-based* safety concerns. In addition, if a product is materially different, FDA requires that the difference be stated on the label but has no requirement that the label state that the difference was produced using in vitro rDNA technology – the safety of the product not the process is the focus of FDA review. By way of illustration, in the not too distant future genetically enhanced oil seeds will be used to produce vegetable oils that are high in heart-healthy ω-3 fatty acids.

These products will carry notice on the label advising consumers of the difference in composition as compared to conventional oils – no doubt a major selling point in this case – however, the label will not be required to state the oil is derived from transgenic or GMO (genetically modified organism) oilseeds in the USA. As we shall see, the requirement, or lack of a requirement, for mandatory labeling is a key difference among regulatory systems around the world (see Sect. 9.1.2).

The first commercially approved transgenic plant, the *Flavr-Savr* tomato, was approved under the Coordinated Framework in 1994 (Chassy 2001). Detailed reviews of the rationale and functionality of the regulatory approach to transgenic crops used in the USA have been published (Chassy 2001; McHughen and Smythe 2008). The US government maintains a website that explains the regulations, roles, and function of biotechnology regulation in the USA (http://usbiotechreg.nbii.gov/, accessed 14 January 2008). The site also hosts a database of regulatory approvals of transgenic crops; according to the database, to date (14 January 2008) 23 transgenic maize varieties have been approved in the USA.

9.1.2 Divergent Regulatory Approaches Around the World

Canada elected a regulatory approach that is similar to that used in the USA by drawing on existing authority previous granted to Health Canada, Environment Canada, and the Food Inspection Agency as found in the Seeds Act, Feeds Act, Fertilizers Act, Food and Drugs Act, Health of Animals Act, or the Canadian Environmental Protection Act (http://www.agbios.com/cstudies.php?book=REG&ev=CAN-USA &chapter=Canada&lang=EN, accessed 14 January 2008). Canada is, however, unique among nations in legislating a national policy that regulates *novel plants* (and novel foods) independent of the technology used in their development.

Other countries have determined that the use of in vitro rDNA technology in variety development is per se sufficient reason to trigger a regulatory review. It should be noted parenthetically that the fundamental problem with this approach is that by regulating process rather than phenotype it is possible to develop and market a variety bearing a novel phenotype that requires no regulatory review, while a similar variety displaying the same phenotype produced though gene transfer technology will be reviewed.

Australia passed the Gene Technology Act in 2000 that created the Office of Gene Technology Regulator (OGTR) which is housed in the Australian Government Department of Health and Ageing. The OGTR coordinates the activities of several government agencies (http://www.ogtr.gov.au/index.htm). Australia and New Zealand have adopted a mandatory GM-labeling policy. Argentina created the Biotechnology Office in the Department of Agriculture, Livestock, Fisheries and Foods (http://www.sagpya.mecon.gov.ar/new/0-0/programas/biotecnologia/index_en.php, accessed 14 January 2008) and has chosen not to require mandatory GMO labels. Neighboring Brazil has, on the other hand, created a national oversight body (National Technical Commission on Biotechnology;

http://www.ctnbio.gov.br/, accessed 14 January 2008) and Brazilian law requires that products containing transgenic content be labeled with a standardized logo containing the letter “T” to denote transgenic content.

The regulation of transgenic crops in the EU parallels changing public attitudes. The original Council Directive on release of genetically modified organisms into the environment (Council Directive 90/220/EEC) has been repealed and replaced by Directive 2001/18/EC. EC Regulation 258/97 concerning novel foods and novel food ingredients and EC Regulation 1829–2003 concerning GM food and feed also regulate GM foods and require mandatory labeling. The European Food Safety Authority (EFSA) has responsibility for food and environmental safety associated with transgenic crops (http://www.efsa.europa.eu/EFSA/KeyTopics/efsa_locale-1178620753812_GMO.htm, accessed 14 January 2008).

The foregoing paragraphs serve to illustrate that governments have chosen to organize the regulation of transgenic crops in a variety of different ways that no doubt arose from the existing organizational arrangement and authority as well as the political and social climate surrounding the introduction of a regulatory system. It is virtually impossible to describe the regulatory system applied to transgenic crops at a global level since almost every nation has put different regulatory frameworks in place (and more than 120 nations have not done so at all). The major functional difference between regulatory systems is the requirement for mandatory labeling of transgenic content. The majority of nations have opted for mandatory labeling of foods and feeds containing ingredients derived from transgenic plants (see Sect. 9.3). It can also be said that no nation has a truly science-based policy since, without exception – Canada and the USA notwithstanding – transgenic crops are singled out for special review by *all* nations.

A point that is beyond the scope of this chapter is the contrast between regulatory systems that are predicated on risk–benefit analysis (which is said to be precautionary in nature) versus those based on the more formal precautionary principle (PP). The PP seeks to avoid new risks and strives for a higher level of safety for consumers and the environment. Suffice it to say that the USA and Argentina – two countries that have embraced transgenic crops – do not claim to be adherents to the PP, while adherence to the PP is mandated by the EU charter. The misinterpretation of the PP as a demand for absolute safety (Hathcock 2000) and the explanation of the fundamental difference between the USA and EU in regulation of transgenic crops arising from the PP (Kalaitzandonakes 2000) have been analyzed.

9.2 Scientific Assessment of Risks Associated with Transgenic Maize

While regulatory systems may vary from nation to nation, the underlying scientific principles of safety assessment of transgenic crops are almost universally accepted within the scientific community. Regulators review the potential agricultural, environmental, and food safety hazards (hazard = potential to do harm) associated with each novel variety and attempt to evaluate which of them will or could do harm (harm × exposure = risk). The task of the risk assessor is to characterize what is likely or unlikely to happen, determine how often any potential harm will occur, and compute the potential damage and cost associated with real risks – if indeed any are identified. The risk assessor may also compute the damage or harm caused by present practices and technology as well as that associated with proposed alternatives, and can compute potential agricultural, environmental, and social benefits of adopting the new technology. This must be done on a case-by-case basis as each trait and each application is different. The analysis is also region and ecosystem specific. Note, however, that opposition to transgenic technology is always against *any* application of the technology since it is an objection to the *process* used to develop the technology.

For each new transgenic crop reviewed, regulatory agencies typically publish the safety assessment data along with the agency's interpretation of the data and conclusions (see, for example, http://www.cfsan.fda.gov/~lrd/biocon.html). It is the risk manager's task to weigh the benefits, costs, and risks and issue an approval or disapproval. It is not an unfair generalization to state that there will often be agreement by risk assessors in most nations regarding scientific risk assessment of the safety of a novel transgenic variety, while approval by the national risk managers may be immediate, slow in coming, or permanently delayed. These national differences most often relate to policy rather than divergent scientific interpretation. For example, more than 160 UN member states of Codex Alimentarius have cooperated in the publication of international guidelines for the food safety assessment of transgenic crops that are followed by national regulators around the globe (Codex Alimentarius 2003). Science-based assessment of hazards of novel transgenic crops is discussed in the following sub-sections.

More than 14,126 field trials have been conducted in the USA with transgenic plants; approximately 43% (5,973) of these were conducted with maize (http://www.isb.vt.edu/cfdocs/fieldtests1.cfm, accessed 14 January 2008). The ratio of commercialization of products to field tests is small. About 43 (0.7% of USA field tests) transgenic maize varieties have been approved around the globe (www.agbios.com, accessed 14 January 2008). With the exception of high lysine maize, these varieties are designed to improve agronomic performance (Table 9.1). The reasons for the disparity between the number of field tests and successful commercial introductions are no doubt numerous and complex. Some field tests are repetitions using the same variety. In other cases the inserted phenotype does not perform well, the event itself does not perform well, or the trait does not confer a marketable advantage. Lurking behind failure to commercialize due to performance and market considerations is another very practical reality. It requires 5–10 years to obtain regulatory approval for a new variety and the direct cost of tests required by regulatory agencies will be at least $7–16 million (Kalaitzandonakes et al. 2007). It is noteworthy that although maize field evaluations have been conducted by numerous public-sector researchers, regulatory approvals have been granted exclusively to products developed by large private sector corporations. Transgenic papaya is the

Table 9.1 Partial list of approved traits in maize[a,b,c]

	Country													
Trait	USA	Argentina	Canada	EU	Australia	Japan	China	Korea	Mexico	Philippines	S. Africa	Switzerland	Taiwan	Uruguay
Cry1Ab	X	X	X	X	X	X	X	X	X	X	X	X	X	X
Cry1AC	X	X	X		X	X		X		X			X	
CryiF	X	X	X	X	X	X	X	X	X	X	X		X	
Cry3A	X		X		X	X		X	X					
Cry3Bb1	X		X	X	X	X	X	X	X	X				
Cry9C	X													
Cry34Ab1	X		X	X	X	X		X	X	X				
Cry35Ab1	X		X	X	X	X		X	X	X				
EPSPS[d]	X	X	X	X	X	X	X	X	X	X	X		X	
PAT[e]	X	X	X	X	X	X	X	X	X	X	X	X	X	X
Male sterility[f]	X		X											
DAM[g]	X													
cDHDPS[h]	X		X		X	X			X	X				

[a]Based on data available in the Agbios database (http://www.agbios.com/dbase.php, accessed 14 January 2008)
[b]The list of traits excludes selective markers. Commercial varieties may contain two or more approved traits ("stacked")
[c]Brazil, the Czech Republic, and India have not approved transgenic maize traits
[d]DNA adenine methylase
[e]Phosphinothricin acetyltransferase
[f] 5-enolpyruvyl shikimate-3-phosphate synthase
[g]Dihydrodipicolinate synthase
[h]Barnase ribonuclease

only public sector transgenic product that has received regulatory approval to date (Fuchs and Gonsalves 2007).

9.2.1 Description of the Event and Organisms

The development of a transgenic crop begins with the identification of a gene or genes that have the potential to improve the agronomic or end-product quality characteristics of the crop. The potential safety of the introduced gene(s) and/or the changes gene introduction might bring about are considered before development is undertaken. This a priori safety assessment is intended to ensure safety as well as to avoid time and resources wasted on the introduction of genes that cause harm. Homology searches (BLAST, FASTA) of genetic and protein databases are performed with candidate genes in order to assure that the introduced genes bear no relationship to known toxicants, anti-nutrients, or allergens (Chassy 2002, 2004; Konig et al. 2004; Goodman et al. 2008).

Subsequent to insertion of genes via transformation, molecular methods are used to characterize the number of inserts and the size(s) of the inserted DNA. Newly produced proteins are characterized to assure identity to the proteins produced in the donor organism; the potential production of unintended protein products is also evaluated. The level and stability of expression in various tissues is determined. These data provide insight into the potential effectiveness of the event and are also useful in calculation of the exposure of humans and/or animals to the novel proteins (see Sect. 9.2.4.7).

9.2.2 Evaluation of Agricultural Hazards

Extensive field tests that are subject to regulatory agency notification and/or permit requirements are used to evaluate composition, seed germination/dormancy, seed bank longevity, crop growth and reproduction, potential for out-crossing, and an overall assessment of fitness. The agricultural evaluation establishes that the trait is stable and efficacious – which are more performance than safety concerns – and provides evidence that no adverse agricultural properties or impacts are associated with the crop (Chassy 2001; McHughen and Smythe 2008). A second important aspect of field trials and other pre-market studies is to provide data for the development of a resistance management plan, where appropriate. Field tests also provide material for food safety assessment and animal performance studies.

9.2.3 Evaluation of Environmental Hazards

The environmental safety assessment evaluates the potential occurrence and consequences of gene transfer to related plants, the potential occurrence and consequences of horizontal gene transfer to unrelated species, potential for weediness, and potential for effects on non-target organisms. Typically, the potential impacts of the crop on avian species (quail), aquatic species (catfish and daphnia), soil organisms (springtails and earthworms), and beneficial insects (honeybee, parasitic wasp, green lacewing, and ladybird beetle) are investigated. Environmental fate studies may also be conducted (Chassy 2001; McHughen and Smythe 2008).

In 1999, a letter published in *Nature* reported that feeding Bt-pollen to Monarch butterfly larvae in a laboratory setting produced greater larvae fatality than feeding conventional pollen (Losey et al. 1999). While it should not have been surprising to observe that pollen that could (Bt was not measured in the study) contain a Bt-protein that is toxic to insects of the order Lepidoptera killed the Monarch butterfly larvae (members of the Lepidoptera), the observations were announced in the media and heralded by opponents of genetic engineering as proof of the dangers of transgenic technology; transgenic maize was (and is today) portrayed as a threat to the survival of the butterfly. There was little opportunity given in the media for scientists to point out that it was an unreasonable stretch of logic to extrapolate from the laboratory results to the agricultural field, that there were major experimental flaws in the work as reported, and that there are many sound reasons to believe that Bt maize poses no threat to Monarchs. For example, since Bt-maize only affects insects that ingest parts of the maize plant it should not affect butterflies such as Monarchs since they do not eat maize plants – certainly Bt maize would do less harm to non-target insects than would chemical pesticides. The findings were thoroughly evaluated in a series of carefully designed and executed studies that were reported in *Proceedings of the National Academy of Sciences* (*PNAS*) in 2001 which concluded that Bt-pollen poses negligible risk to Monarchs (Sears et al. 2001). These studies vindicated Bt-maize but were not mentioned by the media.

Recently is has been claimed that run-off from agricultural fields that contain Bt-maize debris could adversely affect aquatic ecosystems (Rosi-Marshall et al. 2007). The paper described in vitro experiments in which Caddisfly larvae were exposed to maize leaves and concluded that maize-associated Cry1Ab caused toxic effects. These small in vitro toxic effects are then extrapolated to natural headwater ecosystems, and it is concluded that these are threatened. The reported research suffers from serious methodological short-comings – so much so that it is astounding that the paper appeared in *PNAS* (USA) and that the research sponsor, the National Science Foundation (NSF), posted a web page devoted to the proud announcement of the work. The report describes an in vitro study that never evaluated whether Cry1Ab flows from maize fields to aquatic headwaters: in fact, Cry1Ab was not identified or measured in the experiments. The first rule of toxicology is to identify the toxin and quantify the dose. Moreover, non-isogenic varieties of maize were compared without providing data on the quantity used ("Leaves were added to aquaria as needed"), thus rendering repetition of the work impossible. The authors ignore more than 50 years of published research that details the exquisite biological specificity of Cry proteins. Published research suggests that it is highly unlikely that Cry1Ab (anti-Lepidoptera) will adversely affect Caddisfly (Trichoptera); the authors fail to cite reports that Cry proteins degrade rapidly in aquatic ecosystems.

The authors discuss Cry1Ab and ignore other Cry proteins that are not only used in transgenic maize and present in agricultural fields, but also applied directly to aquatic ecosystems for purposes of pest control – it would have been a far more interesting question to ask if direct application of these other Cry proteins to aquatic ecosystems has an adverse effect on Caddisfly larvae. EFSA recently considered this work and concluded: "In summary, the conclusions of the paper Rosi-Marshall et al. (2007) are not supported by the data presented in this paper. The GMO Panel is of the opinion that based on the available information such a low level of exposure to Trichoptera in aquatic ecosystems is unlikely to cause a toxic effect (EFSA 2007)."

9.2.4 Evaluation of Food Safety Hazards

The safety assessment of a whole food is difficult since plant foods can contain a large number of compounds (Konig et al. 2004; Cellini et al. 2004). The concept of substantial equivalence (SE) was developed in order to cope with the complexity of whole foods (Kok and Kuiper 2003). The SE paradigm asserts that if the compositions of two foods are compared, differences in safety can only be associated with differences in composition. Obviously, the choice of an appropriate comparator is essential; most often a near-isogenic strain and several related commercial varieties are selected for comparison. Observed difference(s) in composition is a starting point that guides the safety assessor to changes that might require further analysis. In practice, SE has sometimes been misinterpreted as a conclusion that safety has been demonstrated since two foods are said to be virtually identical in composition. For that reason it has been suggested that the SE paradigm should be referred to as a *comparative safety assessment* (Kok and Kuiper 2003). It is worth reiterating that SE is a starting point and not a conclusion. Comparative analysis demonstrates that one food is as safe as another food on a case-by-case basis; however, it cannot prove absolute safety since no food is absolutely safe under all circumstances.

Three fundamental safety questions need to be evaluated: (1) does the newly inserted DNA pose novel risks; (2) do the product(s) encoded by the inserted gene(s) pose novel risks; and (3) do both intended and unintended changes in composition pose novel risks? These general questions can be answered by consideration of the following points (Chassy 2002; Codex Alimentarius 2003):

- history of use and safety of the donor and recipient organism or food;
- safety of DNA ingestion and selective marker;
- potential for increased toxicity (protein product);
- potential for increased allergenicity (protein product);
- retention or improvement of nutritional value and animal studies;
- equivalence of composition other than intended changes and absence of adverse unintended effects;
- estimation of dietary intake (exposure).

Each of these factors will be discussed briefly in the following section on transgenic maize.

9.2.4.1 History of Use and Safety of the Donor and Recipient Organism or Food

Maize is one of the world's leading cereal grain crops perhaps because it provides excellent nutritional value and has a long history of safe use as food and feed (OECD 2002). Adverse reactions or allergy to maize grain are virtually unknown. To date, all of the genes inserted into transgenic maize varieties have been isolated from non-pathogenic bacteria that are not normally consumed by humans or animals and for which there is no history of safe use or known adverse effect. From a scientific perspective, however, the safety of the donor organism is irrelevant as long as care is taken to not insert genes that encode toxic proteins or allergens.

9.2.4.2 Safety of DNA Ingestion and Selective Marker

DNA is generally regarded safe to eat since no toxic, mutagenic, teratogenic, or carcinogenic effects have been attributed to DNA ingestion (van den Eede et al. 2004). Moreover, DNA is often already extensively degraded when ingested and will be rapidly further degraded in the digestive system. Gene transfer from plants to bacteria or animals has not been demonstrated (van den Eede et al. 2004), rendering the possibility of horizontal gene flow highly unlikely. Although public concerns have been raised about the safety of antibiotic resistance genes being used as selective markers in plant transformations, and alternative non-antibiotic-associated marker systems developed, from a risk perspective such concerns are not justified. Consider for the moment that 90% of human stool samples in Mexico test positive for Ampicillin resistance genes (Calva et al. 1996) or that the average French person deposits between 2.5×10^8 and 2.5×10^9 Ampicillin-resistant *Escherichia coli* into the environment every day (Berche 1998).

9.2.4.3 Potential for Increased Toxicity (Protein Product)

Millions of proteins are described in genetic databases. Of these, an exceedingly small number display toxic or anti-nutrient effects when ingested by humans or animals (Chassy et al. 2008). As mentioned previously, computer programs that compare selected protein sequences with sequences found in protein databases (which include known toxicants, anti-nutrient proteins, and allergens) such as BLAST or FASTA can be used to ensure that potentially hazardous gene products will not be produced in the desired transgenic plant. Proteins are often partially degraded or denatured by natural means as well as processing operations prior to consumption, and most are rapidly degraded in the human or animal gastrointestinal (GI) system. For that reason, inserted proteins are subjected to an in vitro assay using simulated gastric fluid that provides insight into digestibility (Chassy et al. 2008). Digestibility is taken as an indicator that the protein will be innocuous.

The claim that scientific methods can be used to ensure that transgenic foods will not be toxic to animals and humans has been challenged by opponents of GMOs. Ewen and Pusztai reported in a BBC television broadcast in 1999 that feeding rats transgenic potatoes into which had been inserted a gene encoding a lectin resulted in thickening and inflammation of the GI epithelium when compared to controls. Later in the year *Lancet* decided, against the recommendations of the reviewers, to publish a paper by Ewen and Pusztai 1999), noting that by publication *Lancet* "aims to make constructive progress in the debate between scientists, the media, and the general public about the safety of GM food." In the same issue *Lancet* published a critique of the research reported by Ewen and Pusztai that clearly delineated numerous faults and shortcomings of the research (Kuiper et al. 1999). A subsequent Royal Society-UK review concluded that the experiments were improperly designed, and that no meaningful scientific conclusion should be drawn (Royal Society 1999, http://www.royalsoc.ac.uk/displaypagedoc.asp?id=6170, accessed 14 January 2008). The "Pusztai" episode was a tipping point that produced widespread rejection of transgenic crops by food retailers, food processors, the media, and, ultimately, many consumers. The scientific conclusion that the study was fatally flawed went largely unnoticed by the media.

9.2.4.4 Potential for Increased Allergenicity (Protein Product)

A very small number of proteins can produce food allergy in a small percentage of the population (Goodman et al. 2008). A revised scheme for the assessment of allergenic potential that includes bioinformatics analysis, digestibility testing, and, if necessary, human serum screening, as well as whole food challenges in human subjects has been proposed (Codex Alimentarius 2003). Bioinformatic screening is particularly informative since virtually all plant allergens bear a close structural relationship to proteins encoded by one of three gene families (Breiteneder and Mills 2005). Several of the procedures used in the allergy assessment paradigm have recently been questioned as unnecessary and/or invalid (Goodman et al. 2008). Not only do commonly encountered foods (e.g., eggs, milk, soybeans, nuts, ground nuts, wheat, fish, mollusks, crustaceans, and sesame) that cause human allergies continue to be marketed without regulation, new foods such as kiwi can be introduced into the market without any requirement for pre-market safety review, and can subsequently cause food allergy in some subjects. The allergy remedy for conventional foods is to ask allergy-sufferers to avoid consumption of the offending food.

One of the most publicized horror stories about transgenic maize involves the infamous StarLink corn recall by the FDA (Taylor and Hefle 2001). StarLink corn contains a gene that expresses a Bt toxin named Cry9C. The protein was only slowly digested during in vitro digestibility tests, leading to concerns that it could become an allergen. Accordingly, the EPA approved StarLink on the condition that it be used only for animal feed. StarLink corn was marketed and eventually StarLink-encoding DNA was detected in several whole maize-containing food products. The FDA ordered a total mandatory recall of all StarLink-containing products – no matter how small the content of Cry9C. The cost of the recall, penalties, lost markets, and testing for the presence of Cry9C that continues to this day can be estimated in hundreds of millions of US dollars. It is difficult to determine if the greatest damage associated with the StarLink affair is the perception held by some consumers that transgenic foods might cause allergies, or the damage that the affair did to regulatory science. It should not be forgotten that from a scientific perspective StarLink posed no threat of causing food allergy (Taylor and Hefle 2001) for a number of reasons: (1) the human exposure was below the minimal threshold required to sensitize and subsequently elicit a human response; (2) Cry9C was found to be virtually completely degraded by food processing operations and the protein itself could not be measured in the foods in which the inserted StarLink DNA was found; (3) the in vitro digestibility test does not predict allergenicity, as it is one of several indicators that are weighed in the overall decision-making and, in addition, indigestible proteins are only infrequently allergens and digestible proteins can be allergens as well (Goodman et al. 2008); and, perhaps most importantly, (4) Cry9C bears no structural resemblance to any known allergen and therefore would have been highly unlikely to cause allergy. The underlying problem here is an unscientific attempt to demand absolute safety and zero risk. Digestibility is but one of several indicators that should have been used in decision-making. These policies unfairly discriminate against transgenic crops simply because they are transgenic. Recall that numerous foods that cause allergies are sold in markets around the world with no regulatory control.

9.2.4.5 Retention or Improvement of Nutritional Value and Animal Studies

Transgenic crops, as well as food and feeds derived from them, can be tested in animals to ensure that they provide either equivalent or improved nutritional value; however, as noted previously, animal studies are not particularly useful in the safety assessment of whole foods (Chassy et al. 2008). High lysine maize is an example of a nutritionally enhanced crop that improves performance in animal studies (Chassy et al. 2008). New transgenic varieties are routinely evaluated in several production animal species such as broilers, swine, and cattle; aquatic species such as catfish and tilapia can also be evaluated.

There are good reasons to avoid the use of animal studies as a safety assessment tool. Nonetheless, opponents of GMOs have used animal studies similar to those reported by Ewen and Pusztai (1999) in attempts to demonstrate that transgenic crops are inherently more dangerous than their conventional counterparts. In the interests of brevity, the reader is referred to several references that demonstrate the misuse of animal studies. EFSA has reviewed and rejected the claims that rootworm-resistant maize can cause kidney and liver abnormalities (Séralini et al. 2007; http://

www.efsa.europa.eu/EFSA/Statement/GMO_statement_MON863,0.pdf, accessed 14 January 2008). The claims by Ermakova that feeding soyflour prepared from herbicide-tolerant soybeans to rats caused infant mortality and stunted growth of rat pups have been found to be based on improper experimental design, poor animal stewardship, and flawed data analysis (Marshall 2007).

9.2.4.6 Equivalence of Composition Other than Intended Changes and Absence of Adverse Unintended Effects

A key component of safety assessment is examination of the composition of the transgenic variety in comparison to a near isogenic variety and other similar commercial varieties (Chassy et al. 2008; Cellini et al. 2004). Not only do these studies demonstrate the nutritional value of the crop, if no changes other than those intended are observed, but also it is taken as strong evidence that no unintended adverse changes have occurred. Comparative assessment is not intended to prove that two varieties have equivalent composition, nor does it claim that two varieties must have equivalent composition in order to be equally safe. The comparison seeks only to identify differences that are of nutritional and/or safety concern. Statistically significant differences in composition are often observed between varieties owing to natural genetic variation as well as cultural and environmental effects. Comparative analysis is performed to determine whether any *biologically significant* differences exist.

One of the major points that opponents of transgenic crops attempt to make to consumers is that unknown and uncharacterized unintended effects may take place in the process of gene insertion into plants. While this claim is plausible, it overlooks the fact that in the process of conventional breeding unintended effects are also introduced into crop plants (Cellini et al. 2004; Parrott 2005; Bradford et al. 2005). Recently, it has been demonstrated that transgenic plants can be more similar at the proteomic and metabolomic levels to the variety from which they are derived than that variety is to other varieties of the same crop (Shewry et al. 2007; Lehesranta et al. 2005; Catchpole et al. 2005). These findings illustrate both the precision of gene-insertion techniques and the random nature of unintended changes that accumulate in conventional breeding programs (Cellini et al. 2004; Parrott 2005). Genomic analysis has also provided striking insights into the differences in genetic content (e.g., mutations, insertions, deletions, rearrangements, and repetitions) that can be found in varieties of the same crop (Parrott 2005). The important learning here is that unintended effects occur in plant breeding; however, after much selection and culling, new varieties are subjected to a thorough safety analysis that demonstrates no unintended effects that might do harm have occurred. Genetic engineering may be viewed as the more precise and better-defined technology and is thus no more likely to introduce unintended effects than conventional breeding.

9.2.4.7 Estimation of Dietary Intake (Exposure)

In order to perform a meaningful safety assessment of potential toxicological and nutritional impacts, the total exposure (dietary intake) of the food or feed must be estimated (Chassy et al. 2008). Very few transgenic crops are consumed whole; for example, transgenic maize finds its way into the human diet in the form of small amounts of corn meal and flour and, more frequently, as corn starch and corn oil. A large number of processed food products may contain ingredients derived from transgenic canola, maize, or soybeans. Based on quantitative analysis of content in food products, the consumption frequency, and the consumed quantity, it is possible to predict dietary intake. In practice, exposure estimates are for the most part ignored by regulators. Considering the global concerns about the safety of GMOs, it is unlikely that any transgenic product about which there is any residual concern will be approved by regulators on the basis that consumption is sufficiently low to ensure that no harm will occur. Even though safety conclusions based on exposure calculations that predict acceptable safe intakes are the cornerstone of toxicology, transgenic plants are forced to conform to a zero risk standard (see the discussion on StarLink in Sect. 9.2.4.4)

9.3 Discussion and Conclusions

Setting aside the fact that there was no reason to believe that transgenic crops present any new or unusual risks (NAS 1987; Parrott 2005; Bradford et al. 2005; McHughen and Smythe 2008; Kok et al. 2008), it is possible to design a regulatory paradigm that will ensure that transgenic crops are as safe as crops produced by any other technology. Such a paradigm requires evidence that the introduced trait is safe for the environment and safe for humans and animals to consume. Phenotypic behavior and plant composition data can also be compared with that obtained with conventional varieties. While at the moment only transgenic crops are singled out for regulatory scrutiny, a safety analysis should probably be either performed on all new crop varieties or not required of any crop variety. As Cellini et al. (2004) concluded:

"The safety assessment of GM crops should focus primarily on the intended novel traits (target gene(s) and product(s)). Unintended effects occur in both GM and non-GM crops; however, GM crops are better characterised. It may be suggested that the two should be treated the same in safety assessments, bearing in mind that safety assessments are not required for non-GM crops."

Unfortunately, for a variety of reasons (precaution, opposition pressure, adverse publicity, uncertainty, trade advantages, prejudices, emotions, lack of science literacy, and politics, among others) governments around the world have put in place stringent regulatory systems that require a demonstration of virtually zero risk before the approval of a novel transgenic variety. These regulatory requirements far exceed those placed on plants having similar phenotypes – and thus presenting similar risk – produced using alternative technologies. Some of these alternative technologies such as chemical and radiation mutagenesis are surely more likely to produce unknown and unintended effects than precise insertion of well-characterized genes.

In spite of the unreasonable challenges, the regulatory system described in the foregoing sections has been successfully applied to crop approval in a few jurisdictions. Maize is an outstanding example of the application of genetic engineering to strain improvement, with dozens of approved varieties being grown around the globe. One could of course argue that being cautious with a new technology is justified until it is better understood. This overlooks the fact that moving genes between plants is nothing new and fails to account for the cost of the precaution.

It has been calculated that transgenic crops have been worth $US 28 billion globally to farmers in the first 10 years of their use; that chemical use in agriculture has been reduced 14%, and that the reduction in greenhouse gas emissions associated with adoption of transgenic crops is the equivalent of removing 4 million cars from the road (Brookes and Barfoot 2006). Real benefits are lost and real environmental damage continues to be done while the world hesitates to embrace transgenic technology – it is amazing that groups that claim to defend the environment also uniformly oppose GMOs. Over the course of the first 10 years of planting transgenic crops, the forecasted horror stories propagated by opponents have not come to pass (Chassy et al. 2005).

Transgenic technology holds the greatest promise for the world's poorest farmers, but access to that technology is denied to them in part because of the high regulatory barriers placed before novel transgenic crops (Delmer 2005). It is of particular concern that nutritionally enhanced transgenic crops that are designed to help alleviate global malnutrition that seriously affects billions of humans will face an unscientific regulatory barrier not faced by conventionally bred crops (Kok et al. 2008). The same situation prevails for other crops that are not planted as widely as rice, corn, cotton, canola and soybeans, since the high cost of regulation makes it uneconomic to introduce transgenic varieties even though they have been developed (Kalaitzandonakes et al. 2007; McHughen and Smythe 2008). Negative consumer perceptions and trade barriers further inhibit the use of transgenic technology.

Nowhere is the unscientific discrimination against transgenic crops more obvious and more unjustified than the requirement for mandatory labeling imposed by many nations that claim that the public has a right to know and a right to choose. Proponents of labeling even ask supporters of GM crops "if they are safe as you say, why are you afraid to label them?" The label identifies a product that has been unscientifically demonized by opponents so that consumers can decline to buy it. Labels are traditionally used to provide safety and other *meaningful* information to consumers; one of the major points of this chapter has been to make the case that the technology used to produce the crop is not meaningful. More importantly, consumer choice can be protected by a voluntary labeling system such as that used for halal, kosher, and organic foods.

Although they both preserve choice and grant a consumer the right-to-know, the differences between these two approaches (mandatory versus voluntary labels) is not at all trivial. With a voluntary labeling system those consumers desiring the special property, in this case GM-free foods, pay for the cost. A mandatory labeling system means that every consumer bears the cost of repeatedly testing every batch of every raw material, ingredient, intermediate, and final product from the farm, to the elevator, to the factory, to the warehouse, and to the supermarket to ensure GM-free status. This must be done for all foods under a mandatory labeling system, thereby adding billions of dollars each year to the global food bill. This useless testing represents an investment of resources that buys the consumer no added safety – the products would not be on the market if they were not safe. These are resources that could better be spent improving public health, saving the environment, or even feeding the world's hungry.

Two questions remain: (1) why are we over-regulating transgenic crops? and (2) what can the reader do about it? The answers are beyond the scope of this chapter. Suffice it to say in closing that the full benefits of transgenic technology will not be harvested until the discriminatory regulatory barriers are lowered and the discriminatory labels removed.

References

Berche P (1998) Les plantes transgéniques et la resistance aux antibiotiques. Méd Thérap 4:709–719

Bradford KJ, Van Deynze A, Gutterson N, Parrott W, Strauss SH (2005) Regulating transgenic crops sensibly: lessons from plant breeding, biotechnology and genomics. Nat Biotechnol 23:439–444

Breiteneder H, Mills CEN (2005) Plant food allergens – structural and functional aspects of allergenicity. Biotechnol Adv 23:395–399

Brookes G, Barfoot P (2006) Global impact of biotech crops: socio-economic and environmental effects in the first ten years of commercial use. AgBioForum 9:139–151. http://www.agbioforum.org. Accessed 14 January 2007

Calva, JJ, Sifuentes-Osornio J, Cerón C (1996) Antimicrobial resistance in fecal flora: longitudinal community-based surveillance of children from urban Mexico. Antimicrob Agents Chemother 40:1699–1702

Catchpole GS, Beckmann M, Enot DP, Mondhe M, Zywicki B, Taylor J, Hardy N, Smith A, King RD, Kell DB, Fiehn O, Draper J (2005) Hierarchical metabolomics demonstrates substantial compositional similarity between genetically modified and conventional potato crops. Proc Natl Acad Sci USA 102:14458–14462

Cellini F, Chesson A, Colquhoun I, Constable A, Davies HV, Engel KH, Gatehouse AM, Karenlampi S, Kok EJ, Leguay JJ, Lehesranta S, Noteborn HP, Pedersen J, Smith M (2004) Unintended effects and their detection in genetically modified crops. Food Chem Toxicol 42: 1089–1125

Chassy B (2002) Safety and regulatory assessment of crops developed via biotechnology. J Am Coll Nutr 21:166S–173S

Chassy BM (2007) The history and future of GMOs in food and agriculture. Cereal Foods World 52:169–173

Chassy B, Abramson SH, Bridges A, Dyer WE, Faust MA, Harlander SK, Hefle SL, Munro IC, Rice ME (2001) Evaluation of the U.S. regulatory process for crops developed through biotechnology. Council for Agricultural Science and Technology, Ames

Chassy B, Hlywka JJ, Kleter GA, Kok EJ, Kuiper HA, McGloughlin M, Munro IC, Phipps RH, Reid JE (2004) Nutritional and safety assessments of foods and feeds nutritionally improved through biotechnology. Comp Rev Food Sci Food Saf 3:35–104

Chassy B, Parrott W, Roush R (2005) Crop biotechnology and the future of food: a scientific assessment. Council for Agricultural Science and Technology, Ames

Chassy B, Egnin M, Gao Y, Glenn K, Kleter GA, Nestel P, Newell-McGloughlin M, Phipps RH, Shillito R (2008) Nutritional and safety assessments of foods and feeds nutritionally improved through biotechnology: case studies. Comp Rev Food Sci Food Saf (in press)

Codex Alimentarius (2003) Principles for risk analysis and guidelines for safety assessment of foods derived from modern biotechnology. Codex Alimentarius, Rome. http://www.codexalimentarius.net/web/publications.jsp?lang=en. Accessed 10 January 2008

Delmer DP (2005) Inaugural article: agriculture in the developing world: connecting innovations in plant research to downstream applications. Proc Natl Acad Sci USA 102:15739–15746

EFSA (2007) Minutes of the 37th plenary meeting of the scientific panel on genetically modified organisms held on 22–23 November 2007 in Brussels (adopted on 18 December 2007 as "Annex: Analysis by the GMO panel of the PNAS publication of Rosi-Marshall et al. 2007. Toxins in transgenic crop byproducts may affect headwater stream ecosystems." http://www.efsa.europa.eu/EFSA/Event_Meeting/GMO_Minutes_37th_plenmeet,1.pdf. Accessed 14 January 2008

Ewen SW, Pusztai A (1999) Effect of diets containing genetically modified potatoes expressing *Galanthus nivalis* lectin on rat small intestine. Lancet 354:1353–1354

FDA (US Food and Drug Administration) (1992) Statement of policy: foods derived from new plant varieties. Notice. US Federal Register 57:22,984–23,005

Fuchs M, Gonsalves D (2007) Safety of virus-resistant transgenic plants two decades after their introduction – lessons from realistic field risk assessment studies. Ann Rev Phytopath 45:173–202

Goodman RE, Vieths S, Sampson HA, Hill D, Ebisawa M, Taylor, SL, van Ree R (2008) Allergenicity assessment of genetically modified crops: what makes sense? Nat Biotech 26:73–81

Hathcock JN (2000) The precautionary principle – an impossible burden of proof for new products. AgBioForum 3:255–258. http://www.agbioforum.org. Accessed 14 January 2008

James C (2007) Executive summary ISAAA brief 35: global status of commercialized biotech/GM crops: 2006. ISAAA, Ithaca. http://www.isaaa.org/resources/publications/briefs/35/executivesummary/default.html. Accessed 14 Jan 2008

Kalaitzandonakes N (2000) Why does biotech regulation differ so much between the US and EU? AgBioForum 3:75–76. http://www.agbioforum.org. Accessed 14 January 2008

Kalaitzandonakes N, Alston J, Bradford K (2007) Compliance costs for regulatory approval of new biotech crops. Nat Biotech 25:509–511

Kok EJ, Kuiper HA (2003) Comparative safety assessment for biotech crops. Trends Biotechnol 21:439–444

Kok EJ, Keijer J, Kleter GA, Kuiper HA (2008) Comparative safety assessment of plant-derived foods. Regul Toxicol Pharmacol 50:98–113

Konig A, Cockburn A, Crevel RWR, Debruyne E, Grafstroem R, Hammerling U, Kimber I, Knudsen I, Kuiper HA, Peijnenburg AACM, Penninks AH, Poulsen M, Schauzu M, Wal JM (2004) Assessment of the safety of foods derived from genetically modified (GM) crops. Food Chem Toxicol 42:1047–1088

Kuiper HA, Noteborn HP, Peijnenburg AA (1999) Adequacy of methods for testing the safety of genetically modified foods. Lancet 354:1315–1316

Lehesranta SJ, Davies HV, Shepherd LV, Nunan N, McNicol JW, Auriola S, Koistinen KM, Suomalainen S, Kokko HI, Karenlampi SO (2005) Comparison of tuber proteomes of potato varieties, landraces, and genetically modified lines. Plant Physiol 138:1690–1699

Losey JE, Rayor LS, Carter ME (1999) Transgenic pollen harms monarch larvae. Nature 399:214

Marshall A (2007) GM soybeans and health safety – a controversy reexamined. Nat Biotech 25:981–987

McHughen A, Smythe S (2008) US regulatory system for genetically modified [genetically modified organism (GMO), rDNA or transgenic] crop cultivars. Plant Biotechnol J 6:2–12

NAS (National Academy of Science, USA) (1987) Introduction of recombinant DNA-engineered organisms into the environment: key issues. National Academies Press, Washington, DC

OECD (2002) Environmental health and safety publications. Series on the safety of novel foods and feeds. No. 6. Consensus document on compositional considerations for new varieties of maize (*Zea mays*): key food and feed nutrients, anti-nutrients and secondary plant metabolites. Environment Directorate. OECD, Paris. http://www.olis.oecd.org/olis/2002doc.nsf/LinkTo/NT00002F66/$FILE/JT00130429.pdf. Accessed 12 January 2008

Parrott W (2005) The nature of change: towards sensible regulation of transgenic crops based on lessons from plant breeding, biotechnology and genomics. Proc 17th Meeting National Agricultural Biotechnology Council, Nashville, June 27–29, pp 209–220. NABC, Ithaca

Rosi-Marshall EJ, Tank JL, Royer TV, Whiles MR, Evans-White M, Chambers C, Griffiths NA, Pokelsek J, Stephen ML (2007) Toxins in transgenic crop byproducts may affect headwater stream ecosystems. Proc Natl Acad Sci USA 104:16204–16208

Sears MK, Hellmich RL, Stanley-Horn DE, Oberhauser KS, Pleasants JM, Mattila HR, Siegfried BD, Dively GP (2001) Impact of *Bt* corn pollen on monarch butterfly populations: a risk assessment. Proc Natl Acad Sci USA 98:11937–11942

Séralini G-E, Cellier D, de Vendomois J (2007) New analysis of a rat feeding study with a genetically modified maize reveals signs of hepatorenal toxicity. Arch Environ Contam Toxicol 52:596–602

Shewry PR, Baudo M, Lovegrove A, Powers S, Napier JA, Ward JL, Baker JM, Beale MH (2007) Are GM and conventionally bred cereals really different? Trends Food Sci Technol 18:201–209

Taylor SL, Hefle SL (2001) Will genetically modified foods be allergenic? J Allergy Clin Immunol 107:765–771

van den Eede G, Aarts H, Buhk HJ, Corthier G, Flint HJ, Hammes W, Jacobsen B, Midtvedt T, van der Vossen J, von Wright A, Wackernagel W, Wilcks A (2004) The relevance of gene transfer to the safety of food and feed derived from genetically modified (GM) plants. Food Chem Toxicol 42:1127–1156

Part III
Breeding and Genetics

Chapter 10
Doubled Haploids

Ming-Tang Chang and Edward H. Coe, Jr

10.1 Introduction

Maize doubled haploid (DH) technology provides fixed, pure lines from a donor parent. Protocols for breeding of DH lines are available for over 250 crop species, and over 300 DH-derived cultivars have been developed in 12 species worldwide (Forster and Thomas, 2005). In maize, methods for inducing, selecting and doubling haploid plants are advanced and are in widespread use.

In a haploid plant, expression of positive or deleterious effects of genes for seed development, plant growth and function is unmasked, and plants that function effectively will have a better chance to grow to maturity and set seeds. Haploid plants that show good vigor in a natural environment will usually perform well as DH progenies under environmental stress. When doubled and brought to normal genetic balance, DH lines can be selected for agronomic traits, and testing can more accurately estimate yield potential and yield stability under different environments. The DH genome with its pure genetic makeup may still be challenged by environment in the absence of prior selective pressure. The characteristics of DH lines in theory are fixed and stable, and no further inbreeding depression should be observed from generation to generation, although spontaneous mutation or genomic changes caused by transposable elements cannot be avoided (Stadler 1951; Messing 2005).

Ming-Tang Chang
BASF Plant Science LLC, ExSeed Genetics, 2901 South Loop Dr. Bldg 3 Suite 3800, Ames, IA 50010, USA
Current address: 5333 Cervantes Dr., Ames, IA 50014-6914, USA
e-mail: mtcchang@hotmail.com

Edward H. Coe, Jr
University of Missouri, Curtis Hall, Columbia, MO 65211-7020, USA
e-mail: coee@missouri.edu

A.L. Kriz, B.A. Larkins (eds.), *Molecular Genetic Approaches to Maize Improvement* 127
Biotechnology in Agriculture and Forestry, Vol. 63

10.2 History

Maize haploid studies started in the 1920s. Spontaneous parthenogenetic or androgenetic haploids are rare and are generally not noticeable in genetic and breeding studies (Randolph 1938, 1940; Randolph and Fisher 1939). In an article on genetic variation, Randolph (1932) notes the following: "Haploidy in maize was first reported by L.J. Stadler, and the writer, in papers presented before section 0, Amer. Assoc. Adv. Sci., Des Moines, Iowa, 1929." Accordingly, Stadler and Randolph were the first to describe maize haploids (Rober et al. 2005). East (1930) mentions studies by Stadler and unpublished studies by R.A. Emerson, and points out that homozygotes would arise through parthenogenesis and that DHs could eliminate the labor of producing homozygotes through long periods of self-fertilization. There were limited studies on maize haploidy until Chase began to develop pure elite DH lines for commercial hybrid application (Chase 1947, 1949b). He demonstrated practical application of DHs in maize breeding (Chase 1951), and in his breeding career developed very useful DH lines for commercial hybrid application. His first significant haploid-parented hybrid was DeKalb 640 (Forster and Thomas 2005), a double cross (B14×H2167/H2386×H2389) with three DH lines and a station inbred. H2167 was derived from the first cycle Iowa Stiff Stalk Synthetic of Sprague. H2386 and H2389 were sister lines, second cycle DH lines derived from W22xH225 cross. H225 was the ninth, H2167 the 127th, H2386 the 190th and H2389 the 192nd, DH lines produced. It was important in the eastern US market (Pennsylvania) for a number of years and was the first high-density tolerant hybrid of wide acceptance. DeKalb 640 also had a good market in Europe, southern France, northern Italy and the Danube Basin, under different designations. An interesting point is that Marcus Zuber once told Chase that when he was testing Mo17, the high performing single hybrid was Mo17 × N22. N22 was one of the Stiff Stalk Synthetic DH lines out of Chase's Ames program, and this line went out to Nebraska and was released by that station. This performance alerted Zuber to the potency of Mo17 with SSS lines, and when B73 came along, that line took the place of N22 (S.S. Chase, personal communication). Chase did breed many other useful lines for commercial hybrid production, such as popular hybrids DeKalb XL66 (1.9 million seed units) in 1967 and DeKalb XL64 (3.8 million units) in 1970 (Troyer 2004). Chase carried out many firsts in maize haploid breeding, including (1) the first successful selfing of a maize haploid (giving rise to HD-1, a sweet corn line out of Golden Cross Bantam) (Chase 1949b), (2) recognizing that even parthenogenetic rates as low as 1 per 1000 or 2000 were practical if good genetic screening stocks were used, (3) producing DH lines in quantity, (4) showing that the pollinator stock affected parthenogenetic rates (Chase 1952), (5) determining that chromosome doubling treatments were not necessary for success as the rate of natural doubling was high, and (6) suggesting that androgenesis can be used for direct transfer of maize cytoplasm (Chase 1963).

The Northrup King Seed Company had a late Mexican meal corn with the unique genetic traits of red collar, white chalky endosperm and purple-seeded floury flint type. This line had no practical use for Northrup King, but it was good genetic material, and they gave it to Dr. Charles R. Burnham in 1941 for research purposes. Coe, a graduate student with Burnham, received the seeds for a study of pigments. For identification purposes, Coe designated it as Stock 6 in 1950. Based on subsequent studies, self-pollinated plants of Stock 6 yielded 1.97% haploids and Stock 6 selfs (haploid by sib) yielded 2.86% haploids, with an average haploid induction rate of 2.52% (Coe 1959). Coe developed several marker systems, such as *C1-I* and *R1-nj*, which his student, K.R. Sarkar, applied to facilitate identification of haploid seeds (Coe and Sarkar 1964; Sarkar and Coe 1966, 1971). Sarkar advanced studies of maize haploid induction after returning to India (Sarkar et al. 1972; Sarkar 1974; Mathur et al. 1976; Aman and Sarkar 1978). Kato (2002) derived DH progeny from four inbred lines and four hybrids crossed by Stock 6 marked with *R1-scm2*, doubled by the use of nitrous oxide gas.

A mutant gene, *ig1* or indeterminate gametophyte, can increase the frequency of androgenetic haploids in its progeny (Kermicle 1969, 1971; Lin 1981). The effect of *ig1* on seed development is a failure of normal differentiation of nuclei and cells of the female gametophyte, in which the number of cells that function as eggs is indeterminate. In addition to producing 6% poly-embryony, 7% hetero-fertilization, 45% elevated ploidy level of the endosperm and other rare anomalies, the *ig1* stock produces about 3% paternal origin haploid seeds (Kermicle 1969). Androgenesis in higher plants is the development of offspring with paternal chromosomes only, affording breeders and geneticists a means for direct transfer of cytoplasm from one strain to another (Chase 1951, 1963). The practical application of *ig1/ig1* stock in corn breeding is the conversion of an inbred line to its cytoplasmic male sterile form. The homozygous *ig1/ig1* cms W22 stock is used as female and is crossed by a normal inbred. The doubled androgenetic haploid plants are isogenic with the normal except that they carry male sterile cytoplasm.

Study of maize haploids at Krasnodar Agricultural Research Institute in 1969 in Russia was begun by M.V. Chumak and was continued by V.S. Shcherbak, O.A. Shatskaya and E.R. Zabirova. They used Chase PEM (purple embryo marker), Coe Stock 6 and several stocks received in 1982 from V.S. Tyrnov and A.N. Zavalishina from Saratov University as source materials for creating a new haploid inducer. From crosses among those materials and individual selection in their progenies, they were able to generate several new inducers under the general name EMK (Embryo Marker Krasnodarsky) or ZMK (Zarodyshevy Marker Krasnodarsky). EMK-1, with an induction rate of haploids from 6% to over 10%, was produced in 1991 (Shatskaya 2004). In Russia, many registered hybrids use DH lines as one of their parents. Hybrids using DH line Kr716, developed by Chumak, are Krasnodarsky 383MV (1998–2005), Krasnodarsky 384MV (2000–2005), Krasnodarsky 382MV (1992 to current) and ROSS 387MV (1994–2004). A hybrid using DH line Kr503-1 is Krasnodarsky 599MV (2006 to current). Hybrids using DH line Kr640-3 can yield very highly productive hybrids, including Krasnodarsky 290MV (2004 to current), Krasnodarsky 385MV (2005 to current), Krasnodarsky 291MV (2006 to current), Intercras 375 (2006 to current) and Intercras 405 (2006 to current). Several hybrids with Kr640-3 were registered this year in the Ukraine (E.R. Zabirova, personal communication). Line AT-1, developed in 1982 at Saratov University (Tyrnov 1997), gave maternal haploids with a frequency of 90–100%. The frequency of haploids could be higher than 100% because of the occurrence of haploid twins and triplets. During the first 4 years there was a constant threat of losing this line because of its haploid nature, and diploid seeds were rare. In addition, this line was seriously affected by *Ustilago maydis* and could not be practically used. It was converted to a resistant form, in which the haploid induction rate of the newly selected line is about 2–3% (Tyrnov 1997). Further study revealed that early pollination of the AT-1 and AT-3 haploids by normal pollen grains resulted in 3.6% and 2.7% haploids, respectively, and late pollinations resulted in 78% and 75% haploids, respectively (Smolkina and Tyrnov 2003), showing the significant effects of handling on haploid frequency. Chalyk et al. (1994) created a new inducer line MHI (Moldovian Haploid Inducer) from parents of KMS (Korichnevy Marker Saratovsky) and ZMS. This inducer line has an induction rate of 6.5% on average (Chalyk 1999; Eder and Chalyk 2002). The Krasnodar Embryo Marker Synthetic or KEMS (possibly EMK-1) and the French induction line WS14 (W23ig/Stock 6) are the parents of the new induction line RWS (Rober 1999; Rober et al. 2005). Many other inducer lines have been created in different countries, including CAU (China Agricultural University) inducer 1 (Liu and Song 2000a), which was developed by selection of progenies from crosses between Stock 6 and BHO (Beijing High Oil Population) and has an induction rate of 5–6%, and UH400 (University of Hohenheim 400) (Melchinger et al. 2005). UH400 is an inbred line derived from KEMS by W. Schipprack in Melchinger's group at Hohenheim (D. Geiger, personal communication).

10.3 Methods

In corn, haploid methods have been well developed over the last 60 years. Currently there are several types of haploid induction methods, briefly described below.

10.3.1 Spontaneous Haploids

The frequency of spontaneous maize haploids is 0.05–0.1%. The majority are maternal in origin (parthenogenesis). Androgenetic haploids are rare and the rate is about 1 in every 100,000 seeds (Randolph and Fisher 1939; Chase 1949a). Frequencies of both vary according to background.

10.3.2 Genetic Induction

Certain genetic stocks or some unique genes can produce higher percentages of haploid seeds when used as either the male or female, such as A385, 38-11 (Chase 1949a), Stock 6 (Coe 1959), *ig1* (indeterminate gametophyte) gene (Kermicle 1969), and advanced strains derived in part from them, described above.

The mechanism for haploid induction is still not fully understood, although there is some evidence that chromosome elimination may be involved (Rober et al. 2005). In most crosses, Stock 6 can produce 2–3% maternal haploids, Krasnodar marker 6–8%, and MHI 5–6%. Homozygous *ig1* plants used as female can induce from 1–15% paternal haploids (Shatskaya et al. 1994a). Color marker genes have been incorporated into male inducer lines for purple leaf, sheath and plants (with dominant *A1*, *A2*, *B1* and *Pl1*), and for purple endosperm crown and purple plumule color (with dominant *A1*, *A2*, *Bz1*, *Bz2*, *C1*, *C2* and *R1-nj*), to facilitate identification of haploid seeds at the ear level. *R1-nj* is particularly useful because it provides both recessive and dominant marking when crossed as a male on an *r*1 parent. Hybrid seeds have purple endosperm crown and purple plumule, while haploid seeds have purple crown but no plumule color and can be clearly identified by eye. Morphological differences associated with haploidy include fewer, narrower and stiffer leaves with occasional white sectors; smaller plant with slower growth rate; smaller cell size; and smaller guard cells (Chase 1947, 1969; Coe and Sarkar 1964; Greenblatt and Bock 1967; Dankov et al. 1990; Han et al. 2006). The xenia effect from a high oil inducer line can immediately identify haploid seeds with 90% accuracy based on their oil content or embryo size (Chen 2003). The size of haploid embryos in this cross is much smaller than the hybrid seeds. On average from non-destructive single seed NMR (Nuclear Magnetic Resonance) measurement, the oil content of hybrid seeds is 5.26%, the oil content of self-contaminated seeds is 3.86% and the oil content of haploid seeds is 3.42% (Chen 2003). Toward selecting for higher rates from an inducer line, recessive seedling markers such as glossy or liguleless can be reliable aids for the determination of haploid frequencies. A female line carrying glossy or liguleless is crossed with a haploid inducer, and the seeds are planted in a sand bench to be screened for recessive seedlings. The percentage of recessives represents the haploid induction rate, so long as there is no self-contamination or chromosome loss.

10.3.3 Modifications in Handling

Haploid induction rate is affected mainly by genetics, but some studies have shown that changes of environments or handling affect rate of haploid induction. Factors such as delaying pollination to the afternoon, aging of silks, heat, etc. may change the haploidy rate (Rober et al. 2005; Zaharova 1955; Aman et al. 1981; Mathur et al. 1980; Smolkina and Tyrnov 2003).

10.3.4 Artificial Induction

Certain chemicals and radiation can induce haploid formation, such as maleic hydrazide (MH), 2,4-D, NAA-Na, GA_3, IAA, colchicine (Deanon 1957; Zhao and Gu 1984, 1988), trifluralin (Kato 1997), radiation (Mathur et al. 1976), Basagran and other herbicides (Dankov et al. 1997; Wan et al. 1991; Hansen and Andersen 1998).

Kato treated pre-flowering tassels with trifluralin to inhibit the second microspore mitotic division. Zhao and Gu (1988, 1984), Tu et al. (1994) and others used injection to deliver 40 mg/L MH + 2% DMSO + 0.1% colchicine solution into the cob of unfertilized ears, and were successful in obtaining DH lines.

10.3.5 Anther Culture, Embryo Culture and Microspore Culture

Anther culture, embryo culture and microspore culture have been used to generate haploid plantlets (Petolino and Jones 1986; Wan et al. 1991; Aulinger et al. 2003; Zheng et al. 2003; Barnabas 2003; Armstrong et al. 2004), but application is limited because it has low efficiency, is genotype dependent, is time consuming and involves technical demands. The Laboratory of Plant Cell and Tissue Culture (1975) of the Institute of Genetics, Academia Sinica, first reported their success in maize anther culture. Green shoots, leaves and roots emerged from callus and developed into plantlets. Examination of chromosome number in root tip cells showed that they were haploids with 10 chromosomes.

10.3.6 Wide Crosses and Chromosome Elimination

Remote or wide hybridization and chromosome elimination can generate DHs in some species. For example, maize pollen applied to wheat, oat or rice can induce unfertilized haploid embryo development (Zhou et al. 1979; Zenkteler and Nitzsche 1984; Laurie and Bennett 1986; Matzk and Mahn 1994; Bains et al. 1995; Berzonsky et al. 2003; Inagaki 2003; Rines 2003). This is a quite useful technique to develop doubled haploid lines. Because of incompatibility the maize chromosomes are rapidly eliminated during cell mitosis and leave only a haploid genome from the original parent. So far, there is no successful record to indicate that this method is applicable to maize, even though maize can be crossed with closely related relatives, such as gama grass and teosintes.

10.3.7 Apomixis (Parthenogenesis and/or Androgenesis)

Apomixis is a process of regenerating seeds or plantlets from unfertilized gametophytes. It has a genetic basis, and genes responsible for apomictic responses can be identified by monosomic or segment translocation methods. It is possible to introduce apomictic genes from gama grass (*Tripsacum*) into maize (Sokolov et al. 1998; Kindiger 1997, 1998, 2006; Kindiger and Sokolov 1998). Current study has identified a small fragment from gama grass chromosome 16L, transferred to maize chromosome 6L, as being responsible for apomixis.

10.4 Chromosome Doubling

Production of doubled haploids requires that progeny be derived from selected haploid plants. Doubling can occur spontaneously, and its rate can be enhanced by selection. Methods for artificial induction of doubling have been developed.

10.4.1 Spontaneous Doubling

Spontaneous doubling in haploid tassels produces fertile diploid sectors. It is evident that a small proportion of somatic haploid cells of a haploid plant are doubled spontaneously through somatic cell fusion, endoreduplication, endomitosis or some other mechanism (Jensen 1974; Testillano et al. 2004). Doubling at PMC (pollen mother cell) stage just before meiosis can yield a quartet with four normal pollen grains. Staining pollen grains from intact haploid anthers with iodine solution often displays a small proportion of blue round, light blue round and many transparent irregular aborted pollen grains. The dark blue round pollen grains presumably are starch-filled, normal pollen grains, but they are accompanied by much non-fertile pollen. Such anthers are not able to open naturally, and the few normal pollen grains are not able to release and to serve their normal function. Larger amounts of normal pollen grains inside the anther will help the anther to function, split and release pollen grains. It is possible to cut anthers in half or squeeze anthers to force normal pollen grains out to fertilize the silks. This method is tedious and laborious and it is not recommended as a practical exercise for large-scale breeding processes. The percentage of haploid tassels that shed normal pollen grains varies significantly, in the range 2.8–46%, and is genetically dependent (Shatskaya et al. 1994b; Liu and Song 2000b; Wei and Chen 2006; Han et al. 2006). Generally, many haploid tassels have only a few florets that can shed normal pollen grains, and only about 54% fertile tassels have shown a large sector of normal anthers. Spontaneous fertility restoration of the female inflorescence or the ears is in the range 25–94% (Chalyk et al. 1994; Liu and Song 2000b; Han et al. 2006), which is much higher than for the haploid tassel. Therefore, tassel fertility is the limiting factor for application in a DH breeding program. Using pollen grains collected from normal diploid plants to pollinate haploid ears, it is possible to determine the frequency of fertility restorations, seed set, size of doubling tissue and seed distribution of the haploid ears. The average seed set is about 25–30 seeds per ear, and seed distribution on the ear is either from a large cluster in a specific area or randomly distributed (Liu and Song 2000b; Han et al. 2006). This implies that spontaneous doubling events can be either single events leading to a large chimerical sector on the ear or multiple random events that form scattered seed set. Any normal egg with normal silks can be fertilized by a normal pollen grain and develops to a normal mature seed. The time and rate of spontaneous doubling of haploid cells are background dependent. Haploid plants that are derived from certain parent materials show a few early doubling events during mitosis and produce a large fertile sector or sectors with a fair amount of normally shed pollen grains or a large cluster of seeds on the ear. Those materials in general have a higher rate of fertility restoration. Other parent materials show many late doubling events and produce many scattered normal pollen grains within underdeveloped anthers. Those anthers cannot open and release pollen grains naturally, but can be released by cutting or mechanical methods. Those materials in general have a lower rate of fertility restoration. Furthermore, haploid plants that are derived from some parent materials are quite stable and do not show any doubling events during mitosis. The latter require chemical treatment to induce chromosome doubling artificially (Sect. 10.4.3).

10.4.2 Selection for Spontaneous Doubling

Recycling of doubled haploid lines by using DH lines as source parents in the next cycle of selection will increase the frequency of spontaneous diploidization of the haploid tassel. Studies have shown that by recycling DHs for production of haploids, the fertility restoration rate increased from 9.4% to 33% (Chase 1952; Zabirova et al. 1993, 1996) or even to 43% (Shatskaya et al. 1994b).

10.4.3 Artificial Doubling

It is not necessary to apply chemical treatment to double the chromosome number if the spontaneous restoring rate of tassel fertility is over 20%. Generally, it is possible to increase the frequency of doubling in haploid tassels to 20–50% by treating haploid seedlings with 0.06–0.5% colchicine solution (Han et al. 2006). Colchicine response is genotype dependent. Nitrous oxide gas was applied by Kato (2002) to double haploids and derive progeny from haploids from four inbred lines and four hybrids.

10.5 Advantages

10.5.1 Genetic Homozygosity

DH lines provide genetic homozygosity in one generation. Because haploids carry only a single copy of every gene, any gene or genes that have deleterious effects for seed or plant development will have immediate genetic effects to depress or inhibit normal seed or plant development so these plants will be quickly eliminated at the haploid stage. This provides an efficient tool to eliminate unfavorable genes and to enrich favorable genes to improve the genetic pool rapidly. This resembles the process of natural selection but in a very rapid way to quickly fix favorable genomic combinations, conserving many useful genes from a breeding perspective. Doubling of those favorable haploids will generate a DH line with 100% genetic homozygosity. This overcomes the slow process of continuous selfing over many generations to reach almost genetic homozygosity in a conventional breeding program. These DH lines will not show either genetic segregation or inbreeding depression in the following generations, except for the influence from spontaneous gene mutations or transpositions that may cause certain deleterious influences and segregation.

10.5.2 Genetic Enrichment

Studies show that recycling DH lines can quickly improve haploid frequency and fertility restoration (Chase 1952; Zabirova et al. 1993; Shatskaya et al. 1994a, b; Liu and Song 2000a). According to Chase (1952), the original Stiff Stalk Synthetic materials yielded 0.13% haploids and the haploid-derived DHs yielded 0.43% haploids. The haploid fertility restoration of the original Stiff Stalk Synthetic was 9.4% and the DH-derived haploids have increased frequency to 33%. It appears that selection favors genetic or germplasm enrichment for production of haploids and fertility restoration. If that is the case, then germplasm enrichment for yield, general vigor and agronomy of a corn plant can be achieved by applying random mating of high yielding DH lines as source materials for the next cycle of haploid selection. Recycling of selected DH lines through recurrent selection or any other breeding scheme is a fast and powerful way to achieve genetic enrichment of the inbred carrying more favorable alleles for yield, pest resistance, stress tolerance and general agronomic traits (Griffing 1975; Gallais 1988, 1989; Dietzmann and Wehr 1996; Bouchez and Gallais 2000; Chalyk and Rotarenco 2001).

10.5.3 Gamete Selection

Gamete selection (Stadler 1944) is a simple and powerful tool, the potential of which can be realized by DH technology applied to breeding. Selection at the gamete or haploid level is more effective than at the diploid level because the probability of obtaining any genotype that carries n favorable genes is 1 per 2^n individuals for DHs, and the chance is much higher than 1 per 4^n individuals for diploids (Schlegel 2003). Selection is performed at the gamete level, and DHs in a sense are derived from gametes. Using the DH method a gamete with excellent genetic make-up can quickly be fixed to become a homozygous individual. The formation of haploids is a random event, based on isozyme, recessive marker genes and genetic similarity studies (Chang 1992; Chalyk and Chebotar 2000; Seitz 2005). If the number of DH lines is high enough, then results from selection can be effective. DHs have maximum genetic variance in line per se and test-cross trials (Rober et al. 2005). Studies of inbreeding cereals show that DH does not lead to any bias of genotypes in populations, and random DHs were even found to be comparable to selected lines produced by pedigree selection (Forster and Thomas 2005). The DH lines are good source materials for testing of their hybrid yield potential, level of genetic heterosis and yield stability. Results are more reliable than materials with various degrees of genetic segregation or levels of inbreeding. Generally, selected DH lines maintain high yield and outstanding agronomic traits constantly from generation to generation. In other words, their early selection results show high repeatability.

10.5.4 Gene Mutation

Maize haploids are good source materials for mutation study. It is possible to use haploids to estimate the spontaneous mutation rate of a specific gene locus. Since haploids only carry a single genome, the estimation of mutation rate should be more accurate and straightforward. In a haploid field with 50,000–100,000 haploids, it is not unusual to observe certain seedling mutants, such as liguleless, glossy, dwarf, brown midrib, albino and tassel seeds. Kernel mutants are more difficult to study, and can be enumerated only if the number of doubled haploids is high enough, such as 20,000—50,000 DH lines. Occasionally, a DH ear shows all waxy, or sugary or opaque seeds, proving that they are actually homozygous mutants. In microspore culture, it is very effective to generate mutants by treating microspores with chemical mutagens at the uninucleate stage, and this will generate pure elite mutant inbred lines (Szarejko 2003). Another application of the DH method is forward breeding to create new homozygous mutant lines in place of backcross conversion. For example, a waxy inbred can be crossed with a normal yellow dent inbred, and the hybrid seeds can then be crossed with an inducer. In theory, 50% of the DHs will be homozygous waxy inbred lines and 50% will be homozygous yellow dent lines. They are completely new lines, and may or may not be better than the original lines, but they are an alternative to converting traits by backcrossing.

10.5.5 Molecular Mapping Applications

Currently DH lines are routinely used to produce mapping populations for mapping simple genetic traits of agronomic importance, such as disease resistance and plant stature. DHs are ideal mapping materials for constructing a genetic linkage map and can be reconstructed and repeatedly sampled from time to time. Map construction is much easier using DH lines derived from hybrids of two pure homozygous parents. Using DH materials, genetic maps of barley, rice, wheat, rapeseed and pepper have been constructed. It is also possible to use marker-assisted selection (MAS) to identify the most promising DH for commercial line development and application.

DH lines are very effective tools for QTL (quantitative trait locus) analysis. Since DH lines are homozygous, data collected from multiple sites, seasons and years in replicated trials can be pooled and assessed for QTL analysis, such as yield, yield potential, general stress responses and yield stability under different environments (Jansen et al. 2003; Forster and Thomas 2005).

10.6 Future Perspectives

So far, the most effective method of haploid induction in maize is genetic induction. Most of the corn seed companies now have well-established haploid breeding programs and DH lines are routinely produced in reasonable numbers. Yet, the efficiency of certain steps for haploid seed production is so low that application of the DH method is limited: (1) the 3% haploid frequency is too low for efficient application, (2) marker systems are not always precise and efficient to screen for haploids, and (3) the chromosome doubling frequency is low, at about 10%. If haploid frequency could be increased to 12%, color marker intensity enhanced and doubling frequency increased to 30%, the efficiency of the DH method would be improved significantly. Then it would become a still more powerful tool to speed up the breeding program.

DHs are a short-cut way to generate immediate homozygous pure inbred lines. Currently, more than 20 crops are using DH methods to produce new cultivars or pure lines. Anther culture and microspore culture methods are very successful in tobacco, rapeseed, cauliflower, broccoli, Brussels sprouts, cabbage, barley, wheat, triticale, turnip, rape, mustard, flax and apple. Anther culture only is quite promising in eggplant, rice, pepper, rye, oat, ryegrass, potato, cork oak and poplar. The wide crossing method is applicable in wheat, oat, rice, barley, triticale and potato. Gynogenesis (ovary and flower culture) is routinely used in cucumber, onion, sugar beet and citrus breeding. Maize is the only crop that is using the genetic method for mass quantity haploid production. The major problems that limit application of the DH method in other species are technical difficulties, time consumption and high cost. Techniques are genotype dependent in most cases and restrict the application to only a few limited genotypes. In addition, there are many other problems that need to be overcome, such as inbreeding depression, embryo germination, chromosome doubling, polyploids, albinism, physiological weakness, regeneration and fertility (Forster and Thomas 2005).

Technique difficulty in chromosome doubling of haploids is one of the limiting factors that restrict the general application of the DH method to crop breeding. In maize, colchicine and those herbicides inhibiting microtubule formation in mitosis appear to work well in doubling the chromosome number (Wan et al. 1991). Colchicine is a toxic chemical and may enter the environment. Therefore, it may not be a good choice to use for a large-scale application. The herbicides are a better choice because they are degradable in the soil. There are many different methods to apply chemical treatments, including soaking, injecting, dripping and spraying chemical solutions to cultured cells, tissue, seeds, plantlets, seedlings or plants (Jensen 1974). The results are quite different due to differences in procedures and methodologies. In maize, spontaneous fertility restoration of the tassel is useful but is variable. The rate of fertility restoration of the female inflorescence is high enough and doubling chromosome number of the female is less important. Chromosomal doubling is required only if the spontaneous doubling rate of the tassel is less than 20%. A doubling rate of the tassel higher than 20% is acceptable for practical and economic breeding application. Recycling selected DHs as source materials for next cycle haploid production increases the frequency of spontaneous doubling of the tassel and fertility restoration. In addition, recycling of selected DH lines through recurrent selection or other breeding schemes can satisfy objectives for maize population improvement.

The DHs have many advantages for application in basic genetic research, molecular studies and practical applications in plant breeding. Maize has a unique genetic system to generate routinely and randomly a large quantity of haploids. A reasonable amount of DHs are generated from the haploids and yield potential is evaluated in test crosses. Results indicate that the DH lines represent a random sample of gametes of the initial breeding population (Chang 1992; Seitz 2005). Grain yield comparison across all sets, years and locations of S2 lines, S3 lines and DH lines shows no significant advantage of any one method. In contrast, ranges of test-cross means were larger for the DH lines in all ten sets and the top yielding lines were obtained by the DH method in seven out of ten sets (Seitz 2005). This may not be a generally applicable rule, but it does indicate the possible advantages for DH application in maize breeding. Currently, most of the larger maize seed companies have a DH program and produce DH lines routinely for future commercial product development and applications. The forward breeding method can produce improved new DH lines carrying unique genes or genes of commercial importance. The future is filled with challenges and opportunities for the use of dihaploid technology in maize breeding.

References

Aman MA, Sarkar KR (1978) Selection for haploidy inducing potential in maize. Indian J Genet Plant Breed 38:452–457

Aman MA, Mathur DS, Sarkar KR (1981) Effect of pollen and silk age on maternal haploid frequencies in maize. Indian J Genet 41:362–365

Armstrong CL, Behr CF, Brar GS, Duncan DR, Foley T, Marshall LC (2004) A novel method for production of transformed dihaploid corn plants. US Patent 2004/0210959 A1

Aulinger I, Peter S, Schmid J, Stamp P (2003) Rapid attainment of a doubled haploid line from transgenic maize (*Zea mays* L.) plants by means of anther culture. In Vitro Cell Dev Biol-Plant 39:165–170

Bains N, Singh J, Gosal S (1995) Production of wheat haploids through embryo rescue from wheat × maize crosses. Curr Sci 69:621–623

Barnabas B (2003) Anther culture of maize (*Zea mays* L.). In: Maluszynski M, Kasha KJ, Forster BP, Szarejko I (eds) Doubled haploid production in crop plants, a manual. Kluwer, Dordrecht, pp103–108

Berzonsky WA, Kleven SL, Leach GD (2003) The effects of parthenogenesis on wheat embryo formation and haploid production with and without maize pollination. Euphytica 133:285–290

Bouchez A, Gallais A (2000) Efficiency of the use of doubled-haploids in recurrent selection for combining ability. Crop Sci 40:23–29

Chalyk ST (1994) Properties of maternal haploid maize plants and potential application to maize breeding. Euphytica 79:13–18

Chalyk ST (1999) Creating new haploid-inducing lines of maize. Maize Genet Coop Newsl 73:53–54

Chalyk ST, Chebotar OD (2000) Regular segregation of four recessive marker genes among maternal haploids in maize. Plant Breed 119:363–364

Chalyk ST, Rotarenco V (2001) The use of matroclinous maize haploids for recurrent selection. Russ J Genet 37:1382–1387

Chalyk ST, Bylich VG, Chebotar OD (1994) Transgressive segregation in the progeny of a cross between two inducers of maize maternal haploids. Maize Genet Coop Newsl 68:47

Chang MT (1992) Stock 6 induced double haploidy is random. Maize Genet Coop Newsl 66:98

Chase SS (1947) Techniques for isolating monoploid maize plants. Am J Bot 34:582

Chase SS (1949a) Monoploid frequencies in a commercial double cross hybrid maize, and in its component single cross hybrids and inbred lines. Genetics 34:328–332

Chase SS (1949b) The reproductive success of monoploid maize. Am Jour Bot 36:795–796

Chase SS (1951) Efficient methods of developing and improving inbred lines. The monoploid method of developing inbred lines. Report of 6th Hybrid Corn Industry Research Conference, pp 29–34

Chase SS (1952) Selection for parthenogenesis and monoploid fertility in maize. Genetics 37:573–574

Chase SS (1963) Androgenesis – its use for transfer of maize cytoplasm. J Heredity 54:152–158

Chase SS (1969) Monoploids and monoploid-derivatives of maize (*Zea mays* L.). Bot Rev 35:117–167

Chen SJ, Song TM (2003) Identification haploid with high oil xenia effect in maize. Acta Agron Sin 29(4):587–590

Coe EH Jr (1959) A line of maize with high haploid frequency. Am Nat 93(873):381–382

Coe EH Jr, Sarkar KR (1964) The detection of haploids in maize. J Hered 55(5):231–233

Dankov T, Kruleva M, Todorova L (1990) New embryo markers in maize. C R Acad Bulg Sci 43:83–85

Dankov T, Kruleva M, Dimitro B, Krapchev B (1997) Increase the percentage of maternal haploids by effect of herbicide basagran. Maize Genet Coop Newsl 71:78–79

Deanon JR (1957) Treatment of sweet corn silks with maleic hydrazide and colchicines as means of increasing the frequency of monoploids. Phillip Agric 41:364–377

Dletzmann E, Foroughi-Wehr B (1996) Combination of resistance to barley yellow mosaic virus and *Rhynchospordium secales* by recurrent selection with repeated haploid steps. Plant Breed 115:179–182

East EM (1930) The production of homozygotes through induced parthenogenesis. Science 72:148–149

Eder J, Chalyk ST (2002) In vivo haploid induction in maize. Theor Appl Genet 104:703–708

Forster BP, Thomas WTB (2005) Doubled haploids in genetics and plant breeding. In: Janick J (ed) Plant Breed Rev 25:57–88

Gallais A (1988) A method of line development using double haploids: the single doubled haploid descent recurrent selection. Theor Appl Genet 75:330–332

Gallais A (1989) Optimization of recurrent selection on the phenotypic value of doubled haploid lines. Theor Appl Genet 77:501–504

Greenblatt IM, Bock M (1967) A commercially desirable procedure for detection of monoploids in maize. J Hered 58:9–13

Griffing B (1975) Efficiency changes due to use of doubled-haploids in recurrent selection methods. Theor Appl Genet 46:367–368

Han X, Tang Q, Cao M, Rong T (2006) Study on identifying methods of maize haploids induced by Stock 6. J Maize Sci 14(1):64–66

Hansen NJP, Andersen SB (1998) Efficient production of doubled haploid wheat by in vitro treatment of microspores with trifluralin or APM. Plant Breed 117:401–405

Inagaki MN (2003) Doubled haploid production in wheat through wide hybridization. In: Maluszynski M, Kasha KJ, Forster BP, Szarejko I (eds) Doubled haploid production in crop plants, a manual. Kluwer, Dordrecht, pp 53–58

Jansen RC, Jannink J-L,Beavis WD (2003) Mapping quantitative trait loci in plant breeding populations: use of parental haplotype sharing. Crop Sci 43:829–834

Jensen CJ (1974) Chromosome doubling techniques in haploids. In: Kasha KJ (ed) Haploids in higher plants, advances and potential. Proc 1st In Symp University of Guelph, Guelph, pp 153–190

Kato A (1997) Induced single fertilization in maize. Sex Plant Reprod 10:96–100

Kato A (2002) Chromosome doubling of haploid maize seedlings using nitrous oxide gas at the flower primordial stage. Plant Breed, 121(5):370–377

Kermicle JL (1969) Androgenesis conditioned by a mutation in maize. Science 164:1422–1424

Kermicle JL (1971) Pleiotropic effects on seed development of the indeterminate gametophyte gene in maize. Am J Bot 58(1):1–7

Kindiger B (1997) Apomictic corn: progress, potential and opportunity. 52nd Annual Corn and Sorghum Research Conference, Chicago, pp 260–270

Kindiger B (1998) Progress in the development of an apomictic corn. 34th Annual Illinois Corn Breeders School, Urbana, pp 13–20

Kindiger B (2006) Maize and *Tripsacum* hybridization and the transfer of apomixis: In: Acquaah EG (ed) Principles of plant genetics and breeding, Chap 4. Plant reproductive systems. Blackwell, Ames, pp 64–67

Kindiger B, Sokolov V (1998) Apomictic maize. US Patent 5,710,367

Laboratory of Plant Cell and Tissue Culture, Institute of Genetics, Academia Sinica (1975) Primary study on induction of pollen plants of *Zea mays*. Acta Genet Sin 2:138–143 [Chinese]

Laurie DA, Bennett MD (1986) Wheat × maize hybridization. Can J Genet And Cytol 28:313–316

Lin B-Y (1981) Megagametogenetic alternations associated with the indeterminate gametophyte (ig) mutation in maize. Rev Bras Biol 41:557–564

Liu ZZ, Song TM (2000a) The breeding and identification of haploid inducer with high frequency parthenogenesis in maize. Acta Agron Sin 26(5):570–574

Liu ZZ, Song TM (2000b) Fertility spontaneously restoring of inflorescence and chromosome doubling by chemical treatment in maize haploid. Acta Agron Sin 26(6):947–952

Mathur DS, Sachan JKS, Sarkar KR (1976) Radiation induced haploidy and hetero-fertilization in maize. J Nucl Agric Biol 5:76–77

Mathur DS, Aman M, Sarkar KR (1980) Induction of maternal haploids in maize through heat treatment of pollen. Curr Sci India 49:744–746

Matzk F, Mahn A (1994) Improved techniques for haploid production in wheat using chromosome elimination. Plant Breed 113:125–129

Melchinger AE, Longin CF, Utz HF, Reif JC (2005) Hybrid maize breeding with doubled haploid lines: quantitative genetic and selection theory for optimum allocation of resources. Illinois Corn Breeders' School, Urbana, pp 8–21

Messing J (2005) The maize genome. Maydica 50:377–386

Petolino JF, Jones AM (1986) Anther culture of elite genotypes of maize. Crop Sci 26:1072–1074

Randolph LF (1932) Some effects of high temperature on polyploidy and other variations in maize. Proc Natl Acad Sci USA 18:222–229

Randolph LF (1938) Note on haploid frequencies. Maize Genet Coop Newsl 12:12

Randolph LF (1940) Note on haploid frequencies. Maize Genet Coop Newsl 14:23–24

Randolph LF, Fisher HE (1939) The occurrence of parthenogenetic diploids in tetraploid maize. Proc Natl Acad Sci USA 25:161–164

Rines HW (2003) Oat haploids from wide hybridization. In: Maluszynski M, Kasha KJ, Forster BP, Szarejko I (eds) Doubled haploid production in crop plants, a manual. Kluwer, Dordrecht, pp 155–159

Rober FK (1999) Fortpflanzungsbiologische und genetische Untersuchungen mit RFLP-Merkern zur in-vivo-Haploideninduktion bei Mais. Dissertation, University of Hohenhein, Grauer, Stuttgart

Rober FK, Gordillo GA, Geiger HH (2005) In vivo haploid induction in maize – performance of new inducers and significance of doubled haploid lines in hybrid breeding. Maydica 50:275–283

Sarkar KR (1974) Genetic selection techniques for production of haploids in higher plants. In: Kasha KJ (ed) Haploids in higher plants. Advances and potential. University of Guelph, Guelph, pp 33–41

Sarkar KR, Coe EH Jr (1966) A genetic analysis of the origin of maternal haploids in maize. Genetics 54:453–464

Sarkar KR, Coe EH Jr (1971) Origin of parthenogenetic diploids in maize and its implications for the production of homozygous lines. Crop Sci 11:543–544

Sarkar KR, Panke S, Sachan JKS (1972) Development of maternal-haploidy-inducer lines in maize (*Zea mays* L.). Indian J Agric Sci 42:781–786

Schlegel RHJ (2003) Encyclopedic dictionary of plant breeding and related subjects. Food Products Press and Heworth Reference Press, New York

Seitz G (2005) The use of doubled haploids in corn breeding. Illinois Corn Breeders' School, Urbana, pp 1–7

Shatskaya OA, Zabirova ER, Shcherbak VS, Chumak MV (1994a) Mass induction of maternal haploids in corn. Maize Genet Coop Newsl 68:51

Shatskaya O, Zabirova ER, Shcherbak VS (1994b) Autodiploid lines as sources of haploid spontaneous diploidization. Maize Genet Coop Newsl 68:51–52

Shatskaya OA (2004) Frequency increase of maize matroclinous haploid induction by individual selection of pollinators. Collected articles of Krasnodar Scientific Research Instutute of Agriculture – Evolution of Scientific Technology in Plant Growing, vol 2, pp 322–331

Smolkina YV, Tyrnov VS (2003) Development of haploids of parthenogenetical maize lines in crosses n × 2n by different pollen delay terms. Maize Genet Coop Newsl 77:65

Sokolov VA, Kindiger B, Khatipova IV (1998) Investigation of apomictic maize-Tripsacum hybrids. Russ J Genet 34:392–398

Stadler LJ (1944) Gamete selection in corn breeding. J Am Soc Agron 36:988–989

Stadler LJ (1951) Spontaneous mutation in maize. Cold Spring Harbor Symp Quant Biol 16:49–63

Szarejko I (2003) Doubled haploid mutant production. In: Maluszynski M, Kasha KJ, Forster BP, Szarejko I (eds) Doubled haploid production in crop plants, a manual. Kluwer, Dordrecht, pp 351–361

Testillano P, Georgiev S, Mogensen HL, Coronado MJ, Dumas C, Risueno MC, Matthys-Rochon E (2004) Spontaneous chromosome doubling results from nuclear fusion during in vitro maize induced microspore embryogenesis. Chromosoma 112:342–349

Troyer AF (2004) Persistent and popular germplasm in seventy centuries of corn evolution. In: Smith CW, Betran J, Runge ECA (eds) Corn: origin, history, technology and production. Wiley, Hoboken, pp 133–232

Tu SB, Hu YY, Pang KT, Nie S (1994) Study of chemical induced maize parthenogenesis. J Sichuan Agric Univ 12(3):423–430

Tyrnov VS (1997) Producing of parthenogenetic forms of maize. Maize Genet Coop Newsl 71:73

Wan Y, Duncan DR, Rayburn AL, Petolino JF, Widholm JM (1991) The use of antimicrotubule herbicides for the production of doubled haploid plants from anther-derived maize callus. Theor Appl Genet 81:205–211

Wei JJ, Chen MX (2006) Primary study on the natural fertility of maize haploids. J Maize Sci 14(2):24–26

Zabirova ER, Shatskaya OA, Shcherbak VS (1993) Line 613/2 as a source of a high frequency of spontaneous diploidization in corn. Maize Genet Coop Newsl 67:67

Zabirova ER, Chumak MV, Shatskaya OA, Shcherbak VS (1996) Technology of the mass accelerated production of homozygous lines. Kukuruza Sorgo N4:17–19

Zaharova GM (1955) The inheritance of characters in maize and tomato plants pollinated with mixed pollen. Izv Akad Nauk SSSR 1:32–44

Zenkteler M, Nitzsche W (1984) Wide hybridization experiments in cereals. Theor Appl Genet 68:311–315

Zhao ZY, Gu MG (1984) Production of dihaploid pure line of maize through parthenogenesis induced by chemicals. Acta Genet Sin 11:39–46

Zhao ZY, Gu MG (1988) Cytogenetic studies of chemically induced parthenogenesis maize plants. Acta Genet Sin 15:89–94

Zheng MY, Weng Y, Sahibzada R, Konzak CF (2003) Isolated microspore culture in maize (*Zea mays* L.), production of doubled-haploids via induced androgenesis. In: Maluszynski M, Kasha KJ, Forster BP, Szarejko I (eds) Doubled haploid production in crop plants, a manual. Kluwer, Dordrecht, pp 95–102

Zhou GY, Gong ZZ, Wang ZF (1979) The molecular basis of remote hybridization. Acta Genet Sin 6:405–413

Chapter 11
Transposon Tagging and Reverse Genetics

A. Mark Settles

11.1 Introduction

Transposons are mobile genetic elements that can amplify themselves in a genome. Transposable elements were first discovered in maize (McClintock 1948) and have been found to exist in all organisms. Transposons have multiple modes of movement or transposition, which is used to group the elements into two major classes, based on having an RNA or a DNA intermediate during transposition (reviewed in Hua-Van et al. 2005). The maize genome contains examples of most known transposon families including long terminal repeat (LTR) and non-LTR retrotransposons (class I), which use RNA intermediates, as well as class II DNA elements (Bruggmann et al. 2006). Maize class II elements tend to insert near or within genes (Bureau and Wessler 1992; Cowperthwaite et al. 2002; Fernandes et al. 2004; Kolkman et al. 2005; Kumar et al. 2005; McCarty et al. 2005; Settles et al. 2004). Transposition into genes can cause mutant phenotypes, and transposons are used as endogenous mutagens. This chapter focuses on the use of maize DNA transposons in molecular genetics and functional genomics studies.

With the exception of helitrons, DNA transposons share some common molecular and genetic properties. DNA elements have terminal inverted repeats as well as autonomous and non-autonomous transposons (Hua-Van et al. 2005). Autonomous elements encode the genes required for transposition. Non-autonomous transposons contain sequences recognized by transposase proteins and can move only in the presence of an autonomous element. Non-autonomous elements either have mutations in transposase genes or have replaced them with other sequences. DNA transposons also create target site duplications at the site of insertion. The length of the duplication is specific to each family of element. The major families of

A. Mark Settles
Horticultural Sciences Department and Plant Molecular and Cellular Biology Program, University of Florida, Gainesville, FL 32611-0690, USA
e-mail: settles@ufl.edu

A.L. Kriz, B.A. Larkins (eds.), *Molecular Genetic Approaches to Maize Improvement*
Biotechnology in Agriculture and Forestry, Vol. 63

maize transposons that have been used as mutagens include *Activator* and *Dissociation* (*Ac/Ds*), *Enhancer/Suppressor-mutator* (*En/Spm*), and *Robertson's Mutator* (*Mu*). These families were identified because they cause unstable mutations (McClintock 1950, 1954; Peterson 1960; Robertson 1978). The instability of many transposon-induced alleles is due to excision events or epigenetic regulation of the transposons. The structures, mechanisms of transposition, and epigenetic regulation of *Ac/Ds*, *En/Spm*, and *Mu* are discussed in greater detail in multiple reviews (see Kunze and Weil 2002; Lisch 2002; Walbot and Rudenko 2002).

11.2 General Strategies for Transposon Tagging

Transposon mutagenesis is a central tool for current research in maize molecular genetics. A non-exhaustive database search for recent mutant gene cloning reports identified 20 papers (see Table 11.1). Transposon-tagged alleles played a central role in the proof of cloning for 90% of these studies, and conventional transposon-tagging strategies were used in 60% of the reports. Transposon tagging is a gene-cloning strategy that relies on the transposon to provide a DNA "tag" with a known sequence (see Fig. 11.1). The transposon sequence is used to identify DNA sequences adjacent to the transposable element. The strategy was initially developed to clone the *Drosophila white* locus (Bingham et al. 1981). Fedoroff et al. (1984) adapted the method for *Ac* and cloned a tagged allele of the *bronze1* (*bz1*) locus. To identify DNA adjacent to a transposon, the initial approach was to isolate λ phage clones that included the transposon tag. Phage libraries have continued to be useful for cloning transposon-tagged genes (e.g. McSteen et al. 2007; Vollbrecht et al. 2005), but PCR-based methods for amplifying sequences adjacent to endogenous maize transposons have been developed as well (Earp et al. 1990; Frey et al. 1998; Settles et al. 2004).

Transposable elements are useful tags only if the sequences of the elements are known. The *Ac*, *Ds*, *En/Spm*, and *Mu* sequences were cloned using a method that is sometimes referred to as transposon trapping (Barker et al. 1984; Chomet et al. 1991; Fedoroff et al. 1983; Pohlman et al. 1984; Schwarz-Sommer et al. 1984). Previously cloned genes are used as "traps" to isolate transposon-tagged, mutant alleles. The known gene sequence is used to identify clones containing a mutant allele containing the transposon. Through transposon-induced alleles of cloned loci researchers have continued to identify new classes of transposons in the maize genome. For example, the first miniature inverted repeat transposable element and the first helitron insertions in maize were identified through mutants in the *waxy* (*wx*) and *shrunken2* (*sh2*) loci, respectively (Bureau and Wessler 1992; Lal et al. 2003).

Ac/Ds, *En/Spm*, and *Mu* exist in multiple copies within the maize genome. Moreover, it is possible for transposons to induce mutations without tagging the locus of interest (e.g. Satoh-Nagasawa et al. 2006). Thus, it is imperative to collect additional data about putatively tagged mutants to ensure that the cloned locus represents

Table 11.1 Summary of mutant cloning papers since 2005

Locus	Cloning approach	Role for transposon mutagenesis	Reference
Conventional tagging			
delayed flowering1 (dlf1)	Transposon tagging	Directed tagging	Muszynski et al. 2006
glossy1 (gl1)	Transposon tagging	Directed tagging	Sturaro et al. 2005
ramosa1 (ra1)	Transposon tagging	Directed tagging	Vollbrecht et al. 2005
thick tassel dwarf1 (td1)	Transposon tagging	Directed and non-directed tagging	Bommert et al. 2005
barren infloresence2 (bif2)	Transposon tagging	Non-directed mutagenesis	McSteen et al. 2007
empty pericarp4 (emp4)	Transposon tagging	Non-directed mutagenesis	Gutierrez-Marcos et al. 2007
low phytic acid3 (lpa3)	Transposon tagging	Non-directed mutagenesis	Shi et al. 2005
Roothairless1 (rth1)	Transposon tagging	Non-directed mutagenesis	Wen et al. 2005
viviparous10 (vp10)	Transposon tagging	Non-directed mutagenesis	Porch et al. 2006
viviparous15 (vp15)	Transposon tagging	Non-directed mutagenesis	Suzuki et al. 2006
Zea mays shugoshin1 (zmsgo1)	Transposon tagging	Non-directed mutagenesis	Hamant et al. 2005
Zea mays smu2 (zmsmu2)	Transposon tagging	Non-directed mutagenesis	Chung et al. 2007
Confirmed with transposon-tagged alleles			
Corngrass1 (Cg1)	Activation tagging	Reference allele was transposon-tagged	Chuck et al. 2007
Indeterminate gametophyte1 (ig1)	Map-based	Confirmed with directed tagging	Evans 2007
rootless for crown and seminal roots (rtcs)	Map-based	Confirmed with directed tagging	Taramino et al. 2007
ramosa2 (ra2)	Map-based	Confirmed with non-directed and directed tagging	Bortiri et al. 2006b
brittle stalk2 (bk2)	Candidate gene	Confirming alleles from *Mu* reverse genetics	Ching et al. 2006
Oil yellow1 (Oy1)	Candidate gene	Confirming alleles from *Mu* reverse genetics	Sawers et al. 2006
No transposon-tagged alleles			
ramosa3 (ra3)	Map-based	Only non-tagged alleles identified	Satoh-Nagasawa et al. 2006
Mucronate (Mc)	Candidate gene	None – transgenic approach	Kim et al. 2006

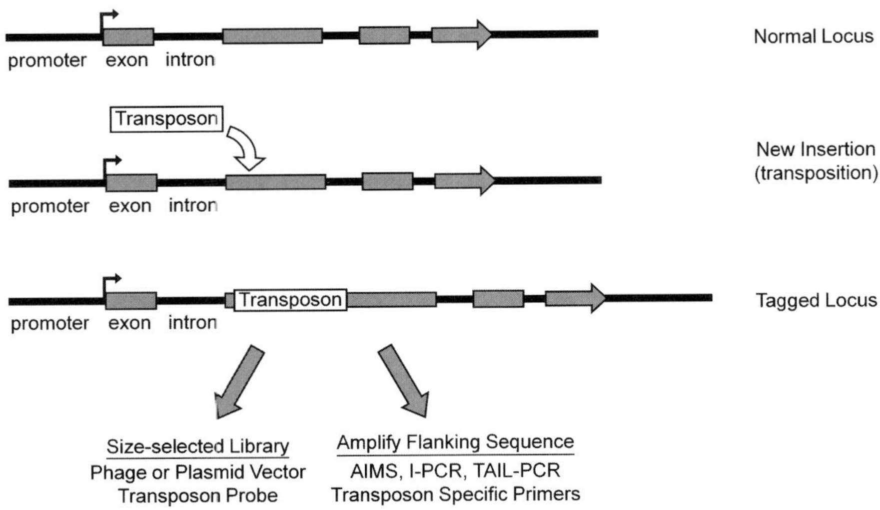

Fig. 11.1 Schematic of a transposon-tagged mutation. DNA transposable elements insert into or near genes with a "cut and paste" type of mechanism. The resulting mutant is a fusion of the normal gene with the transposon sequence. The transposon sequence can then be used as a probe to screen phage or plasmid libraries. More frequently, the transposon sequence is used to anneal specific primers for I-PCR, AIMS, or TAIL-PCR

the mutation of interest (reviewed in .Walbot 1992). Generally accepted criteria for proof of cloning include two types of data. First, the transposon tag and mutant phenotype need to be linked genetically. Second, either the mutant needs to be complemented with a transgene, or multiple mutant and/or revertant alleles of the locus need to be isolated and sequenced. In the past 10 years, there has been significant advancement in genetic resources and technologies, which has simplified cloning transposon-tagged mutants. These efforts have focused on three major areas: (1) directed tagging, (2) non-directed, saturation mutagenesis, and (3) reverse genetics resources.

11.3 Directed Tagging

Directed transposon tagging recovers mutations at a specific locus of interest. The conventional strategy for directed tagging is illustrated well by the cloning of *opaque2* (*o*2) (Schmidt et al. 1987). Plants carrying active transposons are crossed with homozygous mutants for a reference, non-tagged allele (Fig. 11.2A). The F_1 progeny are screened for mutant phenotypes. Obtaining multiple alleles of a locus requires a near-saturation level of mutagenesis. A near-saturation screen will require between 30,000 and 500,000 progeny from tagging crosses, depending on the type of transposon used for mutagenesis. Novel alleles that show unstable or mutable phenotypes are considered the best candidates for being transposon-tagged.

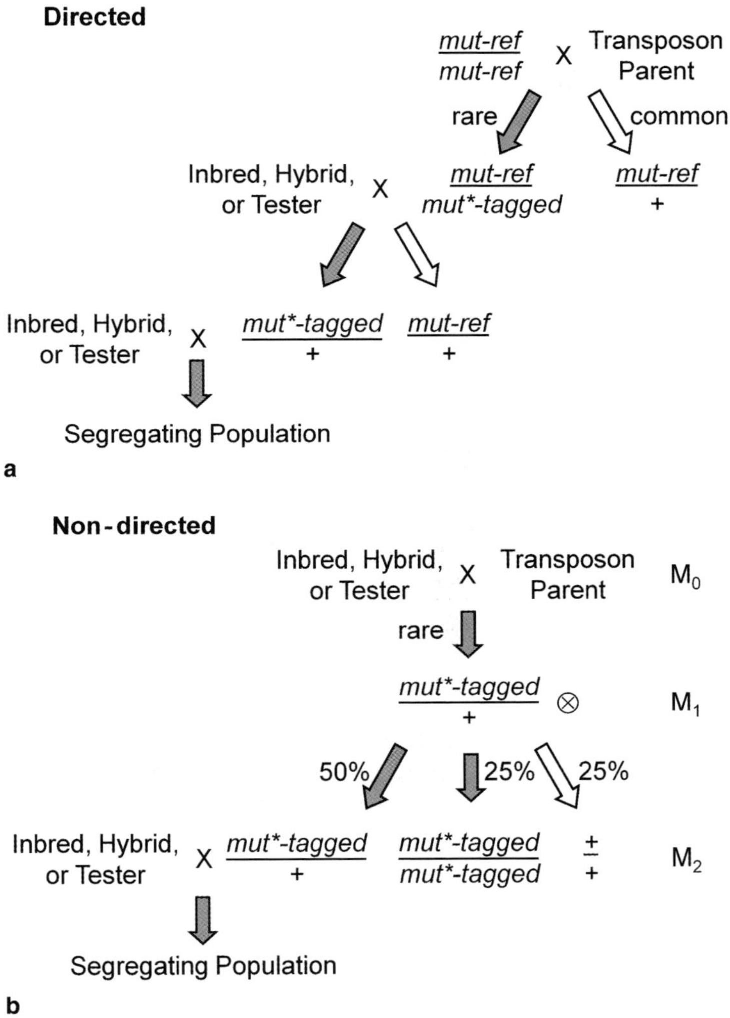

Fig. 11.2 Schematic of transposon mutagenesis strategies. **a** Directed tagging identifies transposon-induced alleles by crossing transposon-active plants with a reference allele of the mutation. The mutable alleles are separated from the reference allele by crossing the F_1 to a standard line (hybrid, inbred, or tester). To identify a co-segregating transposon, the mutable allele is backcrossed into the standard line, and the backcrossed progeny are self-pollinated. **b** For non-directed tagging, transposon-active stocks are generally crossed to a standard line and the resulting progeny are self-pollinated. The self-pollinated families are screened for recessive mutants and segregating populations are generated by backcrossing into the standard line. Although both schematics show the transposon parent as a pollen parent, transposon mutagenesis can be completed with either a male or a female transposon parent

These alleles are crossed to a standard inbred or hybrid to separate the mutable alleles from the reference allele. Segregating populations of the mutable alleles are then screened by DNA gel blot or with PCR methods to identify transposon insertions that co-segregate with the mutable phenotype. The most common PCR method for co-segregation analysis is the amplification of insertion mutagenized sites (AIMS) protocol (Frey et al. 1998). AIMS is a modified amplified fragment length polymorphism protocol that is specific for DNA adjacent to *Mu* transposons. Thermal asymmetric interlaced PCR (TAIL-PCR) has also been used to identify co-segregating *Mu* insertions (e.g. Chung et al. 2007; Porch et al. 2006). The co-segregating insertion is cloned by generating a size-selected phage/plasmid library or by a flanking sequence PCR method such as inverse-PCR (I-PCR) (Earp et al. 1990) or TAIL-PCR (Liu et al. 1995).

Ac/Ds, *En/Spm*, and *Mu* have all been employed successfully in directed tagging experiments. These elements have different transposition characteristics. *Ac/Ds* elements are known to have a low mutation rate and an insertion preference for genetically linked sites (Dooner and Belachew 1989; Greenblatt 1984). *En/Spm* and *Mu* are generally used for tagging loci with unknown map locations. *Mu* has a high mutation rate associated with a high copy number of *Mu* elements (Robertson 1978; Walbot and Warren 1988). A comparison of early directed tagging experiments with *Mu* and *En/Spm* lines suggested that *Mu* lines give a 10-fold higher rate of novel alleles (Walbot 1992). The high-copy nature of *Mu* lines makes subsequent segregation analysis more complicated. Also, *Mu* elements show very low rates of germinal excision events and revertant alleles are difficult to recover (Brown et al. 1989; Levy et al. 1989; Schnable et al. 1989).

Ac/Ds elements are useful for directed tagging when a mutant maps close to a characterized element. The propensity of *Ac/Ds* elements to transpose to linked sites enables the construction of a large allelic series once an *Ac* or *Ds* is positioned close to a locus of interest (Athma et al. 1992; Moreno et al. 1992). In addition, *Ac/Ds* tagged alleles give rise to germinal excision events that are frequently imprecise, leaving partial target site duplications. These "footprints" can generate weak alleles and even proteins with enhanced functional properties (Giroux et al. 1996; Wessler et al. 1986). Footprints are also used as supporting evidence that a tagged locus causes the mutant phenotype of interest. The historical limitation to utilizing *Ac/Ds* for directed tagging has been that only a small number of the transposons were at known map locations.

Several research groups focused on developing *Ac* resources for directed tagging. Currently, there are nearly 170 *Ac* elements that are mapped to distributed locations throughout the genome (Auger and Sheridan 1999; Cowperthwaite et al. 2002; Dooner et al. 1994; Kolkman et al. 2005). These *Ac* stocks are in four collections generated by different genetic strategies. Each collection is propagated and monitored with distinct genetic markers. Dooner et al. (1994) mapped unlinked transpositions from *bz1-m2* using *wx* reciprocal translocations. These stocks are propagated with *wx* translocations to ensure that the *Ac* elements remain at the expected locus. Auger and Sheridan (1999) converted a series of inversion and balanced translocations with a *Pericarp color1* allele tagged by *Ac*, *P1-vv*. The conversions are recombinants of non-tagged alleles of *P1* with *P1-vv*. The translocated or inverted DNA in each stock places the *P1-vv* allele (and thus the *Ac* element) in a linked position to sites throughout the genome. More recently, Cowperthwaite et al. (2002) and Kolkman et al. (2005) generated a series of transpositions originating from *Ac* elements on chromosomes 1, 5, and 9. Flanking DNA from the transposed *Ac* elements was amplified and sequenced using I-PCR. The flanking sequences can be used to confirm that the *Ac* has not moved during the propagation of the stocks. In addition, these stocks have *Ds* markers to monitor *Ac* transposition. To use the *Ac* stocks for conventional tagging, the locus of interest needs to be mapped. One or more *Ac*'s that map close to the mutant can be ordered and crossed to a homozygous reference allele to screen for novel tagged alleles.

11.4 Non-directed Tagging

Directed tagging is limited to mutants that are non-essential for a plant to complete its life cycle. The general approach outlined above requires a homozygous mutant tester and detects tagged alleles via a mutant phenotype in the F_1. If a mutant is lethal or infertile, a heterozygous tester could be generated. Tagged-alleles would be recovered at half the frequency due to the segregation of the mutation in the tester. However, most mutants would be lost in the first generation after the directed tagging crosses. Many lethal or infertile mutants have been cloned by transposon tagging. The tagged alleles of these mutants were identified from non-directed mutagenesis.

Non-directed transposon-induced mutants are generated in a way that is similar to other conventional mutagenesis approaches. Plants with active transposons are maintained through crosses to a reference genotype such as a hybrid, inbred, or transposon-activity tester (Fig. 11.2B). Progeny from these crosses contain a mutagenized gamete and are equivalent to the M_1 generation in conventional mutagenesis. Self-pollinations of the M_1 yield M_2 families segregating for recessive mutants. The recessive phenotypes can be identified through any screening approach practical for maize. Infertile and lethal mutations are propagated with heterozygous siblings from the specific M_2 family. Co-segregating insertions are identified and cloned using the same general strategies as described for directed tagging in Section 11.3. A near-saturation mutagenesis population requires a similar number of mutagenized gametes as in a directed tagging experiment, i.e. 30,000 (for highly active *Mu*) to 500,000 (for *En/Spm*) M_2 families.

M_2 families are more laborious to generate and to screen than progeny of directed-tagging crosses. Consequently, the early focus of non-directed tagging experiments was on mutant classes that occur at a high frequency, such as seed and seedling lethal phenotypes (Cook and Miles 1988; Scanlon et al. 1994; Taylor et al. 1987). Relatively small transposon-tagging populations generally recover single alleles of individual loci. A common approach for cloning "orphan" isolates is to screen for a tightly linked transposon insertion from the individual allele. To confirm that a tagged locus causes the mutant phenotype, reverse genetics screens are used to recover additional alleles (e.g.Gutierrez-Marcos et al. 2007; Hamant et al. 2005; Wen et al. 2005). However, the single allele strategy carries the risk that the mutant may not be tagged.

There are several approaches to ensure that a mutant locus identified in a non-directed tagging experiment can be cloned. First, near-saturation *Mu* populations have been developed for multiple genomics projects (Bensen et al. 1995; Fernandes et al. 2004; Martin et al. 2006; May et al. 2003; McCarty et al. 2005). If a mutant locus has a distinctive phenotype, forward genetic screens of these populations can identify several alleles to reduce the risk of recovering non-tagged alleles (e.g. Lid et al. 2002; Suzuki et al. 2006). Moreover, two functional genomics projects have generated near-saturation collections of non-photosynthetic and seed mutants. Both the photosynthesis mutant library (PML) and the UniformMu seed mutant collection contain thousands of visible mutants that represent hundreds of loci (McCarty et al. 2005; Stern et al. 2004). However, many seed and seedling lethal loci have similar phenotypes. Identifying alleles of the same locus within these collections using conventional genetics requires secondary phenotypic screens, mapping the mutants, and extensive complementation tests (e.g. Scanlon et al. 1994).

Second, there are several recent examples of a hybrid transposon-tagging and map-based approach to clone mutant loci (see Table 11.1). Map-based cloning is practical for maize as the physical map, genome sequence, and synteny relationships to rice have become better understood (reviewed in Bortiri et al. 2006a). A general strategy for this hybrid approach is to generate a segregating population by crossing the transposon-induced allele to a divergent inbred. The segregating population can be screened initially for a co-segregating transposon insertion. If a linked transposon is not identified, the same DNA samples can be used to map the mutant with molecular markers. Once an approximate map position is determined, the population is expanded to ∼1,000 meiotic products for fine mapping.

Third, lethal mutations can be targeted by using regional tagging with *Ac*/*Ds* elements. An example of this approach is the tagging of the *pink scuttellum1* (*ps1*) locus (Singh et al. 2003). An *Ac* element linked to a normal *Ps1* allele was used to generate a non-directed tagging population. A *Ds* insertion in the *R* locus was used to report *Ac* dosage and select 400 linked transpositions from ∼50,000 M_1 seed. Seven alleles of *ps1* were found after self-pollinating the selected transpositions.

The optimal tagging approach will be determined by several factors. Is the phenotype sensitive to genetic background modifiers? Mutants that are sensitive to genetic background effects may be more difficult to map using a map-based cloning strategy. Is the mutant lethal or viable? Directed tagging can be employed for viable mutants. For lethals, are there many or just a few loci that give rise to similar phenotypes? It is easier to screen a near-saturation *Mu* population when the mutant phenotype is simple to score, and relatively rare mutants are more straight-forward to analyze with genetics. Finally, is the map position of the locus known? Tagging using *Ac/Ds* is practical only when one of the elements is at a closely linked site to the locus of interest.

11.5 Reverse Genetics Resources

Conventional transposon tagging is a forward genetics approach. Mutants are characterized due to their phenotypes, and the purpose of identifying tagged alleles is to understand the molecular cause of the phenotype. Transposons can also be used for reverse genetics screens, in which mutations are identified affecting a sequence using PCR. These mutants are analyzed for altered phenotypes to gain insight into the function of the sequence of interest. The maize research community has developed a myriad of reverse genetics resources including multiple transposon-tagged collections. This section will focus on the transposon-tagging populations to discuss the advantages as well as the challenges that come with each resource. Chapter 12 discusses reverse genetics using chemical mutagenesis populations.

A reverse genetics resource begins with a near-saturation collection of mutagenized plants. A large population is necessary to ensure a reasonable chance that a mutation in any given sequence will be present. DNA is sampled from all of the mutagenized individuals for PCR. The specific mutations within the population can be identified on a locus-by-locus basis or systematically depending on the anticipated demand for the particular resource. For the locus-by-locus approach, the DNA samples are pooled, typically in grids, for efficient screening of the population. To identify a mutant in a specific locus, the pooled DNA samples are screened by PCR with a primer for the gene of interest and a primer specific to the transposable element. Amplification products indicate an insertion. Corresponding row and column amplifications identify the plant that has the insertion allele. Plants harboring active transposons will also have somatic insertions in small sectors of the tissue sampled for DNA extractions. Somatic insertions are not inherited and lead to false positive amplification during PCR screening. To limit the impact of somatic insertions, some reverse genetics projects sample from different leaves for row and column pools (Bensen et al. 1995; Fernandes et al. 2004). Other projects have used genetic inhibitors or genetic markers for transposon activity to select against somatic transposition within the plants sampled for DNA (May et al. 2003; McCarty et al. 2005).

11.5.1 Single-Gene Screening Resources

Single-locus screens can be completed either by service facilities or by the individual laboratory. Service facilities include the Trait Utility System for Corn (TUSC), Maize Targeted Mutagenesis (MTM), Biogemma's *Mu* population, and the *Mu* resources at the University of Bristol. The TUSC facility is operated by Pioneer Hi-Bred International, Inc., and academic researchers need to establish a collaboration with the company to complete a screen. These collaborations are relatively straightforward to develop, as evidenced by the steady rate of around two publications per year that report mutants identified using the TUSC service (e.g. Ching et al. 2006; Chung et al. 2007; Golubovskaya et al. 2006; Li et al. 2007).

There are several issues to consider related to choice of a service facility. For example, mutants identified from the TUSC collection require a material transfer agreement (MTA) for distribution, while the MTM population was established as a public screening service to identify mutants that can be freely distributed (May et al. 2003). MTM screens are completed for a user fee, which can become expensive to an individual program when screening for mutations in multiple loci. Also, MTM is not a near-saturation mutagenesis collection and has a lower likelihood of recovering mutations than the TUSC population. The initial MTM screens found mutants for only 42% of the genes screened (May et al. 2003). Moreover, positive MTM screens typically recover a single allele (Martin et al. 2006; Sheehan et al. 2007). In contrast, TUSC screens generally recover two to three alleles.

Biogemma has developed reverse genetics resources that were used in two recent mutant gene cloning reports (Gutierrez-Marcos et al. 2007; Martin et al. 2006). Similar to TUSC, a collaborative agreement is required for a screen and an MTA is required for distributing the mutant seed (Pascual Perez, personal communication). Although the Biogemma population is smaller than the TUSC and MTM populations, it has a similar rate of success as MTM for finding at least one allele of the locus of interest ($\sim$50%). Unlike MTM, the two published reports using Biogemma's population recovered multiple alleles for each locus.

The Functional Genomics group at the University of Bristol has a free screening service (www.cerealsdb.uk.net). This population consists of 5,000 *Mu*-active plants arrayed into a grid. The one report that used this service recovered three insertion alleles for a K^+-channel gene (Philippar et al. 2006).

For researchers who would like to complete single-locus screens in their own laboratories, the Maize Gene Discovery Project developed a *Mu* population with a transgenic *Mu* element, *RescueMu* (Raizada et al. 2001). *RescueMu* contains a plasmid vector, and flanking DNA from the transposon can be recovered using plasmid rescue. *RescueMu* insertions can be screened by ordering plasmid libraries recovered from grids that contain $\sim$27,500 germinal transpositions (Fernandes et al. 2004). Due to the low number of transpositions, there is a significant chance that a mutation will not be found after completing a PCR screen. Although the *RescueMu* population is a challenge to use as a reverse genetics resource, the population is in an active *Mu* genetic background. The non-transgenic, endogenous *Mu* elements are useful for conventional transposon tagging (McSteen et al. 2007). However, it is necessary to obtain appropriate movement and release permits, as well as any specific institutional authorizations, to propagate *RescueMu* lines due to the transgenic *Mu* elements in the population.

11.5.2 Flanking Sequence Tags and Reverse Genetics

Single-gene reverse genetics screens are equivalent to directed-tagging experiments. Mutations in a specific gene of interest can be recovered efficiently. However, each gene that researchers are interested in analyzing requires a separate screen. If a reverse genetics resource will be used to analyze thousands of genes, it becomes more cost effective and faster to identify transposon-induced mutations using a systematic approach. In Arabidopsis and rice, the approach has been to index insertion mutants using flanking sequence tags (FSTs). FSTs are sequenced DNA adjacent to a transposon or T-DNA tag. The sequence anchors the insertion site to a reference genome, allowing gene disruptions to be identified in silico. Sequencing from several hundred thousand insertion sites recovers mutations in the vast majority of genes in the genome (Alonso et al. 2003; Rosso et al. 2003; Samson et al. 2004; Sessions et al. 2002).

Maize FSTs are not as developed as Arabidopsis and rice resources. However, many of the same genetic resources discussed above have also been used to generate FSTs. Most maize insertional mutagenesis resources utilize native transposable element systems. These transposons exist as part of the genome, and plants contain a mix of somatic, novel, and parental insertions. Thus, recovering unique, germinal insertions is more challenging from maize transposon-tagging populations than it is from rice or Arabidopsis T-DNA tagging populations. Amplifying native transposon insertion sites leads to redundant products that represent parental insertions shared by many plants within a mutagenized population. Also, plants with active transposons will have somatic, non-heritable insertions.

Several groups have sequenced random samples of parental, novel, and somatic insertions. The functional genomics group at the University of Bristol amplified transposon insertions from *Mu*-active plants using a modified AFLP method (Hanley et al. 2000). Seven hundred and fifty FSTs resulted in 450 unique insertion sites. Only a small number of these insertions were tested for inheritance, leaving the specific fraction of somatic insertions unknown. MuTAIL-PCR has also been used to amplify insertions for 99 FSTs from a *Mu* population developed in China (Liu et al. 2006). These sequences identified 59 non-redundant insertion sites that were not tested for heritability. The Maize Gene Discovery Project took a large-scale shotgun approach and sequenced 191,717 *RescueMu* FSTs, resulting in 14,887 non-redundant *Mu* FSTs (Fernandes et al. 2004). Only 528 of the insertion sites were identified as germinal, based on the criteria that the insertion site was recovered from two independent DNA samples from the same plant.

The Maize Endosperm Genomics Project used a somatic activity marker to ensure that germinal FSTs are recovered (McCarty et al. 2005; Settles et al. 2004). A total of 37,595 FSTs were sequenced from the UniformMu population using MuTAIL-PCR to amplify the insertions. These FSTs identify over 1,900 non-redundant insertion sites (www.uniformmu.org). Heritability tests for 106 of the insertions gave no evidence of somatic FSTs and showed at least 89% are inherited, germinal insertions (Settles et al. 2007).

FSTs from germinal *Ac* and *Ds* insertions have also been generated. For *Ac* insertions, DNA gel blots were screened to identify novel, germinal *Ac* insertions. After a subsequent gel extraction step, specific insertion sites were amplified with I-PCR (Cowperthwaite et al. 2002; Kolkman et al. 2005). A total of 115 FSTs have been sequenced from *Ac* and a similar approach is being used to generate FSTs from *Ds* elements with 916 FSTs available currently (see Table 11.2).

Table 11.2 Summary of transposon-tagging reverse genetics resources

Resource	Website/contact
FST sequence databases	
Ac/Ds	http://www.plantgdb.org/prj/AcDsTagging/tool/blast.php
RescueMu and UniformMu	http://www.mutransposon.org/cgi-bin/MuBLAST.cgi
UniformMu	http://currant.hos.ufl.edu/mutail/
University of Bristol	http://www.cerealsdb.uk.net/mudb.htm
PCR screening services	
TUSC	Robert Meeley (bob.meeley@pioneer.com)
MTM	http://mtm.cshl.org
Biogemma	Christophe Tatout (christophe.tatout@biogemma.com)
University of Bristol	http://www.cerealsdb.uk.net/pcrscrn.htm

11.5.3 An Optimal Reverse Genetics Strategy?

With five PCR screening resources and seven FST resources, it is challenging to decide which resources are best to obtain mutations in a gene of interest. Since sequence database searches are fast, BLAST searching the FST resources is an easy starting point. Most of the FSTs can be searched at two websites. All of the *Ds* FSTs and most of the *Ac* FSTs are at PlantGDB, and the UniformMu and *RescueMu* FSTs can be searched at www.mutransposon.org (see Table 11.2). A separate UniformMu FST database includes both a BLAST server and analyzed FSTs to help users find non-redundant insertions and annotations for the insertions. The University of Bristol FSTs can be searched at the resource website. Germinal insertions in the *RescueMu* FSTs can be identified when multiple hits are recovered from columns and rows in the same grid. Both a column and a row hit is necessary to identify a specific *RescueMu* plant. For all of the other FST resources, a single hit is sufficient to identify the plant carrying the insertion.

Combined, the *Mu*, *Ac*, and *Ds* FST resources represent less than 4,000 non-redundant insertion sites that are likely to be germinal. With these current sequences, a researcher will be lucky to find a match in a gene of interest. The ease of database searches makes these searches worth completing prior to initiating other reverse genetics screens. If no FSTs are identified after database screening, the researcher needs to decide whether to collaborate with a company or pay for a public reverse genetics screen. The advantages of collaborating with a company are that there are no user fees and the probability of recovering mutant alleles is higher. However, the company will obtain some intellectual property rights to the biological process under study and generally will require a company review of the manuscript describing the mutant alleles. A factor to consider for public resources is that smaller populations are less likely to identify a mutant in a gene of interest. The "do-it-yourself" *RescueMu* screen contains the fewest number of novel insertions.

11.6 Future Perspectives

Conventional transposon tagging is the predominant approach for current maize genetics research. As the maize genome is sequenced, map-based cloning will be easier and is likely to become more common. However, transposon tagging will still play a central role in molecular genetics. Transposon-induced alleles give a different spectrum of mutant phenotypes than other mutagens and are useful in generating allelic series. Transposons also cause mutable or epigenetically regulated alleles that are useful for generating chimeric plants. Most importantly, transposon reverse genetics resources are likely to be critical for obtaining confirming alleles for forward genetics studies. A key challenge in making transposon reverse genetics resources more efficient is generating a saturated collection of FSTs with corresponding seed. Massively parallel sequencing technologies will help in reducing the cost of generating FSTs. Lower costs per FST should allow sufficient numbers of FSTs to make database searches a standard method for recovering maize mutants in a gene of interest.

Acknowledgements I apologize to the authors of many relevant research articles that I was not able to highlight in this chapter due to space constraints. The author's research is supported by grants from the NSF Plant Genome Research Project (DBI-0606607), the National Research Initiative of the USDA-CSREES (grant no. 2007-35304-18439), the University of Florida Genetics Institute, and the Vasil-Monsanto Endowment.

References

Alonso JM, Stepanova AN, Leisse TJ, Kim CJ, Chen HM, Shinn P, Stevenson DK, Zimmerman J, Barajas P, Cheuk R, Gadrinab C, Heller C, Jeske A, Koesema E, Meyers CC, Parker H, Prednis L, Ansari Y, Choy N, Deen H, Geralt M, Hazari N, Hom E, Karnes M, Mulholland C, Ndubaku R, Schmidt I, Guzman P, Aguilar-Henonin L, Schmid M, Weigel D, Carter DE, Marchand T, Risseeuw E, Brogden D, Zeko A, Crosby WL, Berry CC, Ecker JR (2003) Genome-wide insertional mutagenesis of *Arabidopsis thaliana*. Science 301:653–657

Athma P, Grotewold E, Peterson T (1992) Insertional mutagenesis of the maize *P* gene by intragenic transposition of *Ac*. Genetics 131:199–209

Auger DL, Sheridan WF (1999) Maize stocks modified to enhance the recovery of *Ac*-induced mutations. J Hered 90:453–459

Barker RF, Thompson DV, Talbot DR, Swanson J, Bennetzen JL (1984) Nucleotide sequence of the maize transposable element *Mu1*. Nucleic Acids Res 12:5955–5967

Bensen RJ, Johal GS, Crane VC, Tossberg JT, Schnable PS, Meeley RB, Briggs SP (1995) Cloning and characterization of the maize *An1* gene. Plant Cell 7:75–84

Bingham PM, Levis R, Rubin GM (1981) Cloning of DNA sequences from the *White* locus of *Drosophila melanogaster* by a novel and general method. Cell 25:693–704

Bommert P, Lunde C, Nardmann J, Vollbrecht E, Running M, Jackson D, Hake S, Werr W (2005) *thick tassel dwarf1* encodes a putative maize ortholog of the Arabidopsis CLAVATA1 leucine-rich repeat receptor-like kinase. Development 132:1235–1245

Bortiri E, Jackson D, Hake S (2006a) Advances in maize genomics: the emergence of positional cloning. Curr Opin Plant Biol 9:164–171

Bortiri E, Chuck G, Vollbrecht E, Rocheford T, Martienssen R, Hake S (2006b) *ramosa2* encodes a LATERAL ORGAN BOUNDARY domain protein that determines the fate of stem cells in branch meristems of maize. Plant Cell 18:574–585

Brown WE, Robertson DS, Bennetzen JL (1989) Molecular analysis of multiple *Mutator*-derived alleles of the *Bronze* locus of maize. Genetics 122:439–445

Bruggmann R, Bharti AK, Gundlach H, Lai JS, Young S, Pontaroli AC, Wei FS, Haberer G, Fuks G, Du CG, Raymond C, Estep MC, Liu RY, Bennetzen JL, Chan AP, Rabinowicz PD, Quackenbush J, Barbazuk WB, Wing RA, Birren B, Nusbaum C, Rounsley S, Mayer KFX, Messing J (2006) Uneven chromosome contraction and expansion in the maize genome. Genome Res 16:1241–1251

Bureau TE, Wessler SR (1992) *Tourist*: a large family of small inverted repeat elements frequently associated with maize genes. Plant Cell 4:1283–1294

Ching A, Dhugga KS, Appenzeller L, Meeley R, Bourett TM, Howard RJ, Rafalski A (2006) *Brittle stalk 2* encodes a putative glycosylphosphatidylinositol-anchored protein that affects mechanical strength of maize tissues by altering the composition and structure of secondary cell walls. Planta 224:1174–1184

Chomet P, Lisch D, Hardeman KJ, Chandler VL, Freeling M (1991) Identification of a regulatory transposon that control the *Mutator* transposable element system in maize. Genetics 129:261–270

Chuck G, Cigan AM, Saeteurn K, Hake S (2007) The heterochronic maize mutant *Corngrass1* results from overexpression of a tandem microRNA. Nat Genet 39:544–549

Chung T, Kim CS, Nguyen HN, Meeley RB, Larkins BA (2007) The maize *Zmsmu2* gene encodes a putative RNA-splicing factor that affects protein synthesis and RNA processing during endosperm development. Plant Physiol 144:821–835

Cook WB, Miles D (1988) Transposon mutagenesis of nuclear photosynthetic genes in *Zea mays*. Photosynth Res 18:33–59

Cowperthwaite M, Park W, Xu ZN, Yan XH, Maurais SC, Dooner HK (2002) Use of the transposon *Ac* as a gene-searching engine in the maize genome. Plant Cell 14:713–726

Dooner HK, Belachew A (1989) Transposition pattern of the maize element *Ac* from the *bz-m2(Ac)* allele. Genetics 122:447–457

Dooner HK, Belachew A, Burgess D, Harding S, Ralston M, Ralston E (1994) Distribution of unlinked receptor-sites for transposed *Ac* elements from the *bz-m2(Ac)* allele in maize. Genetics 136:261–279

Earp DJ, Lowe B, Baker B (1990) Amplification of genomic sequences flanking transposable elements in host and heterologous plants – a tool for transposon tagging and genome characterization. Nucleic Acids Res 18:3271–3279

Evans MMS (2007) The *indeterminate gametophyte1* gene of maize encodes a LOB domain protein required for embryo sac and leaf development. Plant Cell 19:46–62

Fedoroff N, Wessler S, Shure M (1983) Isolation of the transposable maize controlling elements *Ac* and *Ds*. Cell 35:235–242

Fedoroff NV, Furtek DB, Nelson OE (1984) Cloning of the *Bronze* locus in maize by a simple and generalizable procedure using the transposable controlling element *Activator* (*Ac*). Proc Natl Acad Sci USA 81:3825–3829

Fernandes J, Dong QF, Schneider B, Morrow DJ, Nan GL, Brendel V, Walbot V (2004) Genome-wide mutagenesis of *Zea mays L.* using *RescueMu* transposons. Genome Biol 5:R82

Frey M, Stettner C, Gierl A (1998) A general method for gene isolation in tagging approaches: amplification of insertion mutagenised sites (AIMS). Plant J 13:717–721

Giroux MJ, Shaw J, Barry G, Cobb BG, Greene T, Okita T, Hannah LC (1996) A single gene mutation that increases maize seed weight. Proc Natl Acad Sci USA 93:5824–5829

Golubovskaya IN, Hamant O, Timofejeva L, Wang C-JR, Braun D, Meeley R, Cande WZ (2006) Alleles of *afd1* dissect REC8 functions during meiotic prophase I. J Cell Sci 119: 3306–3315

Greenblatt IM (1984) A chromosome replication pattern deduced from pericarp phenotypes resulting from movements of the transposable element, *Modulator*, in maize. Genetics 108:471–485

Gutierrez-Marcos JF, Dal Pra M, Giulini A, Costa LM, Gavazzi G, Cordelier S, Sellam O, Tatout C, Paul W, Perez P, Dickinson HG, Consonni G (2007) *empty pericarp4* encodes a mitochondrion-targeted pentatricopeptide repeat protein necessary for seed development and plant growth in maize. Plant Cell 19:196–210

Hamant O, Golubovskaya I, Meeley R, Fiume E, Timofejeva L, Schleiffer A, Nasmyth K, Cande WZ (2005) A REC8-dependent plant *Shugoshin* is required for maintenance of centromeric cohesion during meiosis and has no mitotic functions. Curr Biol 15:948–954

Hanley S, Edwards D, Stevenson D, Haines S, Hegarty M, Schuch W, Edwards KJ (2000) Identification of transposon-tagged genes by the random sequencing of *Mutator*-tagged DNA fragments from *Zea mays*. Plant J 23:557–566

Hua-Van A, Le Rouzic A, Maisonhaute C, Capy P (2005) Abundance, distribution and dynamics of retrotransposable elements and transposons: similarities and differences. Cytogenet Genome Res 110:426–440

Kim CS, Gibbon BC, Gillikin JW, Larkins BA, Boston RS, Jung R (2006) The maize *Mucronate* mutation is a deletion in the 16-kDa gamma-zein gene that induces the unfolded protein response. Plant J 48:440–451

Kolkman JM, Conrad LJ, Farmer PR, Hardeman K, Ahern KR, Lewis PE, Sawers RJH, Lebejko S, Chomet P, Brutnell TP (2005) Distribution of *Activator* (*Ac*) throughout the maize genome for use in regional mutagenesis. Genetics 169:981–995

Kumar CS, Wing RA, Sundaresan V (2005) Efficient insertional mutagenesis in rice using the maize *En/Spm* elements. Plant J 44:879–892

Kunze R, Weil CF (2002) The hAT and CACTA superfamilies of plant transposons. In: Craig Nl, Craigie M, Gellert Am, Lambowitz Am (eds) Mobile DNA II. ASM Press, Washington, DC, pp 533–564

Lal SK, Giroux MJ, Brendel V, Vallejos CE, Hannah LC (2003) The maize genome contains a Helitron insertion. Plant Cell 15:381–391

Levy AA, Britt AB, Luehrsen KR, Chandler VL, Warren C, Walbot V (1989) Developmental and genetic-aspects of *Mutator* excision in maize. Dev Genet 10:520–531

Li J, Harper LC, Golubovskaya I, Wang CR, Weber D, Meeley RB, McElver J, Bowen B, Cande WZ, Schnable PS (2007) Functional analysis of maize RAD51 in meiosis and double-strand break repair. Genetics 176:1469–1482

Lid SE, Gruis D, Jung R, Lorentzen JA, Ananiev E, Chamberlin M, Niu XM, Meeley R, Nichols S, Olsen OA (2002) The *defective kernel 1* (*dek1*) gene required for aleurone cell development in the endosperm of maize grains encodes a membrane protein of the calpain gene superfamily. Proc Natl Acad Sci USA 99:5460–5465

Lisch D (2002) *Mutator* transposons. Trends Plant Sci 7:498–504

Liu WT, Gao YJ, Teng F, Shi Q, Zheng YL (2006) Construction and genetic analysis of *Mutator* insertion mutant population in maize. Chinese Sci Bull 51:2604–2610

Liu YG, Mitsukawa N, Oosumi T, Whittier RF (1995) Efficient isolation and mapping of *Arabidopsis thaliana* T-DNA insert junctions by thermal asymmetric interlaced PCR. Plant J 8: 457–463

Martin A, Lee J, Kichey T, Gerentes D, Zivy M, Tatout C, Dubois F, Balliau T, Valot B, Davanture M, Terce-Laforgue T, Quillere I, Coque M, Gallais A, Gonzalez-Moro MB, Bethencourt L, Habash DZ, Lea PJ, Charcosset A, Perez P, Murigneux A, Sakakibara H, Edwards KJ, Hirel B (2006) Two cytosolic glutamine synthetase isoforms of maize are specifically involved in the control of grain production. Plant Cell 18:3252–3274

May BP, Liu H, Vollbrecht E, Senior L, Rabinowicz PD, Roh D, Pan XK, Stein L, Freeling M, Alexander D, Martienssen R (2003) Maize-targeted mutagenesis: a knockout resource for maize. Proc Natl Acad Sci USA 100:11541–11546

McCarty DR, Settles AM, Suzuki M, Tan BC, Latshaw S, Porch T, Robin K, Baier J, Avigne W, Lai JS, Messing J, Koch KE, Hannah LC (2005) Steady-state transposon mutagenesis in inbred maize. Plant J 44:52–61

McClintock B (1948) Mutable loci in maize. Carnegie Inst Wash Year B 47:155–169

McClintock B (1950) The origin and behavior of mutable loci in maize. Proc Natl Acad Sci USA 36:344–355

McClintock B (1954) Mutations in maize and chromosomal aberrations in Neurospora. Carnegie Inst Wash Year B 56:415–429

McSteen P, Malcomber S, Skirpan A, Lunde C, Wu X, Kellogg E, Hake S (2007) *Barren inflorescence2* encodes a co-ortholog of the PINOID serine/threonine kinase and is required for organogenesis during inflorescence and vegetative development in maize. Plant Physiol 144:1000–1011

Moreno MA, Chen J, Greenblatt I, Dellaporta SL (1992) Reconstitutional mutagenesis of the maize *P* gene by short-range *Ac* transpositions. Genetics 131:939–956

Muszynski MG, Dam T, Li B, Shirbroun DM, Hou ZL, Bruggemann E, Archibald R, Ananiev EV, Danilevskaya ON (2006) *Delayed flowering1* encodes a basic leucine zipper protein that mediates floral inductive signals at the shoot apex in maize. Plant Physiol 142:1523–1536

Peterson PA (1960) The pale green mutable system in maize. Genetics 45:115–133

Philippar K, Buchsenschutz K, Edwards D, Loffler J, Luthen H, Kranz E, Edwards KJ, Hedrich R (2006) The auxin-induced K+ channel gene *Zmk1* in maize functions in coleoptile growth and is required for embryo development. Plant Mol Biol 61:757–768

Pohlman RF, Fedoroff NV, Messing J (1984) The nucleotide-sequence of the maize controlling element *Activator*. Cell 37:635–643

Porch TG, Tseung CW, Schmelz EA, Settles AM (2006) The maize *Viviparous10/Viviparous13* locus encodes the *Cnx1* gene required for molybdenum cofactor biosynthesis. Plant J 45:250–263

Raizada MN, Nan GL, Walbot V (2001) Somatic and germinal mobility of the *RescueMu* transposon in transgenic maize. Plant Cell 13:1587–1608

Robertson DS (1978) Characterization of a mutator system in maize. Mut Res 51:21–28

Rosso MG, Li Y, Strizhov N, Reiss B, Dekker K, Weisshaar B (2003) An *Arabidopsis thaliana* T-DNA mutagenized population (GABI-Kat) for flanking sequence tag-based reverse genetics. Plant Mol Biol 53:247–259

Samson F, Brunaud V, Duchene S, De Oliveira Y, Caboche M, Lecharny A, Aubourg S (2004) FLAGdb(++): a database for the functional analysis of the Arabidopsis genome. Nucleic Acids Res 32:D347–D350

Satoh-Nagasawa N, Nagasawa N, Malcomber S, Sakai H, Jackson D (2006) A trehalose metabolic enzyme controls inflorescence architecture in maize. Nature 441:227–230

Sawers RJH, Viney J, Farmer PR, Bussey RR, Olsefski G, Anufrikova K, Hunter CN, Brutnell TP (2006) The maize *Oil yellow1* (*Oy1*) gene encodes the I subunit of magnesium chelatase. Plant Mol Biol 60:95–106

Scanlon MJ, Stinard PS, James MG, Myers AM, Robertson DS (1994) Genetic analysis of 63 mutations affecting maize kernel development isolated from *Mutator* stocks. Genetics 136:281–294

Schmidt RJ, Burr FA, Burr B (1987) Transposon tagging and molecular analysis of the maize regulatory locus *Opaque-2*. Science 238:960–963

Schnable PS, Peterson PA, Saedler H (1989) The *Bz-Rcy* allele of the *Cy* transposable element system of *Zea mays* contains a *Mu*-like element insertion. Mol Gen Genet 217:459–463

Schwarz-Sommer Z, Gierl A, Klosgen RB, Wienand U, Peterson PA, Saedler H (1984) The *Spm* (*En*) transposable element controls the excision of a 2-Kb DNA insert at the *wx-m8* allele of *Zea mays*. EMBO J 3:1021–1028

Sessions A, Burke E, Presting G, Aux G, McElver J, Patton D, Dietrich B, Ho P, Bacwaden J, Ko C, Clarke JD, Cotton D, Bullis D, Snell J, Miguel T, Hutchison D, Kimmerly B, Mitzel T, Katagiri F, Glazebrook J, Law M, Goff SA (2002) A high-throughput Arabidopsis reverse genetics system. Plant Cell 14:2985–2994

Settles AM, Latshaw S, McCarty DR (2004) Molecular analysis of high-copy insertion sites in maize. Nucleic Acids Res 32:e54

Settles AM, Holding DR, Tan BC, Latshaw SP, Liu J, Suzuki M, Li L, O'Brien BA, Fajardo DS, Wroclawska E, Tseung CW, Lai JS, Hunter CT, Avigne WT, Baier J, Messing J, Hannah LC,

Koch KE, Becraft PW, Larkins BA, McCarty DR (2007) Sequence-indexed mutations in maize using the UniformMu transposon-tagging population. BMC Genomics 8

Sheehan MJ, Kennedy LM, Costich DE, Brutnell TP (2007) Subfunctionalization of *PhyB1* and *PhyB2* in the control of seedling and mature plant traits in maize. Plant J 49:338–353

Shi J, Wang H, Hazebroek J, Ertl DS, Harp T (2005) The maize *low-phytic acid 3* encodes a myo-inositol kinase that plays a role in phytic acid biosynthesis in developing seeds. Plant J 42:708–719

Singh M, Lewis PE, Hardeman K, Bai L, Rose JKC, Mazourek M, Chomet P, Brutnell TP (2003) *Activator* mutagenesis of the *pink scutellum1/viviparous7* locus of maize. Plant Cell 15:874–884

Stern DB, Hanson MR, Barkan A (2004) Genetics and genomics of chloroplast biogenesis: maize as a model system. Trends Plant Sci 9:293–301

Sturaro M, Hartings H, Schmelzer E, Velasco R, Salamini F, Motto M (2005) Cloning and characterization of GLOSSY1, a maize gene involved in cuticle membrane and wax production. Plant Physiol 138:478–489

Suzuki M, Settles AM, Tseung CW, Li QB, Latshaw S, Wu S, Porch TG, Schmelz EA, James MG, McCarty DR (2006) The maize *viviparous15* locus encodes the molybdopterin synthase small subunit. Plant J 45:264–274

Taramino G, Sauer M, Stauffer JL, Multani D, Niu X, Sakai H, Hochholdinger F (2007) The maize (*Zea mays L.*) *RTCS* gene encodes a LOB domain protein that is a key regulator of embryonic seminal and post-embryonic shoot-borne root initiation. Plant J 50:649–659

Taylor WC, Barkan A, Martienssen RA (1987) Use of nuclear mutants in the analysis of chloroplast development. Dev Genet 8:305–320

Vollbrecht E, Springer PS, Goh L, Buckler IV ES, Martienssen R (2005) Architecture of floral branch systems in maize and related grasses. Nature 436:1119–1126

Walbot V (1992) Strategies for mutagenesis and gene cloning using transposon tagging and T-DNA insertional mutagenesis. Annu Rev Plant Physiol Plant Mol Biol 43:49–82

Walbot V, Rudenko GN (2002) *MuDR/Mu* transposable elements in maize. In: Craig Nl, Craigie M, Gellert Am, Lambowitz Am (eds) Mobile DNA II. ASM Press, Washington, DC, pp 533–564

Walbot V, Warren C (1988) Regulation of *Mu*-element copy number in maize lines with an active or inactive *Mutator* transposable element system. Mol Gen Genet 211:27–34

Wen T-J, Hochholdinger F, Sauer M, Bruce W, Schnable PS (2005) The *roothairless1* gene of maize encodes a homolog of *sec3*, which is involved in polar exocytosis. Plant Physiol 138:1637–1643

Wessler SR, Baran G, Varagona M, Dellaporta SL (1986) Excision of *Ds* produces WAXY proteins with a range of enzymatic activities. EMBO J 5:2427–2432

Chapter 12
EMS Mutagenesis and Point Mutation Discovery

Clifford F. Weil and Rita-Ann Monde

12.1 Introduction

One of the ultimate goals of biotechnology is to design beneficial phenotypes and then achieve them by altering gene expression. For any given organism that is a target of biotechnological improvement, an indexed collection of every single base change that created either a non-silent missense mutation or a truncated gene product would be ideal. Such a collection would provide the most flexibility in design of new varieties. While the number of mutant lines for such an idealized scenario is prohibitive, lines with a high density of point mutations can provide the starting material for a similar sort of approach.

The two greatest sources of point mutations are spontaneous mutation and chemical mutagenesis. Both of these sources rely on altering bases in DNA and then allowing the DNA replication system to use the altered base as a template, creating the mutation. In maize, where mutagenesis of pollen is quite straightforward (Neuffer 1994), this has been the method of choice for several decades, primarily because it avoids the creation of chimaeric plants that may or may not transmit induced mutations to progeny. To avoid lysis of the pollen by aqueous solutions, an emulsion of the mutagen is made in paraffin oil, and the pollen mixed in.

12.2 EMS Mutagenesis

Chemicals such as ethyl methane sulfonate (EMS) and nitrosoguanidine (NTG) have been used successfully to introduce single base changes throughout the maize

Clifford F. Weil
Agronomy Department, Purdue University, 915 West State St., West Lafayette, IN 47907, USA
e-mail: cweil@purdue.edu

Rita-Ann Monde
Agronomy Department, Purdue University, 915 West State St., West Lafayette, IN 47907, USA

A.L. Kriz, B.A. Larkins (eds.), *Molecular Genetic Approaches to Maize Improvement*
Biotechnology in Agriculture and Forestry, Vol. 63

genome (Neuffer and Coe 1978). Identification of mutants using forward genetic screens (going from mutant phenotype to affected gene) placed a greater emphasis on being able to identify a single phenotype in a mutant plant. Very large numbers of plants can be screened as a result, each plant carrying a relatively small number of mutations.

With the advent of reverse genetics, there has been renewed interest in understanding the limits to which such mutageneses can be pushed, however. In reverse genetics, one identifies an altered sequence and then determines what, if any, effect the change has on the plant. Efficiently screening the large number of interesting, possible point mutations dictates the use of a relatively small number of lines carrying a large number of mutations per line. The idea, then, is to create the greatest possible mutational loads on each plant without compromising fertility or viability.

Because a combination of temperature and time of treatment as well as concentration of mutagen often dictate the effectiveness of mutagens on maize pollen, results of mutageneses are often quite variable when carried out in the field. Higher temperatures and longer times produce more mutation but also a higher incidence of pollen death. An effective strategy has been to test a range of mutagen concentrations on pollen at some fixed combination of temperature (30–35°C) and time (45–60 min) so that conditions can be reproduced reliably. This is typically achieved using an incubator housed in a building located near the field or greenhouse. There is also a noticeable difference in the response of different inbred lines to EMS treatment, even when the conditions are kept constant. For example, side-by-side treatments of W22 and B73 generally show nearly twice the mutations in W22 as in B73 (C.W., unpublished data). EMS is light sensitive and, even though these treatments are relatively short, they are best kept in the dark, either inside an incubator or, at the least, wrapped in foil.

Empirical tests of EMS treatments are important for several reasons. EMS proves to have a relatively short shelf life for this application. Bottles of EMS should be relatively new (<1 month old) when used and their efficacy should be tested just before large-scale use. This can usually be done the day before treatment by testing a small sample of pollen treated and then placed on germination medium (Schreiber and Dresselhaus 2003). In addition, mixing procedures during the treatment may vary from lab to lab and these need to be kept constant in both the tests and the treatments themselves. For example, Neuffer (1994) recommends that the EMS be added to paraffin oil and mixed thoroughly for an hour prior to being diluted further for the treatment itself. This is typically done on a magnetic stirrer at high speed in a brown glass bottle. The emulsion that forms separates quickly, however, and keeping the mixture stirring rapidly while removing aliquots for dilution into the treatments leads to more consistent results. For similar reasons, it is important to mix the treatment every few minutes as well, particularly to keep the pollen and EMS from separating out of the oil, which will result in pollen lysis. There are reports that a stock solution of 1% EMS/oil (v/v) can be kept indefinitely and diluted as needed, but we find that best results come when this 1% stock is no more than a week old.

In addition, in our hands, pollen viability in paraffin oil begins to decrease rapidly by ~60 min even in the absence of mutagen, and this can vary even between batches of paraffin oil. Pollen collected at different times of the day varies in both viability and response to treatment, and these parameters vary between maize inbreds. For example, greenhouse-grown B73 pollen is best when collected by bagging tassels first thing in the morning and harvesting the pollen before noon. For field-grown material when temperatures are warmer, these times may need to be even earlier in the day.

Treatments that begin to affect pollen germination rates significantly often result in no seed. While in some organisms, treatment levels that target 1–10% survival of the treated organism are common, in this case such strategies prove unwieldy. Instead, the treatments just below the levels where pollen viability begins to decrease are those that prove most effective, typically yielding 3–5 M1 seed per treated ear. Given the sterility and viability issues these seed can have, three to four times the number of desired M1 seed need to be generated. For TILLING populations, for which a target of ~3000–4000 fertile M1 plants is the norm, this means a large number of ears must be pollinated; however, the resulting populations are extremely valuable and can contain upwards of 100 billion induced mutations.

Assessment of a mutagenesis is often done by measuring induced dominant mutations in the M1 plants, such as *Oil Yellow* (photosynthetic aberrations) or dominant, lesion mimic mutations. Where it is possible to do so, an alternative approach is to mutagenize pollen of a plant homozygous for an easily scored, non-mutant seed marker, such as *Bz1* or *C*1, and use the treated pollen to fertilize ears of a line homozygous for a mutant form of the seed marker. M1 seeds are thus reporters for the effectiveness of mutating at least one gene, although this "readout" in the endosperm is not usually reflected in the accompanying embryo. Maize pollen typically has two sperm nuclei already by the time pollen is shed and mutagenesis is therefore on each sperm nucleus independently. Another classic assessment of mutagenesis efficacy is to self M1 lines and score photosynthetic aberrations that segregate among M2 seedlings (albino, yellow-green, virescent, etc. plants). Interestingly, these measures do not always correlate well with mutation density determined by molecular methods. A more reliable M2 phenotypic screen has been to measure embryo/seedling lethality in M2 seedlings grown in sandbenches. Treatments in which 40–60% of M2 families segregate lethality prove highly effective in reverse genetic screens.

The best indicator of all is a direct assessment of mutation density in five to ten genes using a subset (typically either 384 or 768) of M1 plants. These assessments are quite straightforward, using the TILLING procedure itself together with primer pairs that have been demonstrated to be effective already (see below). At present, maize TILLING uses a population of 3072 individuals that have produced 511 induced mutations in 340.7 Mb of sequence screened. Across the entire population, this rate translates into 1 mutation every 217 bp throughout the genome.

Technologies for rapid, de novo identification of point mutations using reverse genetics typically fall into one of two groups, both of which will be reviewed here. The first is an initial screen using enzymatic detection of mismatched base pairs, followed by DNA sequencing. The protocol known as TILLING (Targeting Induced Limited Lesions IN Genomes) has proven highly effective at doing just that in maize. The second approach is simply resequencing many potential mutants to find any with DNA alterations in specific genes of interest. Once regarded as too costly and inefficient, revolutionary changes in massively parallel sequencing technology have made such ideas a reality.

12.3 TILLING

12.3.1 TILLING Mutagenized Lines

Developed in 2000 for Arabidopsis, TILLING detects mutations by PCR amplification of a target gene using pools of templates, then screening for DNA mismatches among the mixture of amplicons (McCallum et al. 2000; Till et al. 2003, 2004b). These mismatches arise only if any member of a given pool has a sequence difference from the other members of the pool (Fig. 12.1). The mismatch is revealed by use of an S1 family endonuclease (typically Cel 1, derived from celery extract) that cleaves the DNA backbone 3' of the mismatched base (Till et al. 2004a).

Using differentially end-labeled PCR primers, the result is doubly end-labeled full-length, amplified molecules and, if a mutation is present in a pool of templates, a pair of singly end-labeled sub-fragments that are complementary in size. The products are then resolved electrophoretically in DNA sequencers. Two-color laser (700 and 800 nm wavelength), slab gel analyzers (Li-COR, Lincoln, Nebraska) are used in the Maize TILLING Project (MTP) because the two-dimensional readout has provided more accurate fragment sizing, increased sensitivity and a more diagnostic readout both for resolving two-dimensional pools of individuals using both lasers and for troubleshooting when required. Other projects have used capillary electrophoresis instruments successfully, which are more conducive to carrying out the entire process robotically (Perry et al. 2003; Cordeiro et al. 2006).

Using two-dimensional pooling strategies, individuals within the pools can be identified, the target gene amplified specifically from those individuals, and the mutant genes sequenced to identify the base change in each. Non-silent mutations (nonsense alleles, splice site alterations and missense alleles) can be identified and studied further; thus far, non-silent mutations comprise 49.7% of the 511 mutations identified by MTP. The best uses of EMS mutagenesis and TILLING are typically to identify alleles with partial rather than null phenotypes. While knockouts of genes are made using EMS, they comprise only $\sim$5% of the total mutation spectrum. Rather, EMS provides alleles that can help identify functionally important amino acids or identify null alleles when such alleles are not among the large transposon insertion collections. In addition, EMS alleles can provide sublethal alleles of essential genes and substerile alleles of fertility genes.

Identifying regions of genes to perform TILLING on and then interpreting the results obtained is an important part of the TILLING procedure. A key component in TILLING has been the bioinformatics tool CODDLe, developed by the proWeb group (http://proweb.org), which is used to select regions of a gene most likely to yield useful mutations. CODDLe identifies the gene region with the highest frequency of potential G to A changes that are likely to produce either truncations of

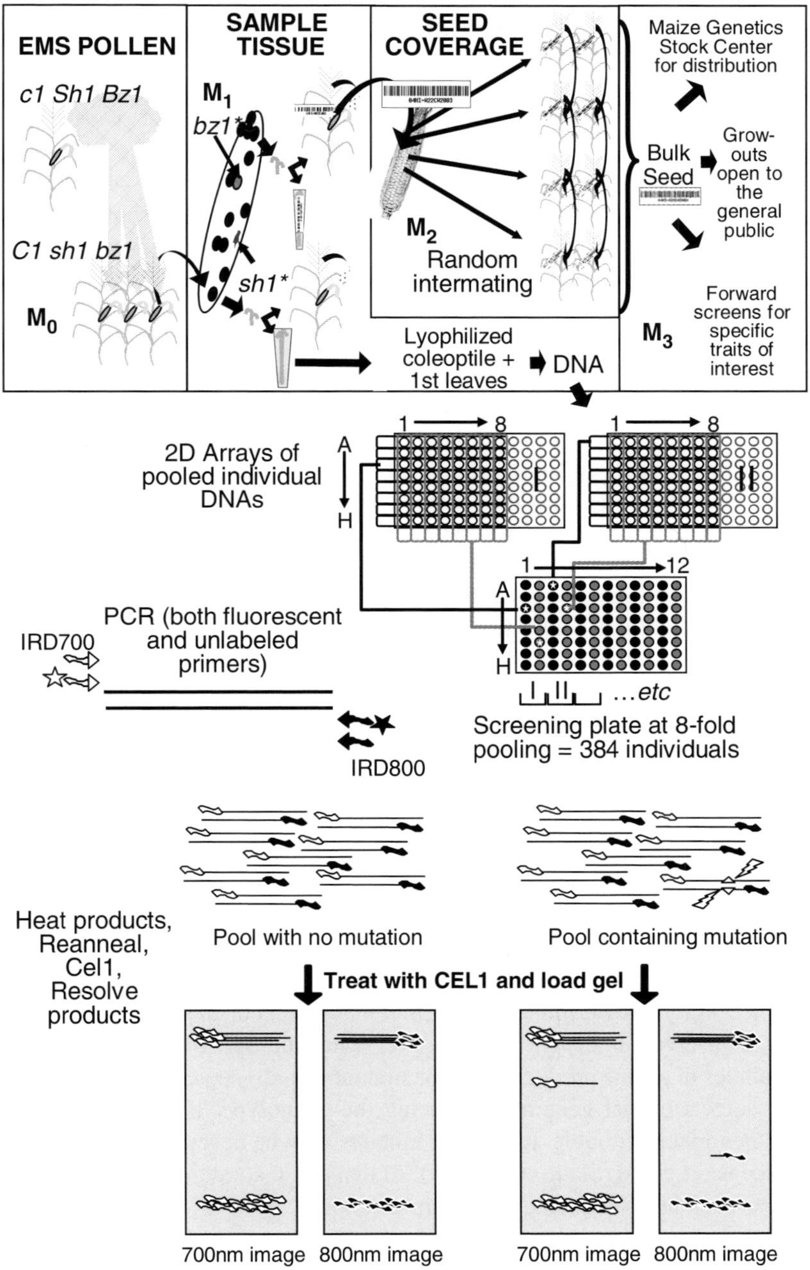

Fig. 12.1 Cel 1 based TILLING. Pollen mutagenesis is followed by growth of the resulting M1 seed, and tissue is harvested from these M1 seedlings for DNA, ensuring that all mutations present are heterozygous. DNA from individuals is then pooled two-dimensionally, PCR amplified using gene-specific primers and then treated with Cel 1. In pools where a mutation is present, Cel 1 cleaves at mismatched bases, producing singly end-labeled sub-fragments. The 2-D pooling allows resolution of which individual contains the mutation and the sizes of the sub-fragments give a good indication of where the mutation is located. Mutations are then verified in individual lines by DNA sequencing

the gene product or missense alleles predicted to be damaging to the gene product. These predictions are based primarily on the assumption that blocks of conserved amino acid sequence are functionally important, and that non-conservative changes within these blocks are more likely to be severely deleterious. It is worth pointing out that, valuable though these suggestions can be, they need to be taken only as suggestions, and CODDLe allows users to scan additional regions of their gene. Individual amino acids important for function may not lie within blocks of highly conserved sequence and conserved regions are not always functionally significant (Consortium 2007). Thus, at least initially, all non-conservative missense alleles should be tested further.

An additional pre-screening procedure has been implemented in maize TILLING that has proven valuable in maize and in other duplicated and polyploid genomes. Precautions are taken to ensure that primers designed for TILLING PCR amplify only one gene, producing a single amplicon that gives a clean Sanger sequence trace. In addition, this prescreening provides a baseline sequence for both B73 and W22, the two inbreds currently used for TILLING, against which mutations can be compared. The detailed PCR protocol used is available at http://genome.purdue.edu/maizetilling to permit users of the service to test primers themselves, which can help speed the prescreening process. Primers that meet these prescreen criteria are then advanced to TILLING.

Once mutations have been identified, a second proWeb algorithm is used to analyze the predicted effects of any mutations found for users of MTP (SIFT) (Ng and Henikoff 2003). Automated scores are assigned to each mutation identified that reflect what change they create (if any) in the amino acid sequence, where those changes are in relation to blocks of known conservation, and the confidence with which the prediction can be made. Screens are continued until either truncation alleles or multiple missense alleles predicted to be damaging are delivered.

Stocks that contain mutations identified by TILLING are made available through the Maize Genetic Stock Center for further study. These stocks have been heavily mutagenized and each stock can carry as many as 4000 individual base changes (Weil and Monde 2007). Thus, backcrossing to isolate a specific mutation of interest may be necessary to determine whether that mutation is responsible for any mutant phenotype observed. In many cases, however, when multiple lines carrying different mutant alleles of a gene produce the same mutant phenotype, it is a strong indication that the defects in that gene are producing the phenotype. If genes are duplicate factors, then making double, triple, etc. mutants may be necessary before a mutant phenotype is observed (Slade et al. 2005). At that point, combining various missense alleles can then be used to characterize interactions.

12.3.2 EcoTILLING

Natural variation in gene sequence is nature's own mutagenesis experiment and the source, together with environmental effects, of the observed phenotypic variation within a species. In maize, it is reported that many inbreds differ from one another in DNA sequence by as much as 1.5% (Liu et al. 2003). While it may be expected that the majority of that variation does not alter protein sequences, a useful fraction of it probably does.

These differences are also amenable to TILLING techniques, comparing one cultivar/accession/ecotype of a species with another. A variation on the protocol, originally devised by Luca Comai and co-workers for Arabidopsis ecotypes and called EcoTILLING (Comai et al. 2004), compares different inbred lines against the reference B73 genome. Taking advantage of the Maize Diversity Lines, a set of inbreds that capture ~80% of the diversity in maize germplasm worldwide (Liu et al. 2003), primers already submitted for TILLING are also used for EcoTILLING to provide a quick, inexpensive look at allelic diversity available for those same gene targets among natural populations, and over 40 gene targets have had this analysis included since we introduced it as part of MTP. The gene model provided as part of submitting genes for TILLING then allows a prediction of whether natural variations observed fall within exons; as expected, more variation is observed among introns. Users of the service get annotated gel images (Fig. 12.2) and can then sequence and analyze alleles from various inbreds as appropriate.

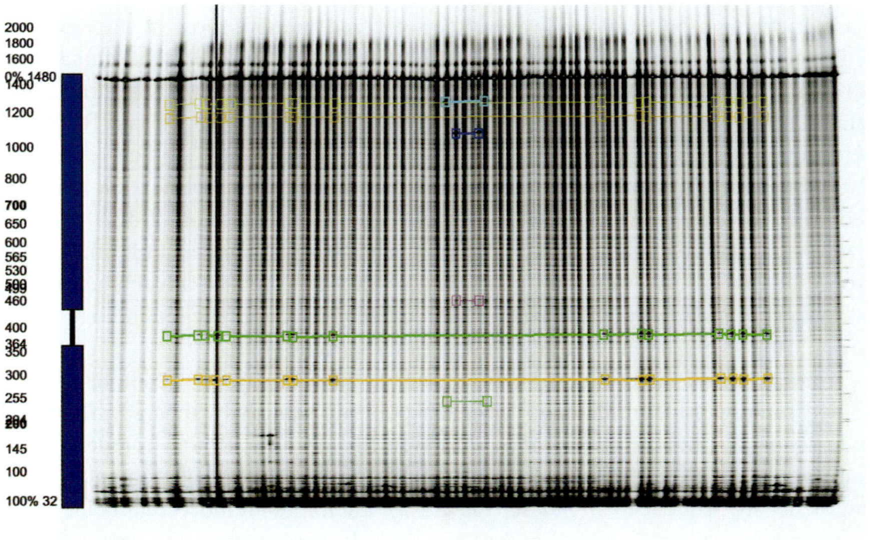

Fig. 12.2 SNPs between various inbred maize lines and B73 detected by EcoTILLING. Pairwise comparisons for a sample gene of the indicated maize inbreds against B73; each comparison is loaded on the gel twice, in mirror image, for confirmation; axis of symmetry denoted by *vertical dashed line*. The gene model, shown *left*, indicates whether the variation detected is within an intron (*black line*) or an exon (*blue box*). Size standards in base pairs are indicated at *far left*. *Boxes* are drawn with *thick lines around bands* representing SNPs. The 800-nm channel gel image is shown; bands of complementary size that appear in the 700-nm channel (not shown) are indicated by *boxes with thin lines*

12.4 Targeted Resequencing Using Massively Parallel Strategies: TRUMPing TILLING

The strength of TILLING approaches has been the ability to assess rapidly and inexpensively (compared to dideoxy sequencing) the presence of a mutation in one individual among several thousand genomes simultaneously. As sequencing technologies have achieved much higher throughput, much higher output and become much less expensive, the notion of resequencing entire mutagenized genomes has become much more feasible. However, even with the best current technologies, the idea of sequencing (as an example) 3200 entire genomes, archiving these sequences as a searchable database and then distributing seed for these lines on request is still a project that would not be cost effective. Even when the goal of a "$1000 genome" is realized, resequencing entire genomes across thousands of mutant lines and to sufficient depth for analysis of point mutations will be an expensive proposition that needs to be repeated with each new inbred screened. An alternative approach, Targeted Resequencing Using Massively Parallel (TRUMP) methodology that focuses on resequencing specific gene targets within those genomes is much more reasonable and can be accomplished right now.

These are resequencing analyses comparing mutant lines against a known "scaffold" developed both from the genome-sequencing project and from MTP prescreening procedures. As a result, technologies with extremely high output and short read lengths can be employed very effectively and without most of the assembly problems faced in de novo sequencing. An additional advantage to these massively parallel technologies is that they allow deeper pooling of samples, with reports of reliable mutation detection among up to 80-fold pools (E. Cuppen and R. Nutter, pers. comm.). Even at 40-fold pooling, already five times better than current TILLING, this aspect dramatically reduces the number of PCR amplification reactions needed to screen a given target. For example, an entire population of 3200 mutagenized lines requires only 160 PCR reactions and two gene targets at a time can be processed on one 384-well thermal cycler. Perhaps the greatest advantage of these resequencing strategies is that each improvement in sequencing technology drives down cost to the user and steadily decreases turnaround time.

Amplifying as large a portion of a gene as possible from two-dimensionally pooled samples, the amplicons from each pool are randomly sheared. Short barcode sequences unique to each pool are then ligated onto all the fragments from that pool, along with the adapters required for the sequencing instrument, and the fragments are sequenced *en masse* (Fig. 12.3). Each mutant individual is identified by a unique combination of two barcodes and all the sequence fragments derived from that mutant have one or the other of these barcodes. In addition, avoiding genes that have similarities over short regions in the same sequencing run, 30–50 different gene targets can be analyzed across the entire mutant population simultaneously. Mutations, always heterozygous in maize screens (Weil et al. 2005), can then be identified as SNPs. False positive identification of mutations is not a problem because of the depth of the sequence coverage (typically approximately

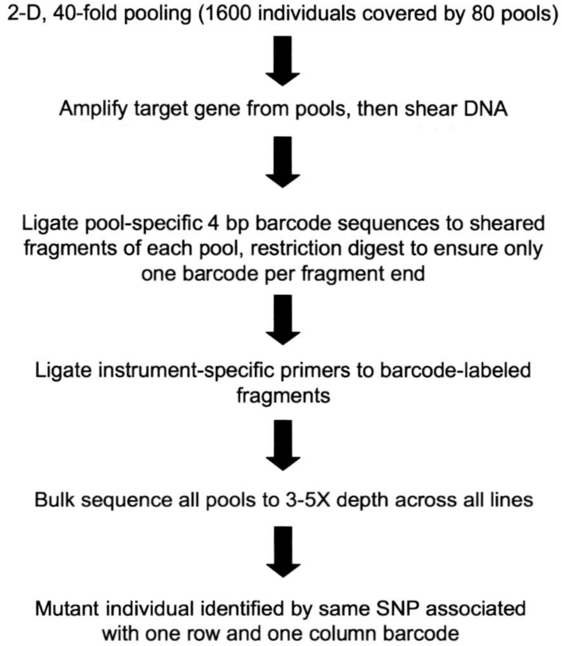

Fig. 12.3 Targeted Resequencing Using Massively Parallel (TRUMP) methodology workflow. Fragments are sequenced in massively parallel instruments such as those made by Illumina/Solexa or ABI SOLiD (Roche 454 sequencing can also be done, with longer read lengths but higher cost). Sequences are then aligned and compared against the "scaffold" of the wild type sequence, and mutant individuals identified on the basis of a unique pair of identifying barcode sequences

three to five times or greater for each individual allele). The false positive rate may be reduced still further by the use of ligase-based sequencing techniques rather than polymerase-based sequencing (Shendure et al. 2004, 2005). Conservatively assuming useable read lengths of 25 bp (not including the barcode sequences), three times coverage of a 2-kb sequence target requires approximately 240 reads, or a total of 1,536,000 reads across 3200 mutant lines with two alleles of every gene in each. In instruments already capable of as many as 50,000,000 reads per run, however, the ability to screen many genes across entire populations with each run of the sequencer provides an extremely inexpensive and effective tool for functional genomics. With today's technology, the capacity of this method is over 1500 genes screened per year per sequencing instrument, and these screens cost users hundreds rather than thousands of dollars. As run times decrease and sequencer output increases, this capacity will only go up and the cost is very likely to come down even further. Alternatively, this approach provides the ability to screen even larger mutagenized populations, increasing the probability that screens will produce an informative and diverse allelic series for each gene.

12.5 Conclusion

Induction and/or detection of small changes in genes is a crucial part of our functional analysis of maize genes. Reverse genetic approaches are particularly important in this effort, particularly with a sequenced genome to explore. Mutation discovery technologies are available today to take immediate advantage of that sequence, and the genetic resources to leverage those technologies are available as well. This combination, together with transposon insertion collections (described elsewhere in this volume), makes understanding the functions of most maize genes a goal that can be realized within the next 10–15 years.

References

Comai L, Young K, Till BJ, Reynolds SH, Greene EA, Codomo CA, Enns LC, Johnson JE, Burtner C, Odden AR, Henikoff S (2004) Efficient discovery of DNA polymorphisms in natural populations by ecotilling. Plant J 37:778–786

Consortium E (2007) Identification and analysis of functional elements in 1% of the human genome by the ENCODE pilot project. Nature 447:799–816

Cordeiro G, Eliott FG, Henry RJ (2006) An optimized ecotilling protocol for polyploids or pooled samples using a capillary electrophoresis system. Anal Biochem 355:145–147

Liu K, Goodman M, Muse S, Smith JS, Buckler E, Doebley J (2003) Genetic structure and diversity among maize inbred lines as inferred from DNA microsatellites. Genetics 165:2117–2128

McCallum CM, Comai L, Greene EA, Henikoff S (2000) Targeting induced local lesions in genomes (TILLING) for plant functional genomics. Plant Physiol 123:439–442

Neuffer MG (1994) Mutagenesis. In: Freeling M, Walbot V (eds) The maize handbook. Springer, New York, pp 212–219

Neuffer MG, Coe EH Jr (1978) Paraffin oil technique for treating mature corn pollen with chemical mutagens. Maydica 23:21–28

Ng PC, Henikoff S (2003) SIFT: predicting amino acid changes that affect protein function. Nucleic Acids Res 31:3812–3814

Perry JA, Wang TL, Welham TJ, Gardner S, Pike JM, Yoshida S, Parniske M (2003) A TILLING reverse genetics tool and a web-accessible collection of mutants of the legume *Lotus japonicus*. Plant Physiol 131:866–871

Schreiber DN, Dresselhaus T (2003) In vitro pollen germination and transient transformation of *Zea mays* and other plant species. Plant Molec Biol Rep 21:31–41

Shendure J, Mitra RD, Varma C, Church GM (2004) Advanced sequencing technologies: methods and goals. Nat Rev Genet 5:335–344

Shendure J, Porreca GJ, Reppas NB, Lin X, McCutcheon JP, Rosenbaum AM, Wang MD, Zhang K, Mitra RD, Church GM (2005) Accurate multiplex polony sequencing of an evolved bacterial genome. Science 309:1728–1732

Slade AJ, Fuerstenberg SI, Loeffler D, Steine MN, Facciotti D (2005) A reverse genetic, nontransgenic approach to wheat crop improvement by TILLING. Nat Biotechnol 23:75–81

Till BJ, Reynolds SH, Greene EA, Codomo CA, Enns LC, Johnson JE, Burtner C, Odden AR, Young K, Taylor NE, Henikoff JG, Comai L, Henikoff S (2003) Large-scale discovery of induced point mutations with high-throughput TILLING. Genome Res 13:524–530

Till BJ, Burtner C, Comai L, Henikoff S (2004a) Mismatch cleavage by single-strand specific nucleases. Nucleic Acids Res 32:2632–2641

Till BJ, Reynolds SH, Weil C, Springer N, Burtner C, Young K, Bowers E, Codomo CA, Enns LC, Odden AR, Greene EA, Comai L, Henikoff S (2004b) Discovery of induced point mutations in maize genes by TILLING. BMC Plant Biol 4:12

Weil CF, Monde R, Till B, Comai L, Henikoff S (2005) Mutagenesis and functional genomics in maize. Maydica 50:415–424

Weil CF, Monde R-A (2007) Getting the point: mutations in maize. Crop Sci 47:S60–S67

Chapter 13
Applications of Linkage Disequilibrium and Association Mapping in Maize

Elhan S. Ersoz, Jianming Yu, and Edward S. Buckler

13.1 Introduction

Association mapping, also known as linkage disequilibrium mapping, is a relatively new and promising genetic method for complex trait dissection. Association mapping has the promise of higher mapping resolution through exploitation of historical recombination events at the population level, that may enable gene level mapping on non-model organisms where linkage-based approaches would not be feasible (Risch and Merikangas 1996; Nordborg and Tavare 2002).

Association mapping utilizes ancestral recombinations and natural genetic diversity within a population to dissect quantitative traits and is built on the basis of the linkage disequilibrium concept (Geiringer 1944; Lewontin and Kojima 1960). One of the working definitions of linkage disequilibrium (which here on will be referred to as LD) is the non-random co-segregation of alleles at two loci.

In contrast to linkage-based studies, LD-based genetic association studies offer a potentially powerful approach for mapping causal genes with modest effects (Hirschhorn and Daly 2005). While linkage analysis is based upon detection of non-random association between a genotype and a phenotype in well-characterized pedigrees, association mapping focuses on associations within populations of *unrelated* individuals. In general, chromosomes sampled from *unrelated* individuals in a population will be much more distantly related than those sampled from members of traditional pedigrees. In other words, the time to most recent common ancestor

Elhan S. Ersoz
Institute for Genomic Diversity, Cornell University, Ithaca, NY 14853, USA

Jianming Yu
Department of Agronomy, Kansas State University, Manhatten, KS 66506, USA

Edward S. Buckler
USDA-ARS, US Plant, Soil and Nutrition Laboratory, Ithaca, NY 14853, USA
Institute for Genomic Diversity, 159 Biotechnology, Cornell University, Ithaca, NY 14853, USA
e-mail: esb33@cornell.edu

A.L. Kriz, B.A. Larkins (eds.), *Molecular Genetic Approaches to Maize Improvement* 173
Biotechnology in Agriculture and Forestry, Vol. 63

(MRCA) of any given two individuals from a population of unrelated individuals would be greater than that of a pedigree population. This is what makes LD mapping suitable for fine-scale mapping: there will have been more opportunities for recombination to take place over several generations, between many alleles, in a species, while there can be only a few generations of recombination present in pedigree populations. Increase in the rate of recombination will lead to reshuffling of the chromosomal segments into smaller pieces. This will lead to reduction of the LD in short distances around loci, and lead to significant co-occurrence (i.e. LD) between only loci physically close, allowing high resolution. Whereas pedigree studies work with recombination events in few generations that enable exchange between chromosomes at the order of megabases, association studies deal with segmental exchanges measured in kilobases (Paterson et al. 1990; Stuber et al. 1992; Thornsberry et al. 2001).

13.2 What is Linkage Disequilibrium and How is it Related to Association Mapping Studies

The term *linkage disequilibrium* was first introduced back in the late 1940s to describe the degree of non-random association between pairs of loci. In the absence of demographic effects that might confound the LD patterns, LD summary statistics such as r^2 can be used to define the level of co-occurrence of alleles at two loci (Hill and Robertson 1968). When r^2 is zero, alleles at two loci do not co-occur more frequently than would be expected under random sampling. r^2 approaches its maximum of 1 as alleles at two loci show more frequent co-occurrence within the population sample examined. There are various other LD statistics that can be used for this purpose (Hedrick 1987) all of which aim to estimate the predictive value of a marker locus on another locus that is displaying non-zero LD with it (if LD statistic is zero, two loci examined have zero predictive value for each other).

Association mapping uses these properties of the measures of pairwise LD statistics to infer the predictive value of a marker locus for the association of the chromosomal region where it resides with the phenotype. The high-LD chromosomal region around a marker locus defines the predictive range of a certain genetic marker. If LD within this genomic range is complete, any polymorphism within this range will have the same predictive value for the association with the phenotype. Hence, as a result of a significant marker–phenotype association, it can be concluded that the causative polymorphism resides within this high LD region around the marker locus.

With respect to association mapping, the most significant aspect of LD is its predictive properties over the haplotype it resides in. However, the extent of LD (in base pairs) within species and even within individual genomes is highly variable, and therefore most reliably estimated empirically (Long and Langley 1999).

Theoretical estimation of the levels of LD for realistic population models that does not satisfy the assumptions of the Wright-Fisher model is complex. The hardship is mostly due to the large number of interrelated factors involved in the formation of patterns of LD, including but not limited to genetic drift, population admixture, and natural selection (Pritchard and Przeworski 2001; Wall and Pritchard 2003).

The statistical power of associations is determined by the extent of LD with the causative polymorphism, as well as sample size used for the study (Long and Langley 1999; Wang and Rannala 2005). If LD decays too fast within a region, a large number of markers would be required to scan target regions of a genome. On the other hand, if LD decays too slowly, the size of the haplotype blocks would be too large to unambiguously reveal underlying causative locus. In other words, the decay of LD over physical distance in the study population determines the marker density required and the level of resolution that may be obtained in an association study.

13.2.1 How to Estimate LD

There are several summary statistics proposed for estimation of LD (Hedrick 1987); however, the most commonly used summary statistic within the association study framework is known as r^2 (Hill and Robertson 1968; Lewontin 1988). Conceptually and mathematically r is the Pearson's (product moment) *correlation coefficient* of the correlation that describes the predictive value of the allelic state at one polymorphic locus on the allelic state at another polymorphic locus, where r^2 is the squared value of correlation coefficient that is also called *coefficient of determination*. r^2 explains the proportion of a sample variance of a response variable that is *explained* by the predictor variables when a linear regression is performed.

Lewontin's D is another summary statistic for LD that is commonly used. D describes the difference between the coupling gamete frequencies and repulsion gamete frequencies at two loci. From D a second measure of LD, that is normalized D', can also be estimated. Even in samples taken from populations at equilibrium under neutrality, variances of LD summary statistics are typically large, but D' has the lowest variance (Hedrick 1987). However, estimation using D' may generate erratic and unreliable results when low frequency alleles or small sample sizes are used for the analysis. It is advisable to collapse the alleles using an allele frequency cut-off prior to estimation of LD statistics D and D'.

Other than these commonly used summary statistics for LD, there are also likelihood-based methods that investigate probability of independence between pairs of sites using two-locus sampling distributions, rather than calculating a summary statistic for LD. These methods, usually referred to as model-based LD estimators, also provide means of estimating population recombination parameter $4Nc$

under a neutral equilibrium model from nucleotide sequence data (Golding 1984; Hudson 1985, 2001) or generating other model-based estimates of LD for comparisons with observed patterns (Mueller 2004) under various population structure and demographic history scenarios. Although the estimation of LD through these methods is more computationally intensive compared to pairwise-LD estimation methods, they are extensively used for evolutionary and population genetic studies as well as investigations into the domestication of various crop plant species (Wright et al. 2005; Wright and Gaut 2005).

13.2.2 Interpretation of LD Data

Estimating LD from empirical data is a straightforward procedure; however, interpretation of the results of LD analysis and extrapolation of this information to the genome may be more complex. It is important to estimate the rate of decay of LD with physical distance to be able to extrapolate information gathered from a small collection of sampled loci to the whole genome investigated. This extrapolation is essential for association mapping study design since it may be used to determine the marker density required for scanning previously unexplored regions of the genome as well as the maximum resolution that can be achieved for genotype–phenotype associations in the study population.

The levels of LD are expected to be highly variable across the genome due to several factors, such as variation in recombination rate and selection. For reliable results, this variation needs to be taken into account when designing experiments to exploit LD. Variation in rate of recombination across the genome is a key factor that contributes to the variance observed in patterns of LD. A number of researchers have focused on the distance at which average r^2 is reduced to 0.10 as a reasonable point to conclude that there is minimal LD to detect associations with complex traits. The reasoning for this r^2cut-off is as follows: in a complex trait a large quantitative trait locus (QTL) may only explain approximately 10% of the phenotypic variation. If a marker only explains 10% of the total QTL variation, then the marker will only explain 1% of the phenotypic variation. Detection of locus effects that cause larger than 1% phenotypic variation requires exponentially increasing population sizes, and therefore such small effects would be considered undetectable in a moderate size study population.

To maintain sufficient power for dissection of complex traits through association studies, the choice of marker density and population size are of importance. Not only high enough marker density to screen and target region(s) at blocks of greater LD (i.e. $r^2 > 0.8$) but also large-sized populations are required in order to achieve sufficient power. Current human genetic studies focus on genome scans aiming for much higher LD (e.g. $r^2 > 0.80$) (Barrett and Cardon 2006), and are developing haplotype-based approaches that can help capture more variants (Pe’er et al. 2006).

13.2.3 LD in Maize

Studies on rates of decay of LD in various plant taxa (Flint-Garcia et al. 2003) such as maize (*Zea mays* ssp. *mays*) (Remington et al. 2001b; Ching et al. 2002; Tenaillon et al. 2002; Palaisa et al. 2003), barley (*Hordeum vulgare*) (Caldwell et al. 2004, 2006), *Arabidopsis thaliana* (Nordborg et al. 2002, 2005), sorghum (*Sorghum bicolor*) (Hamblin et al. 2005) and durum wheat (*Triticum durum*) (Maccaferri et al. 2005) indicate tremendous variation in the extent of LD. This variation is mostly due to founder effect followed by genetic drift that leads to unequal number of effective recombinations in species sub-populations. Selfing also plays an important role (Nordborg 2000).

The population sample effect is clearly observed in maize, where LD decays within 1 kb in land races (Tenaillon et al. 2001), in approximately 2 kb in diverse inbred lines (Remington et al. 2001b) and can extend up to 100–500 kb in commercial elite inbred lines (Ching et al. 2002; Jung et al. 2004). One key issue in comparing distances within genes and between genes is that recombination occurs very rarely outside of genes, so LD can extend for great distances in retroposon regions.

LD decay can also vary considerably from locus to locus. For example, significant LD was observed up to 4 kb for the Y1 locus (encoding phytonene synthase), but was seen at only 1 kb for PSY2 (a putative phytonene synthase) in the same maize population (Palaisa et al. 2003). A more recent study showed that LD for some haplotypes extends over 800 kb around Y1 (Palaisa et al. 2004). The Y1 case is a clear example of strong selection, with a decade-long period tremendously reducing the diversity linked to the key polymorphism, which created very extensive LD.

13.3 Association Populations and Statistics

There are five main stages for association studies: (1) selection of population samples; (2) determination of the level and influence of population structure on the sample; (3) phenotyping the population sample for traits of interest; (4) genotyping the population, for either candidate genes/regions or as a genome-wide scan; and (5) testing the genotypes and phenotypes for their associations (Fig. 13.1).

The choice of association test is the last step of the study and is mostly dependent on the previous steps, according to the characteristics of the population that was used to collect the genotypic and phenotypic data (Lewis 2002; Breseghello and Sorrells 2006a, b). Furthermore, possible complications due to population structure in the study sample may adversely affect the association test results. The influence of population structure on each association study depends on the relatedness between sampled individuals in the studied population (Fig. 13.2, Fig. 13.3). Therefore, the populations amenable for association studies may be classified according to the level of relatedness between the individuals forming the association population.

Preliminary Analysis and Feasibility Study

Population: Small sized diversity sample(s) to be used as a *Discovery Panel.*

Data: Nucleotide sequence, from locus samples with genome-wide coverage from the *Discovery Panel.*

Analysis: Nucleotide diversity (θ), decay of linkage disequilibrium with physical distance (r^2), population recombination rate (ρ), population structure and demography.

Results: Range of diversity to be sampled for association population, marker density required for sufficient coverage of target genomic regions (or the genome) for association, level of population structure that exists within the species, evaluation of genome-wide influence of demography, determination of genomic regions targeted by natural selection and domestication, and number and density of the neutral markers required to evaluate background associations.

Data Collection

Genotype: Select species-wise informative and high-through put genotyping-amenable markers. Choice of genotyping platform is dependent on the size of the population to be studied as well as the number of available markersthereby per marker per individual experimental cost is optimized. In addition, since genotypes from the candidate regions are trait dependent, in order to test the levels of background-stochastic associations, other neutral markers should be genotyped as well.

Phenotype: Phenotypes of interest should be replicated temporally and spatially to increase accuracy and precision of the phenotypic measurements. Quantitative measures of the traits of interest are preferable over categorical phenotyping. Evaluation of the heritability helps define the expectation for the genetic component of the phenotypic variance.

Statistical Association

- Build statistical model(s) for the expectation of phenotypic correlation with environmental and genetic variability ($V_p = V_G + V_E$).
- Evaluate the level of co-variance between the phenotypes, and combine the highly correlated traits in the same model.
- Evaluate co-variance between the neutral marker genotypes and candidate gene genotypes.
- Determine the type I error thresholds according to the number of tests performed and the level of flexibility in the study.
- Determine power and false positive rate expectations for the study.
- Run statistical association tests.

Post-Association Follow-up

Evaluation: The genotypic value of the associated allele should be evaluated on several different genetic backgrounds, for its overall phenotype as well as biochemical and molecular genetic studies for elucidation of structure and function.
Verification: The association reported should be verified either through re-evaluation in an independent population sample or through allelic silencing/knock-outs.
Breeding: The best alleles obtained through the study should be incorporated into breeding programs for integration into elite varieties.

Fig. 13.1 The steps employed during an association study

In the following subsections, we will first discuss the influences of population structure on various association study designs, followed by examples of control on its influences by accounting for the relatedness between individuals forming the association population.

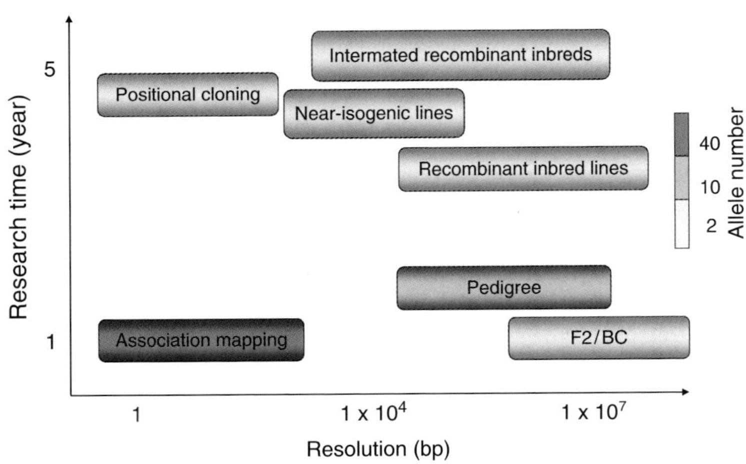

Fig. 13.2 Schematic comparison of various methods for identifying nucleotide polymorphism trait association in terms of resolution, research time and allele number

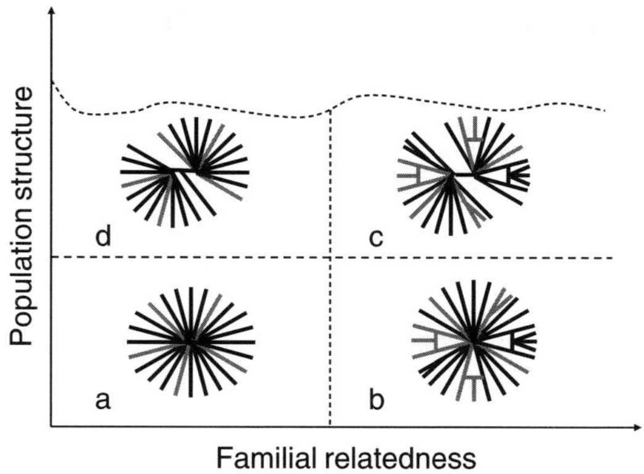

Fig. 13.3 Schematic diagram of the different types of population encountered in association mapping studies. Examples and relevant statistical methods for the analysis of the different population types are described. *a* Ideal sample with subtle population structure and familial relatedness (e.g. F2 population or synthetic), regression and genomic control (GC). *b* Family-based sample (e.g. extended pedigree), transmission disequilibrium test, quantitative transmission disequilibrium test, GC and mixed model (pedigree-based coancestry matrix and relative kinship matrix). *c* Sample with population structure (e.g. maize landraces), structured association (SA) and GC. *d* Sample with both population structure and familial relationships (e.g. maize association panel), SA, GC and mixed model (population structure (Q) plus relative kinship matrix (K))

13.3.1 Population Structure

The most important constraint to the use of association mapping for crop plants is unidentified population substructuring and admixture due to factors such as adaptation or domestication (Thornsberry et al. 2001; Wright and Gaut 2005). Population structure creates genome-wide LD between unlinked loci. When the allele frequencies between sub-populations of a species are significantly different, due to factors such as genetic drift, domestication or background selection, genetic loci that do not have any effect whatsoever on the trait may demonstrate statistical significance for their co-segregations with a trait of interest. Provided that a large number of neutral markers are available for estimation of genome-wide effects of structure, it is possible to statistically account for such effects in association data analysis (Yu et al. 2006b).

In cases where population structuring is mostly due to population stratification (Pritchard 2001; Bamshad et al. 2004), three methods are often acknowledged to be suitable for statistically controlling the effects of population stratification on association tests: (1) genomic control (GC) (Devlin and Roeder 1999; Devlin et al. 2001, 2004); (2) structured association (SA) method, including two extensions that are modified for the type of association study – case control (the SA-model) (Pritchard et al. 2000b) and quantitative trait association study (the Q-model) (Thornsberry et al. 2001; Camus-Kulandaivelu et al. 2006); and (3) the unified mixed model approach (Q+K) (Yu et al. 2006b).

The first method suggested for statistically controlling population structure was GC, which assumes that population structuring has equivalent effects on all loci genome-wide. In the GC method, a small random set of markers (e.g., polymorphisms unlikely to affect the trait of interest) are used to estimate influence of population structure on the association test statistics (*inflation factor*), such that the significance of the association statistic (*P* value) estimated is adjusted to account for population structure. The general principle of GC is to use individual genomes from the sample to estimate the levels of confounding due to substructure and more direct relatedness, such as familial relationship, in the study, and scale the final significance level of the association reported accordingly (Devlin et al. 2001).

Structured association methodology utilizes marker loci unlinked to the candidate genes under investigation to infer *sub-population membership*. The application of structured association to qualitative and quantitative traits is done using the appropriate model, depending on the trait and population type, with either SA or Q models, respectively. In application of SA for quantitative trait association (Q-model), a two-stage procedure is constructed, where for the first stage each subject's probability of membership in each sub-population is estimated (Pritchard et al. 2000a, b), and then in the next stage a test of association is conducted using sub-population membership as a variable for the association model tested (Pritchard et al. 2000b); then, in the next stage, a test of association is conducted using sub-population membership. In case-control studies, the probability of the SNP frequency distribution based on population structure is compared between the case and control samples. For quantitative traits, the population structure estimates are used as co-variates in the regression model that defines the correlation of the genotype with the phenotype (Thornsberry et al. 2001; Camus-Kulandaivelu et al. 2006).

In the unified mixed model approach (aka Q+K model) of Yu et al. (2006b), a large set of random markers that can provide genome-wide coverage are used to estimate population structure (Q) and relative kinship matrix (K), which are fit into a mixed-model framework to test for marker-trait association. In the unified mixed-model approach, each of the factors that may confound association analysis, that is familial relatedness between individuals (K) and relatedness due to population structure (Q), are considered as independent variables within the species population. In order to account for the combined effects of such relatedness factors, they are included as covariates in the regression model that defines the correlation between genotype and phenotype during association testing.

The genetic makeup of the study population that was used to collect genotypic and phenotypic data defines the model and type of association statistics to be used for association tests. This will be discussed further in the next section.

13.3.2 Classic Association Populations

If the individuals forming the study population are *effectively* unrelated, the study population may be considered a random sample of individuals from species populations and is therefore equivalent to any natural population. The relatedness amongst the individuals forming the population can be either estimated using pedigrees (Emik and Terrill 1949) or inferred using molecular markers (Lynch and Ritland 1999; Wang 2002; Blouin 2003; Oliehoek et al. 2006). These individuals can be either selected from originally natural populations or subselected from selections included in breeding programs, to form a classic association population. Selecting individuals from breeding programs offers the advantage of easy incorporation into future breeding programs; however, the number of lineages incorporated in the association study becomes limited (Breseghello and Sorrells 2006a, b).

All the previously mentioned statistical methods for population structure inferences are applicable to the classic association populations; however, the Q+K model has the widest base of applicability across all structured association study designs in natural populations.

In plants, so far the focus has been on quantitative traits in natural populations. In maize, using diverse inbred lines, it was possible to select a sample of 102 lines with relatively few closely related individuals by sampling across the world's breeding programs (Remington et al. 2001b; Thornsberry et al. 2001). However, as larger samples were gathered to increase statistical power to over 300 maize lines it became extremely difficult to find samples that match the structure expected in natural populations (Flint-Garcia et al. 2005). These are the cases where the combined natural and family-based approaches are most powerful (Yu et al. 2006a). In *Arabidopsis* (Nordborg et al. 2005), natural samples were collected from around the world, but because of strong population structure and selfing, these samples in many respects behave more like families for association mapping purposes (Aranzana et al. 2005). Association studies with some tree species are more likely to fall into the model of effectively unrelated individuals (Thumma et al. 2005; Gonzaléz-Martinéz et al. 2006). Most crop plant studies will probably fall on a continuum between natural and family-based association populations.

13.3.3 Family-Based Association Populations

If the association population is a collection of unrelated families, instead of single unrelated individuals, it is possible to perform a joint linkage and association analysis on the population, that potentially can be more informative on the trait of interest than either approach alone (Holte et al. 1997; Karayiorgou et al. 1999). For instance, in human genetics, where the association populations are collections of parent–offspring trios, two types of study design are considered: transmission disequilibrium tests (TDTs) (Spielman et al. 1993; Allison 1997; Rabinowitz 1997; Monks et al. 1998; Fulker et al. 1999) and family-based association tests (FBATs) (Laird et al. 2000; Lake et al. 2000; Horvath et al. 2001; Lange et al. 2003; Herbert et al. 2006; Laird and Lange 2006). Stich et al. (2006) modified the QTDT algorithm to test its applicability to inbred plant populations, and developed a model named the Quantitative Inbred Pedigree Disequilibrium Test (QIPDT), for analysis of joint linkage and association data from crop plant populations. Another family-based population design that was essentially developed for crop and livestock breeding is the Henderson's Mixed Model Approach (Henderson 1975), which is generally known for its applications in best linear unbiased predictors (BLUPs). Family-based association study design investigates co-segregation and linkage simultaneously (Spielman et al. 1994).

A long-standing mixed model method has been used by animal scientists to analyze the data from extended pedigree in dairy and beef cattle breeding programs (Henderson 1975, 1976, 1984). The superiority of the mixed model lies in its incorporation of the phenotypic observations from relatives of an individual in the estimation of the breeding value of that individual. The amount of information that is incorporated depends on the heritability of the trait and the genetic relationships (traditionally defined by pedigree information) among individuals. Naturally, this method has been extended to quantify the single gene effect while accounting for the pedigree relationship (Kennedy et al. 1992), and is applicable to association mapping with family-based association populations. Taking this mixed model framework, Yu et al. (2006b) suggested replacing the pedigree-based co-ancestry with a marker-based relative kinship (K) to account for the relatedness among individuals.

This unified mixed model approach is demonstrated to be the most powerful statistic compared to all the rest of the statistics for the family-based association studies and those studies falling between classical and family-based designs. The flexibility and generality of this approach allow association studies to be carried out on any population without the restriction on the specific family structure.

13.3.4 Special Association Populations

Recently, the field of plant association genetics pioneered the use of a new type of association population, designed to incorporate advantages of both linkage-based and LD-based quantitative trait dissection approaches in association studies, in a stronger design than transmission-disequilibrium test (TDT) design. This builds on some of the joint linkage-association approaches encountered in cattle breeding (Meuwissen and Goddard 1997; Blott et al. 2003). Nested association populations (NAM) are developed through controlled crosses between a diverse selection of unrelated individuals according to a breeding scheme that aims to shuffle alleles in diverse samples either across backgrounds or against a reference background, while keeping track of number and locations of the recombination events that shuffle the parental chromosomes (Yu et al. 2006a). The subsequent generations of progeny of the crosses can then be used as association populations. A population generated according to this described scheme not only provides tremendous power to the statistical tests of association, but also enables the projection of genotype information from the parents to the progeny, optimizing genotyping cost for large studies. The cross design is expected to effectively reduce many of the effects of admixture and population structure on the association population. For such populations, a two-step procedure for associations is suggested.

The two-stage study design of nested association mapping requires deep sequencing or genotyping of the parents for SNP identification across the genome, followed by lower density genotyping in the progeny in order to infer the locations of the recombination breakpoints during the crosses. Once the recombination breakpoints are localized and the recombination blocks are traced back to the contributing parent, the haplotype information from the parents can be directly projected on the progeny genome, without further need for genotyping within these blocks.

This design scheme enables the researcher to utilize the advantages of both linkage-based and LD-based genetic mapping approaches. It provides genome-wide coverage with high resolution and is performed on an experimental cross that is robust to genetic heterogeneity, with representation of several alleles per loci in a large population.

Because of the balanced design, straightforward multiple regression approaches can be applied (Yu et al. 2006a) for association testing. Currently, availability of such nested association populations are reported for maize (Yu et al. 2006a) and loblolly pine (Baltunis et al. 2005; Kayihan et al. 2005; Ersoz 2006). Further statistical methods that are going to utilize and combine information from both parent and progeny generations for NAM-type populations are currently under development.

These mentioned association population structures represent the continuum of LD levels from low in classic association populations towards high in biparental breeding populations. Nested association populations that are similar to heterogenous intermated populations (Niebur et al. 2004) fall in the mid-range of this continuum with moderate levels of LD and linkage.

13.4 False Positives and Power of Association

One of the major concerns of association mapping studies is the statistical power of the association testing, since, as it stands, there is a trade-off between the power and accuracy of reporting associations due to false positives. The major determinant of the levels of false positives and power of associations is the level of population structure in the association population.

A false positive (type I error) occurs when a test incorrectly reports that it has found a positive result where none really exists. The classical definition of type I error is an incorrect rejection of the null hypothesis – accepting the alternative hypothesis even though the null hypothesis was true. The second functional biological definition of false positives is also used in association studies. In this framework, false positives arise not only due to the failure of the statistical test performed, but also in cases where the statistical test is valid and the association exists but it is an association with population structure instead of the trait of interest. Population structure can lead to identification of loci that generate statistically significant but biologically invalid associations solely due to their tight correlation with population structure. However, if the population structure in an association study is properly dealt with, this is not expected to be a source of false positives.

Traditionally, type I error rate (α) for multiple testing is controlled with the Bonferroni correction. The Bonferroni correction in general is conservative and leads to power loss for detection if the polymorphisms are in LD and/or the traits are correlated with one another.

Another statistical method suggested for control of multiple testing is the false discovery rate (FDR) procedure. The FDR is the proportion of positive results that are actually false positives versus the whole set of positive results obtained from a statistical test. The procedure can be used to estimate a cutoff for a particular FDR (Benjamini and Hochberg 1995) or an FDR for a particular cutoff (Storey 2002; Storey and Tibshirani 2003). FDR approaches may be most appropriate when multiple traits are being compared or when the markers are not in extensive LD (Chen and Storey 2006). Essentially based on the relative costs of false positives on further follow-up research, appropriate FDRs should be determined and used.

A third procedure that can be applied for multiple testing correction is the permutation test (Churchill and Doerge 1994; Doerge and Churchill 1996), which controls for the genome-wide error rate (GWER). The permutation test has the ability to estimate effects on significance levels caused by the use of correlated markers as well as correlated traits. In this approach, the trait values are permuted relative to the genotypic data. These permutation approaches are appropriate ways to control the GWER; however, they can be quite conservative if one expects numerous QTLs. Recently, the GWER_k approach of Chen and Storey (2006) incorporating a more liberal balance of true and false positives provides a reasonable avenue.

Other than the statistical methods proposed, it is also possible to non-parametrically estimate the FDR through comparison of distributions of P values against a set of markers of known influence and a set of random markers scored on the same association population, with simulations. The probability of false associations is simply the ratio of the proportion of significant associations detected in the random set to the proportion of significant associations detected in the simulated set of known influence loci. This method provides a fast and rigorous way of estimating FDR if a set of random markers has been scored on the association population. Since random markers are required to estimate population structure, this method should be applicable for association testing in most cases.

The power of a statistical test is the probability that the test will reject a false null hypothesis. Some of the relevant parameters that can affect the power of association studies are, but are not limited to, (1) the type of association test – single marker or haplotype based; (2) the multiplicity control method; (3) the population-structure control method; (4) genetic architecture of the trait; (5) population size; (6) marker density; and (7) type of populations used for associations – family based or effectively unrelated (Long and Langley 1999).

Simulation studies that investigate the power of the association tests for the candidate gene association approach report that 300 individuals in a natural population provide enough power to detect *repeatable* associations when population structure is controlled properly (Long and Langley 1999; Thornsberry et al. 2001; Camus-Kulandaivelu et al. 2006; Yu et al. 2006a). These power estimates are based on candidate gene studies, where there are few SNPs being evaluated relative to the entire genome. Genome scan-type association studies are rapidly becoming feasible, but for such studies the population sample size required to obtain sufficient power will be larger. The exact population size required will depend on the LD structure for the population. Population sizes of 1000 to 5000 genotypes will likely be sufficient in most cases.

The power of association will be low if the trait is highly correlated with population structure. Statistical controls for population structure under such circumstances would result in false negatives. An example of such a case is demonstrated for maize and *Arabidopsis* flowering time traits (Aranzana et al. 2005; Flint-Garcia et al. 2005). The reason for flowering time and population structure to be correlated is that flowering time is an adaptive trait that largely defines the structure. The Q+K model can produce somewhat better results in these situations (Yu et al. 2006b), but in general a different sample or genetic design is required to work with traits that are tightly correlated with population structure. From a study of 60 traits on a maize diversity panel of 302 inbred lines, the only traits that showed strong relationship with structure were two flowering time-related traits.

Three studies using different germplasm have analyzed maize flowering time and the *dwarf8* (*d*8) gene (Thornsberry et al. 2001; Andersen et al. 2005; Camus-Kulandaivelu et al. 2006). These studies highlight the difficulties of studying traits related to population structure. In all three studies, when population structure is ignored, highly significant associations between the traits and polymorphisms in *d*8 are detected that are often much more significant than any of the random markers. It is clear that the putatively functional allele is segregating with a very high allele frequency in some populations, while it is represented at very low frequencies in other populations. This is exactly what would be expected if flowering time is under diversifying selection between the various sub-populations. Furthermore, upon application of standard corrections for managing population structure (Q), the *d*8-flowering time association is significant for some samples but not for others, in all three studies. Essentially, there is low statistical power to evaluate candidate genes that are involved in the clinal adaptation and/or creation of population structure. While empirical significance estimates obtained through contrasting the significances of the candidates with large numbers of random markers, the most effective approach for this type of trait may be specially constructed association populations with balanced designs.

13.5 Phenotyping and Genotyping Strategies for Association Testing

As in all other quantitative genetic studies, the success of an association study is heavily dependent on the accurate evaluation of the phenotype of interest. The within-population variation observed for genotypes and phenotypes for an association is much greater than that found in most bi-parental mapping populations. While greater variation is preferable when aiming for higher resolution and allele mining, it can pose problems for accurate evaluation of this variation in a meaningful way in a single environment.

The inherent variation observed in phenotypic trait measurement, when combined with the substantial genetic variation included in some association studies, requires careful experimental design to acquire quality data. In addition, evaluations in multiple environments with controls and unbalanced designs may be required. In our experience with maize, we found that evaluating the germplasm in short-day environments facilitated some trait evaluation by reducing photoperiod effects between lines. Additionally, we found that evaluating the germplasm in testcrosses (F1 hybrids) has reduced the phenotypic range to a manageable level. Since each of these approaches interact with the genetic architectures of the traits, future studies will be needed to fully understand the tradeoffs of various study design approaches.

In the association study design, genotyping is required for inferences both on the genotype/phenotype associations and on the population structure and demography. The first aim of querying candidate regions for polymorphisms is best achieved by genotyping SNPs within these candidate regions. The second aim of gathering information on population-specific phenomena, such as structure, linkage, demography and kinship, can be achieved through genotyping neutral background markers, such as SNPs on non-coding regions and SSRs (simple sequence repeats) distributed evenly throughout the genome.

All genetic markers can be used for investigating association; however, SNPs potentially have the most utility compared to other genetic markers. Various assays were developed for detection of known and unknown SNPs. Some are relatively easy to implement and low in cost, while others are developed for high-volume screening at substantial cost. As the cost of genotyping diminishes, genome-wide scans of all available polymorphisms in a species' genome are becoming rapidly feasible and preferable over targeted SNP genotyping approaches. SSR markers have historically been useful in association studies and do have high information content, but they may be difficult to find in candidate gene regions and they are several-fold more expensive to score than SNPs.

For the purposes of inferences on the population history, genotype information from a large number of neutral marker loci is required. We are using the term neutral marker loosely here to indicate the non-candidate loci, i.e. the loci that were *not* designated as candidate loci that can putatively influence a trait of interest. The density of the markers required should be scaled to provide genome-wide coverage. Simulation studies suggest 100 SSR or 200 SNP markers would suffice to get a reasonable estimate of population structure and relatedness for most crop plants (J. Yu and E.S. Buckler, unpublished results).

When targeting candidate loci for association studies, the greatest statistical power is achieved when the marker and QTL have equal allele frequencies (Abecasis et al. 2001) in the study population. This is due to the opportunity created for maximal linkage and LD, since robust detection of associations requires that the marker and trait loci are in phase. If there is no knowledge of the QTL frequency distribution a priori, the best alternative is to choose markers with a wide range of allele frequencies that are likely to mimic the QTL mutation rate. Some SSRs probably mutate faster and have a different frequency distribution than QTLs, which may make them less useful for association mapping. SNPs with a wide range of allele frequencies are most likely to be informative. In order to maximize the information content of SNPs, a large number of them can be chosen to scan a particular genomic region, and this can be achieved with numerous algorithms available for choosing SNPs (Daly et al. 2001; Johnson et al. 2001; Patil et al. 2001; Gabriel et al. 2002; Ackerman et al. 2003; Ke and Cardon 2003; Sebastiani et al. 2003; Zhang and Jin 2003; Halldorsson et al. 2004; Forton et al. 2005).

Whether the trait of interest has a binary or quantitative phenotype, it is also of interest for the association study design. When a binary trait is being investigated, case-control-type populations are required for association analysis, where equivalent sized sub-populations of individuals that display the phenotype of interest (cases) and do not display the phenotype of interest (controls) are queried for allelic association of genetic loci with the case and control phenotypes in a statistically significant manner. The statistical test performed is simply an hypothesis test that asks whether or not the allelic frequency distribution of a locus is the same or different for a given locus between the two sub-populations. Bulk segregant analysis (BSA)-type (Michelmore et al. 1991) bulked sample genotype screening methods for all the available marker loci may facilitate candidate gene and association discovery for binary traits (Shaw et al. 1998). The challange of case-control type studies is to make sure that the case and control groups are comparable in terms of their genetic makeup. Most of the statistical methods aim to detect and correct for the effects of population statification and ancestry differences between the case and control groups (Pritchard et al. 2000b; Price et al. 2006).

13.6 Association Mapping in Crop Plants

The motivations for attempting association mapping in different crop plants are highly variable. For historically well-studied crop plants, such as maize and rice, the major motivation for the association approach is dissection of complex traits at very high-level resolution, as well as allele mining from natural genetic diversity resources. For other organisms where there is insufficient or few genetic resources, the major motivation is functional marker development and identification of molecular markers tightly linked to the trait locus for marker assisted selection and breeding practices. Thus, each association study stands alone for its own motivations and should be evaluated for its utility and success based on the initial motivations and aims.

The association mapping approach requires extensive infrastructure development and preliminary studies to determine population structure and LD (Fig. 13.1). Once the preliminary data and infrastructure for association mapping for a species are available, several association studies on various plant taxa report successful results for tests of associations between candidate locus genotypes and various complex phenotypes (Table 13.1).

Table 13.1 Association studies that report significant results. SA Structured association; MLM mixed linear model

Species	Population type	Association method	Trait	References
Zea mays	Diverse inbred lines	SA (Q model)	Flowering time	Thornsberry et al. 2001; Andersen et al. 2005; Camus-Kulandaivelu et al. 2006
		SA (Q model)	Kernel composition Starch pasting properties	Wilson et al. 2004
		SA (Q model)	Maysin synthesis	Szalma et al. 2005
		Case-control	Carotenoid content	Palaisa et al. 2004
		MLM (Q+K model)	Carotenoid content	Harjes et al. 2008
Zea mays	Diverse inbred lines	Haplotype tree scanning	Sweet taste	Tracy et al. 2006

In the model organism *Arabidopsis*, the association mapping practice is mostly motivated by generating proof of concept, identification of QTLs involved in adaptation, and additional alleles to supplement other mutagenesis approaches. The candidate-gene association study at the *CRY2-Cryptochrome2* locus reported diverse functional alleles (Olsen et al. 2004). In their first attempt at a genome-wide association study in *Arabidopsis*, Aranzana et al. (2005) reported identification of previously known flowering time (*FRI* locus) and three known pathogen-resistance genes.

In maize, all reported association studies so far have targeted candidate genes with known mutant phenotypes and are motivated by high resolution mapping and allele mining purposes. For instance, *d8*locus with flowering time (Thornsberry et al. 2001; Andersen et al. 2005; Camus-Kulandaivelu et al. 2006), *bt2* (*brittle2*), *sh1* (*shrunken1*) and *sh2* (*shrunken2*) with kernel composition, *ae1* (*amylose extender1*) and *sh2* (*shrunken2*) with starch pasting properties (Wilson et al. 2004) and sweet taste (Tracy et al. 2006), *a1*(*anthocyaninless1*) and *whp1* (*whitepollen1*) genes with maysin synthesis (Szalma et al. 2005), and *lyc-e* (*lycopene epsilon cyclase*) gene with carotenoid content (Harjes et al. 2008) are studies that report very high resolution associations, as well as localizing the causative polymorphism within 1–2 kb of the marker loci reported. In maize, very little is known about association mapping from a genomic scale, mostly due to incomplete genomic sequence and very rapid decay of LD. At the *Y1* locus a relatively large genomic context was examined. *Y1* is a key gene in carotenoid production in maize (Buckner et al. 1990, 1996), and through an association study (Palaisa et al. 2003) the allelic variation was traced down to multiple independent insertions in the *Y1* promoter region that cause up-regulation of the downstream *Y1* gene. At this locus, associations were also shown to extend to neighboring genes (Palaisa et al. 2004), albeit with weaker significances. This extended LD is mostly the result of breeding efforts in the twentieth century that specifically targeted this simple Mendelian inherited trait. The extended LD at the *Y*1 locus is likely to be one of the most extensive in the maize genome, effective over hundreds of kilobases, while other domestication loci, *tb1*(*teosinte branched 1*) (Lukens and Doebley 2001) and *tga* (*teosinte glume architecture*) (Wang et al. 2005), show LD that extends over tens of kilobases. However, it should be emphasized that *tb1* and *tga* domestication loci demonstrate patterns of reduced diversity as well as extended LD, indicating that the estimates of LD are not as efficient as they are at *Y*1. Furthermore it is plausible to assume that not all of the selection events may have similar LD patterns to that of the *Y*1 locus.

Another motivation for the association approach is the opportunity to unify the elite germplasm resource of an organism through investigation of the breeding material. In an association study, Breseghello and Sorrells (2006b) investigated wheat kernel size and milling quality in an elite germplasm collection of soft-winter wheat from eastern USA. They identified three candidate regions on chromosomes 2D, 5A and 5B that are significantly associated with these traits (Breseghello and Sorrells 2006a). This study clearly demonstrates how results obtained from association mapping-based genetic trait dissection studies can be utilized for marker-assisted selection.

13.7 Conclusions

So far, map-based cloning approaches have been reported to successfully isolate 12 major-effect QTLs and nine small-effect QTLs (Price 2006). The time from QTL mapping to positional cloning is estimated to be between 5 and 10 years, while sufficient marker resolution for QTL cloning through association mapping can be achieved within 2–3 years. Furthermore, there is a substantial lag between QTL discovery and marker assisted crop improvement practices dedicated to verification of the presence and stability of QTL in traditional linkage-based studies. In a well-designed association study, some of the results can be immediately applied to marker-assisted improvement.

The true large-scale applications of association mapping will become apparent as multiple species begin to have marker densities sufficiently high for whole genome scan by association mapping. Currently, several research groups are working on whole genome scan approaches in half a dozen species that have whole genome sequences available, and there are at least 50 more species whose genome sequences will be completed in the near future.

The goal of association mapping in many crop plants is to identify key genes controlling various traits and mine the best alleles from diverse germplasm for incorporation into elite breeding material. Traditionally, genetic markers were mostly used for trait improvement through several breeding-based approaches, such as marker assisted selection (MAS), marker assisted breeding (MAB) and mapping as you go (MAYG) (Podlich et al. 2004), as well as QTL cloning/transformation-based approaches (Remington et al. 2001a). Association mapping has the potential to provide numerous useful alleles to these marker-assisted breeding programs. Marker-assisted breeding programs using association data are now underway in numerous plant breeding companies. In the next few years, we will also witness applications of association mapping and MAS for public breeding programs.

Association mapping holds an important and rapidly expanding niche in quantitative trait mapping studies, along with linkage mapping and positional cloning, and it is likely that this niche will continue to expand over the next decade.

References

Abecasis GR, Cookson WO, Cardon LR (2001) The power to detect linkage disequilibrium with quantitative traits in selected samples. Am J Hum Genet 68:1463–1474

Ackerman H, Usen S, Mott R, Richardson A, Sisay-Joof F, Katundu P, Taylor T, Ward R, Molyneux M, Pinder M, Kwiatkowski DP (2003) Haplotypic analysis of the TNF locus by association efficiency and entropy. Genome Biol 4:R24

Allison DB (1997) Transmission-disequilibrium tests for quantitative traits. Am J Hum Genet 60:676–690

Andersen JR, Schrag T, Melchinger AE, Zein I, Lubberstedt T (2005) Validation of Dwarf8 polymorphisms associated with flowering time in elite European inbred lines of maize (*Zea mays* L.). Theor Appl Genet 111:206–217

Aranzana MJ, Kim S, Zhao K, Bakker E, Horton M, Jakob K, Lister C, Molitor J, Shindo C, Tang C, Toomajian C, Traw B, Zheng H, Bergelson J, Dean C, Marjoram P, Nordborg M (2005) Genome-wide association mapping in Arabidopsis identifies previously known flowering time and pathogen resistance genes. PLoS Genet 1:e60

Baltunis BS, Huber DA, White TL, Golfard B, Stelzer HE (2005) Genetic effects of rooting loblolly pine stem cuttings from a partial diallel mating design. Can J Forest 35:1098–1108

Bamshad M, Wooding S, Salisbury BA, Stephens JC (2004) Deconstructing the relationship between genetics and race. Nat Rev Genet 5:598–609

Barrett JC, Cardon LR (2006) Evaluating coverage of genome-wide association studies. Nat Genet 38:659–662

Benjamini Y, Hochberg Y (1995) Controlling the false discovery rate – a practical and powerful approach to multiple testing. J R Stat Soc Series B 57:289–300

Blott S, Kim JJ, Moisio S, Schmidt-Kuntzel A, Cornet A, Berzi P, Cambisano N, Ford C, Grisart B, Johnson D, Karim L, Simon P, Snell R, Spelman R, Wong J, Vilkki J, Georges M, Farnir F, Coppieters W (2003) Molecular dissection of a quantitative trait locus: a phenylalanine-to-tyrosine substitution in the transmembrane domain of the bovine growth hormone receptor is associated with a major effect on milk yield and composition. Genetics 163:253–266

Blouin JD (2003) DNA-based methods for pedigree reconstruction and kinship analysis in natural populations. Trends Ecol Evol 18:503–511

Breseghello F, Sorrells ME (2006a) Association mapping of kernel size and milling quality in wheat (*Triticum aestivum* L.) cultivars. Genetics 172:1165–1177

Breseghello F, Sorrells M (2006b) Association analysis as a strategy for improvement of quantitative traits in plants. Crop Sci 46:1323–1330

Buckner B, Kelson TL, Robertson DS (1990) Cloning of the y1 locus of maize, a gene involved in the biosynthesis of carotenoids. Plant Cell 2:867–876

Buckner B, Miguel PS, Janick-Buckner D, Bennetzen JL (1996) The y1 gene of maize codes for phytoene synthase. Genetics 143:479–488

Caldwell KS, Langridge P, Powell W (2004) Comparative sequence analysis of the region harboring the hardness locus in barley and its colinear region in rice. Plant Physiol 136: 3177–3190

Caldwell KS, Russell J, Langridge P, Powell W (2006) Extreme population-dependent linkage disequilibrium detected in an inbreeding plant species, *Hordeum vulgare*. Genetics 172:557–567

Camus-Kulandaivelu L, Veyrieras JB, Madur D, Combes V, Fourmann M, Barraud S, Dubreuil P, Gouesnard B, Manicacci D, Charcosset A (2006) Maize adaptation to temperate climate: relationship between population structure and polymorphism in the Dwarf8 gene. Genetics 172:2449–2463

Chen L, Storey JD (2006) Relaxed significance criteria for linkage analysis. Genetics 173:2371–2381

Ching A, Caldwell KS, Jung M, Dolan M, Smith OS, Tingey S, Morgante M, Rafalski AJ (2002) SNP frequency, haplotype structure and linkage disequilibrium in elite maize inbred lines. BMC Genet 3:19

Churchill GA, Doerge RW (1994) Empirical threshold values for quantitative trait mapping. Genetics 138:963–971

Daly MJ, Rioux JD, Schaffner SF, Hudson TJ, Lander ES (2001) High-resolution haplotype structure in the human genome. Nat Genet 29:229–232

Devlin B, Roeder K (1999) Genomic control for association studies. Biometrics 55:997–1004

Devlin B, Roeder K, Wasserman L (2001) Genomic control, a new approach to genetic-based association studies. Theor Popul Biol 60:155–166

Devlin B, Bacanu SA, Roeder K (2004) Genomic control to the extreme. Nat Genet 36:1129–1130; author reply 1131

Doerge RW, Churchill GA (1996) Permutation tests for multiple loci affecting a quantitative character. Genetics 142:285–294

Emik LO, Terrill CE (1949) Systematic procedures for calculating inbreeding coefficients. J Hered 40:51–55

Ersoz ES (2006) Candidate gene-association mapping for dissecting fungal disease resistance in loblolly pine. PhD Dissertation in Genetics, University of California, Davis

Flint-Garcia SA, Thornsberry JM, Buckler ES (2003) Structure of linkage disequilibrium in plants. Annu Rev Plant Biol 54:357–374

Flint-Garcia SA, Thuillet AC, Yu J, Pressoir G, Romero SM, Mitchell SE, Doebley J, Kresovich S, Goodman MM, Buckler ES (2005) Maize association population: a high-resolution platform for quantitative trait locus dissection. Plant J 44:1054–1064

Forton J, Kwiatkowski D, Rockett K, Luoni G, Kimber M, Hull J (2005) Accuracy of haplotype reconstruction from haplotype-tagging single-nucleotide polymorphisms. Am J Hum Genet 76:438–448

Fulker DW, Cherny SS, Sham PC, Hewitt JK (1999) Combined linkage and association sib-pair analysis for quantitative traits. Am J Hum Genet 64:259–267

Gabriel SB, Schaffner SF, Nguyen H, Moore JM, Roy J, Blumenstiel B, Higgins J, DeFelice M, Lochner A, Faggart M, Liu-Cordero SN, Rotimi C, Adeyemo A, Cooper R, Ward R, Lander ES, Daly MJ, Altshuler D (2002) The structure of haplotype blocks in the human genome. Science 296:2225–2229

Geiringer H (1944) On the probability theory of linkage in Mendelian heredity. Ann Math Stat 15(1)25–57

Golding GB (1984) The sampling distribution of linkage disequilibrium. Genetics 108:257–274

Gonzaléz-Martinéz SC, Wheeler N, Ersoz ES, Neale DB (2006) Association genetics in *Pinus taeda* L.I. Wood property traits. Genetics 175:399–409

Halldorsson BV, Bafna V, Lippert R, Schwartz R, De La Vega FM, Clark AG, Istrail S (2004) Optimal haplotype block-free selection of tagging SNPs for genome-wide association studies. Genome Res 14:1633–1640

Hamblin MT, Salas Fernandez MG, Casa AM, Mitchell SE, Paterson AH, Kresovich S (2005) Equilibrium processes cannot explain high levels of short- and medium-range linkage disequilibrium in the domesticated grass *Sorghum* bicolor. Genetics 171:1247–1256

Harjes CE, Rocheford TR, Bai L,, Brutnell TP, Kandianis CB, Sowinski SG, Stapleton AE, Vallabhaneni R, Williams M, Wurtzel ET, Yan J, Buckler ES (2008) Natural genetic variation in lycopene epsilon cyclase tapped for maize biofortification. Science 319(5861):330–333

Hedrick PW (1987) Gametic disequilibrium measures – proceed with caution. Genetics 117:331–341

Henderson CR (1975) Best linear unbiased estimation and prediction under a selection model. Biometrics 31:423–447

Henderson CR (1976) Simple method for computing inverse of a numerator relationship matrix used in prediction of breeding values. Biometrics 32:69–83

Henderson CR (1984) Application of linear models in animal breeding. University of Guelph, Guelph

Herbert A, Gerry NP, McQueen MB, Heid IM, Pfeufer A, Illig T, Wichmann HE, Meitinger T, Hunter D, Hu FB, Colditz G, Hinney A, Hebebrand J, Koberwitz K, Zhu X, Cooper R, Ardlie K, Lyon H, Hirschhorn JN, Laird NM, Lenburg ME, Lange C, Christman MF (2006) A common genetic variant is associated with adult and childhood obesity. Science 312:279–283

Hill WG, Robertson A (1968) Linkage disequilibrium in finite populations. Theor Appl Genet 38:226–231

Hirschhorn JN, Daly MJ (2005) Genome-wide association studies for common diseases and complex traits. Nat Rev Genet 6:95–108

Holte S, Quiaoit F, Hsu L, Davidov O, Zhao LP (1997) A population based family study of a common oligogenic disease – part I: association/aggregation analysis. Genet Epidemiol 14:803–807

Horvath S, Xu X, Laird NM (2001) The family based association test method: strategies for studying general genotype–phenotype associations. Eur J Hum Genet 9:301–306

Hudson RR (1985) The sampling distribution of linkage disequilibrium under an infinite allele model without selection. Genetics 109:611–631

Hudson RR (2001) Two-locus sampling distributions and their application. Genetics 159:1805–1817

Johnson GC, Esposito L, Barratt BJ, Smith AN, Heward J, Di Genova G, Ueda H, Cordell HJ, Eaves IA, Dudbridge F, Twells RC, Payne F, Hughes W, Nutland S, Stevens H, Carr P, Tuomilehto-Wolf E, Tuomilehto J, Gough SC, Clayton DG, Todd JA (2001) Haplotype tagging for the identification of common disease genes. Nat Genet 29:233–237

Jung M, Ching A, Bhattramakki D, Dolan M, Tingey S, Morgante M, Rafalski A (2004) Linkage disequilibrium and sequence diversity in a 500-kbp region around the *adh1* locus in elite maize germplasm. Theor Appl Genet 109:681–689

Karayiorgou M, Sobin C, Blundell ML, Galke BL, Malinova L, Goldberg P, Ott J, Gogos JA (1999) Family-based association studies support a sexually dimorphic effect of COMT and MAOA on genetic susceptibility to obsessive-compulsive disorder. Biol Psychiatry 45:1178–1189

Kayihan GC, Huber DA, Morse AM, White TL, Davis JM (2005) Genetic dissection of fusiform rust and pitch canker disease traits in loblolly pine. Theor Appl Genet 110:948–958

Ke X, Cardon LR (2003) Efficient selective screening of haplotype tag SNPs. Bioinformatics 19:287–288

Kennedy B, Quinton M, Vanarendonk J (1992) Estimation of effects of single genes on quantitative traits. J Anim Sci 70:2000–2012

Laird NM, Lange C (2006) Family-based designs in the age of large-scale gene-association studies. Nat Rev Genet 7:385–394

Laird NM, Horvath S, Xu X (2000) Implementing a unified approach to family-based tests of association. Genet Epidemiol 19:S36–S42

Lake SL, Blacker D, Laird NM (2000) Family-based tests of association in the presence of linkage. Am J Hum Genet 67:1515–1525

Lange C, Lyon H, DeMeo D, Raby B, Silverman EK, Weiss ST (2003) A new powerful nonparametric two-stage approach for testing multiple phenotypes in family-based association studies. Hum Hered 56:10–17

Lewis CM (2002) Genetic association studies: design, analysis and interpretation. Brief Bioinform 3:146–153

Lewontin RC (1988) On measures of gametic disequilibrium. Genetics 120:849–852

Lewontin RC, Kojima K (1960) The evolutionary dynamics of complex polymorphisms. Evolution 14:458–472

Long AD, Langley CH (1999) The power of association studies to detect the contribution of candidate genetic loci to variation in complex traits. Genome Res 9:720–731

Lukens L, Doebley J (2001) Molecular evolution of the teosinte branched gene among maize and related grasses. Mol Biol Evol 18:627–638

Lynch M, Ritland K (1999) Estimation of pairwise relatedness with molecular markers. Genetics 152:1753–1766

Maccaferri M, Sanguineti MC, Noli E, Tuberosa R (2005) Population structure and long-range disequilibrium in a durum wheat elite collection. Mol Breed 15:271–290

Meuwissen TH, Goddard ME (1997) Estimation of effects of quantitative trait loci in large complex pedigrees. Genetics 146:409–416

Michelmore RW, Paran I, Kesseli RV (1991) Identification of markers linked to disease-resistance genes by bulked segregant analysis: a rapid method to detect markers in specific genomic regions by using segregating populations. Proc Natl Acad Sci USA 88:9828–9832

Monks SA, Kaplan NL, Weir BS (1998) A comparative study of sibship tests of linkage and/or association. Am J Hum Genet 63:1507–1516

Mueller J (2004) Linkage disequilibrium for different scales and applications. Brief Bioinfo 5:355–364

Niebur W, Rafalski JA, Smith OS, Cooper M (2004) New directions for a diverse planet. Proceedings of the 4th International Crop Science Congress, Brisbane

Nordborg M (2000) Linkage disequilibrium, gene trees, selfing: an ancestral recombination graph with partial self-fertilization. Genetics 154:923–929

Nordborg M, Tavare S (2002) Linkage disequilibrium: what history has to tell us. Trends Genet 18:83–90

Nordborg M, Borevitz JO, Bergelson J, Berry CC, Chory J, Hagenblad J, Kreitman M, Maloof JN, Noyes T, Oefner PJ, Stahl EA, Weigel D (2002) The extent of linkage disequilibrium in *Arabidopsis thaliana*. Nat Genet 30:190–193

Nordborg M, Hu TT, Ishino Y, Jhaveri J, Toomajian C, Zheng H, Bakker E, Calabrese P, Gladstone J, Goyal R, Jakobsson M, Kim S, Morozov Y, Padhukasahasram B, Plagnol V, Rosenberg NA, Shah C, Wall JD, Wang J, Zhao K, Kalbfleisch T, Schulz V, Kreitman M, Bergelson J (2005) The pattern of polymorphism in *Arabidopsis thaliana*. PLoS Biol 3:e196

Oliehoek PA, Windig JJ, van Arendonk JA, Bijma P (2006) Estimating relatedness between individuals in general populations with a focus on their use in conservation programs. Genetics 173:483–496

Olsen KM, Halldorsdottir SS, Stinchcombe JR, Weinig C, Schmitt J, Purugganan MD (2004) Linkage disequilibrium mapping of Arabidopsis CRY2 flowering time alleles. Genetics 167:1361–1369

Palaisa KA, Morgante M, Williams M, Rafalski A (2003) Contrasting effects of selection on sequence diversity and linkage disequilibrium at two phytoene synthase loci. Plant Cell 15:1795–1806

Palaisa K, Morgante M, Tingey S, Rafalski A (2004) Long-range patterns of diversity and linkage disequilibrium surrounding the maize Y1 gene are indicative of an asymmetric selective sweep. Proc Natl Acad Sci USA 101:9885–9890

Paterson AH, DeVerna JW, Lanini B, Tanksley SD (1990) Fine mapping of quantitative trait loci using selected overlapping recombinant chromosomes, in an interspecies cross of tomato. Genetics 124:735–742

Patil N, Berno AJ, Hinds DA, Barrett WA, Doshi JM, Hacker CR, Kautzer CR, Lee DH, Marjoribanks C, McDonough DP, Nguyen BT, Norris MC, Sheehan JB, Shen N, Stern D, Stokowski RP, Thomas DJ, Trulson MO, Vyas KR, Frazer KA, Fodor SP, Cox DR (2001) Blocks of limited haplotype diversity revealed by high-resolution scanning of human chromosome 21. Science 294:1719–1723

Pe'er I, Chretien YR, de Bakker PI, Barrett JC, Daly MJ, Altshuler DM (2006) Biases and reconciliation in estimates of linkage disequilibrium in the human genome. Am J Hum Genet 78:588–603

Podlich D, Winkler C, Cooper M (2004) Mapping as you go. Crop Sci 44:1560–1571

Price AH (2006) Believe it or not, QTLs are accurate! Trends Plant Sci 11:213–216

Price AL, Patterson NJ, Plenge RM, Weinblatt ME, Shadick NA, Reich D (2006) Principal components analysis corrects for stratification in genome-wide association studies. Nat Genet 38:904–909

Pritchard JK (2001) Deconstructing maize population structure. Nat Genet 28:203–204

Pritchard JK, Przeworski M (2001) Linkage disequilibrium in humans: models and data. Am J Hum Genet 69:1–14

Pritchard JK, Stephens M, Donnelly P (2000a) Inference of population structure using multilocus genotype data. Genetics 155:945–959

Pritchard JK, Stephens M, Rosenberg NA, Donnelly P (2000b) Association mapping in structured populations. Am J Hum Genet 67:170–181

Rabinowitz D (1997) A transmission disequilibrium test for quantitative trait loci. Hum Hered 47:342–350

Remington DL, Ungerer MC, Purugganan MD (2001a) Map-based cloning of quantitative trait loci: progress and prospects. Genet Res 78:213–218

Remington DL, Thornsberry JM, Matsuoka Y, Wilson LM, Whitt SR, Doebley J, Kresovich S, Goodman MM, Buckler ES (2001b) Structure of linkage disequilibrium and phenotypic associations in the maize genome. Proc Natl Acad Sci USA 98:11479–11484

Risch N, Merikangas K (1996) The future of genetic studies of complex human diseases. Science 273:1516–1517

Sebastiani P, Lazarus R, Weiss ST, Kunkel LM, Kohane IS, Ramoni MF (2003) Minimal haplotype tagging. Proc Natl Acad Sci USA 100:9900–9905

Shaw SH, Carrasquillo MM, Kashuk C, Puffenberger EG, Chakravarti A (1998) Allele frequency distributions in pooled DNA samples: applications to mapping complex disease genes. Genome Res 8:111–123

Spielman RS, McGinnis RE, Ewens WJ (1993) Transmission test for linkage disequilibrium: the insulin gene region and insulin-dependent diabetes mellitus (IDDM). Am J Hum Genet 52:506–516

Spielman RS, McGinnis RE, Ewens WJ (1994) The transmission/disequilibrium test detects cosegregation and linkage. Am J Hum Genet 54:559–560; author reply 560–553

Stich B, Melchinger AE, Piepho H-P, Heckenberger M, Maurer HP, Reif JC (2006) New test for family-based association mapping with inbred lines from plant breeding programs. Theor Appl Genet 113(6):1121–1130

Storey JD (2002) A direct approach to false discovery rates. J R Stat Soc Series B 64(3):479–498

Storey JD, Tibshirani R (2003) Statistical significance for genomewide studies. P Natl Acad Sci USA 100:9440–9445

Stuber CW, Lincoln SE, Wolff DW, Helentjaris T, Lander ES (1992) Identification of genetic factors contributing to heterosis in a hybrid from two elite maize inbred lines using molecular markers. Genetics 132:823–839

Szalma SJ, Buckler ES, Snook ME, McMullen MD (2005) Association analysis of candidate genes for maysin and chlorogenic acid accumulation in maize silks. Theor Appl Genet 110:1324–1333

Tenaillon MI, Sawkins MC, Long AD, Gaut RL, Doebley JF, Gaut BS (2001) Patterns of DNA sequence polymorphism along chromosome 1 of maize (*Zea mays* ssp. *mays* L.). Proc Natl Acad Sci USA 98:9161–9166

Tenaillon MI, Sawkins MC, Anderson LK, Stack SM, Doebley J, Gaut BS (2002) Patterns of diversity and recombination along chromosome 1 of maize (*Zea mays* ssp. *mays* L.). Genetics 162:1401–1413

Thornsberry JM, Goodman MM, Doebley J, Kresovich S, Nielsen D, Buckler ES (2001) Dwarf8 polymorphisms associate with variation in flowering time. Nat Genet 28:286–289

Thumma BR, Nolan MF, Evans R, Moran GF (2005) Polymorphisms in cinnamoyl CoA reductase (CCR) are associated with variation in microfibril angle in *Eucalyptus* spp. Genetics 171:1257–1265

Tracy WF, Whitt SR, Buckler ES (2006) Recurrent mutation and genome evolution: example of *Sugary1* and the origin of sweet maize. Crop Sci 46:1–7

Wall JD, Pritchard JK (2003) Haplotype blocks and linkage disequilibrium in the human genome. Nat Rev Genet 4:587–597

Wang H, Nussbaum-Wagler T, Li B, Zhao Q, Vigouroux Y, Faller M, Bomblies K, Lukens L, Doebley JF (2005) The origin of the naked grains of maize. Nature 436:714–719

Wang J (2002) An estimator for pairwise relatedness using molecular markers. Genetics 160:1203–1215

Wang Y, Rannala B (2005) In silico analysis of disease-association mapping strategies using the coalescent process and incorporating ascertainment and selection. Am J Hum Genet 76:1066–1073

Wilson LM, Whitt SR, Ibanez AM, Rocheford TR, Goodman MM, Buckler ES (2004) Dissection of maize kernel composition and starch production by candidate gene association. Plant Cell 16:2719–2733

Wright SI, Gaut BS (2005) Molecular population genetics and the search for adaptive evolution in plants. Mol Biol Evol 22:506–519

Wright SI, Bi IV, Schroeder SG, Yamasaki M, Doebley JF, McMullen MD, Gaut BS (2005) The effects of artificial selection on the maize genome. Science 308:1310–1314

Yu J, Holland JB, McMullen MD, Buckler ES (2006a) Genetic design and statistical power of nested association mapping in maize genetics. Nat Genetics 178:539–551

Yu J, Pressoir G, Briggs WH, Vroh Bi I, Yamasaki M, Doebley JF, McMullen MD, Gaut BS, Nielsen DM, Holland JB, Kresovich S, Buckler ES (2006b) A unified mixed-model method for association mapping that accounts for multiple levels of relatedness. Nat Genet 38:203–208

Zhang K, Jin L (2003) HaploBlockFinder: haplotype block analyses. Bioinformatics 19:1300–1301

Chapter 14
Maize Genetic Resources

Martin M. Sachs

14.1 Introduction

Maize is *the* most diverse crop plant analyzed at both morphological and molecular levels (Anderson and Cutler 1942; Brown 1949, 1985; Buckler et al. 2006; Duvick 1981; Galinat 1961; Goodman 1968; Ho et al. 2005; Iltis 1972; Timothy and Goodman 1979; Vigouroux et al. 2005; Xia et al. 2005). Enormous levels of allelic polymorphism exist in maize (Chin et al. 1996; Guo et al. 2004; Goodman and Stuber 1983; Stuber et al. 1980) and this diversity has allowed for selection of beneficial agronomic traits that have been utilized in breeding over the millennia (Dudley 1988; Pollak 2003; Vigouroux et al. 2002; Yamasaki et al. 2005). Visible phenotypic trait diversity due to natural allelic variation or induced mutation has allowed a greater understanding of maize biology, which can also lead to agronomic improvements (Coe 2001; Peterson and Bianchi 1999; Rhoades 1984; Sachs 2005). Maize germplasm stock centers exist to categorize, preserve, maintain, and distribute this genetic diversity to researchers, breeders, educators, and others who can utilize this variation (De Vincente 2004; Dillmann et al. 1997; Hoisington et al. 1999; Troyer 1990). These genetic resources in maize have proven to be extremely useful and germplasm centers will ensure that they continue to be so.

Maize genetic resources are divided into two major categories: (1) genetic stocks and (2) germplasm accessions (Bird 1982; Bretting and Widrlechner 1995; Bretting and Widrlechner 1995; Brown and Goodman 1977; Crossa et al. 1994; Goodman 1990; Janick 1989; Scholl et al. 2003; Shands 1990, 1995; Shands et al. 1989; Taba et al. 2004; White et al. 1989; Wilson et al. 1985). The Maize Genetics Cooperation Stock Center (MGCSC) specializes in maize genetic stocks. Other types of maize and wild *Zea* germplasm are maintained and distributed by the

Martin M. Sachs
Maize Genetics Cooperation Stock Center, US Department of Agriculture, Agricultural Research Service, Soybean/Maize Germplasm, Pathology and Genetics Research Unit, Department of Crop Sciences, University of Illinois, Urbana, IL 61801, USA
e-mail: msachs@uiuc.edu

A.L. Kriz, B.A. Larkins (eds.), *Molecular Genetic Approaches to Maize Improvement* 197
Biotechnology in Agriculture and Forestry, Vol. 63

North Central Regional Plant Introduction Station (NCRPIS) in Ames, Iowa. Maize germplasm can also be obtained from Centro Internacional de Mejoramiento de Maiz y Trigo (CIMMYT) in Mexico. CIMMYT specializes in tropical germplasm. The stocks of all three collections are backed up at the National Center for Genetic Resources Preservation (NCGRP) in Fort Collins, Colorado.

Maize genetic stocks are focused upon one or a limited number of defined variations or genetic tools. Examples of what genetic stocks contain include: an allele of a specific gene (induced mutation or natural variant), a combination of mutations that give a unique phenotype, a series of mutant alleles of genetically linked genes, a variant cytoplasmic trait, a chromosomal aberration (e.g., translocation or inversion), a trisomic aneuploid, and a tetraploid. Genetic stocks also include tools such as recombinant inbred lines for mapping gene locations and active transposable element lines for generating new mutants.

The other types of maize germplasm available at NCRPIS and CIMMYT are maintained to preserve natural genetic diversity. These accessions include inbred lines, land races, open pollinated varieties, exotic accessions, wild relatives (e.g., teosinte, *Tripsacum*, *Coix*), cultivars, and other breeding stocks. These germplasm accessions also carry undefined variation that may prove to be a valuable resource for breeders and research scientists.

This chapter describes the resources held at the Maize Genetics Cooperation Stock Center in detail and also provides some information about NCRPIS, CIMMYT, and NCGRP, which is the backup facility for the three active collections.

14.2 Genetic Stocks

14.2.1 Maize Genetics Cooperation Stock Center (MGCSC; GSZE)

MGCSC is part of the National Plant Germplasm System (NPGS) and is supported by the US Department of Agriculture, Agricultural Research Service (USDA/ARS). It is located at the University of Illinois in Urbana, Illinois, USA.

Researchers working with maize have been continually sharing resources and tools to further fundamental research efforts in this model biological organism (Neuffer et al. 1997; Freeling and Walbot 1994). This sharing of resources was strengthened when the concept of a maize genetics cooperation was introduced at an evening get-together in Rollins A. Emerson's (Rhoades 1949) hotel room during the 1928 Winter Science meetings in New York. During that meeting a dozen or so maize workers discussed the current state of maize linkage maps (Kass et al. 2005). The formal organization of the Maize Genetics Cooperation occurred in August 1932, when maize geneticists attending the Sixth International Genetics Congress agreed to establish a cooperative enterprise to further the advance of maize genetics. Among the aims of this organization were the collection and dissemination of unpublished data and information to interested workers and the maintenance and distribution of tester stocks. The establishment of the Maize

Genetics Cooperation promoted the sharing of genetic marker stocks that had been developed by maize geneticists over the years in a spirit of cooperation and generosity. A collection of stocks was assembled and maintained at Cornell University, and samples were supplied upon request. Information was exchanged through the medium of an informal Maize Genetics Cooperation Newsletter. Both aspects of this cooperation continue to the present day (http://www.uiuc.edu/ph/www/maize; http://www.maizegdb.org/mnl.php; Coe 2001; Kass et al. 2005; Peterson and Bianchi 1999; Rhoades 1984).

The stock center moved from Ithaca, New York, to Urbana, Illinois, in 1953. During the period from 1953 to 1981, operating funds were provided by grants from the National Science Foundation (NSF). Beginning in 1982, most operating funds were provided by the USDA/ARS. The MGCSC continues to be housed in the University of Illinois Department of Crop Sciences (formerly Agronomy). In 1992, ARS took over the day-to-day operations of the Stock Center and an ARS scientist was appointed as Director. In 1993, an ARS support scientist was appointed to serve as Curator.

14.2.2 The Genetic Stock Collection

In 1953, first plantings at the University of Illinois were made of the Cornell source stocks, together with numerous additional stocks hurriedly solicited and assembled from other sources. Stocks are still being added to the collection each year.

Most of the variants in the MGCSC's collection were identified by maize geneticists and saved by them. Maize breeders and growers also notice mutant traits and submit seed samples for evaluation and potential addition to the collection. Information about maize mutants and chromosomal aberrations as well as their use in biological research has been published over the years (e.g., Coe et al. 1988; Carlson 1988; Freeling and Walbot 1994; Neuffer et al. 1997; Sachs 2005; Sheridan 1982).

Maintenance of maize genetic stocks is very labor intensive. Features of the maize plant that make self- or cross-pollinations simple also make hand pollination mandatory if controlled crosses are to be made. Seed samples are increased by hand pollinations. Ears are shelled individually, and the seed samples from each are stored in packets labeled with pedigrees in the form of genetic symbols. Long-term storage is in a cold room maintained at about 8°C and less than 30% relative humidity. Some samples of good quality seed placed immediately into these storage conditions have shown good viability even after as many as 30 years. However, as a general practice, efforts are made to perpetuate fresh seed stocks within a 10-year period.

When the MGCSC was moved from Cornell University to the University of Illinois in 1953, the collection consisted of only 220 stocks. Today, the total core collection is now over 7,500 genetic stocks represented by approximately 100,000 individually pedigreed samples. The bulk of the current collection consists of several hundred symbolized genes, together with many additional gene combinations and other heritable variants. Many of the stocks are maintained in forms or combinations suitable for specific research uses. Included are about 1,000 chromosome aberrations (e.g., translocations and inversions; Anderson 1948) as well as stocks varying in chromosome number (e.g., trisomic aneuploids; Einset 1943) and complete sets of chromosomes (e.g., tetraploids; Randolph 1932) and mapping tools such as recombinant inbred lines (Lee et al. 2002). Mutant genes are also maintained that generate unbalanced chromosome complements. There are also over 80,000 stocks that have been incorporated from recent NSF-funded plant genome projects (Cook 1998; Lee 1998). These are mostly transposable element- and EMS-induced mutants.

The Plant Genome Initiative (Cook 1998; National Research Council of the National Academies 2002) has led to the support of numerous projects involved in creating genomic resources that will further advance plant biological and agronomic research. Maize research benefited greatly from this support through the National Science Foundation (NSF) Plant Genome Research Program as well as from recent projects funded by the USDA/ARS, USDA National Research Initiative (USDA/NRI), National Institutes of Health (NIH), the Department of Energy (DOE), and other agencies.

The NSF-funded Maize Gene Discovery, Sequencing, and Phenotypic Analysis project (Lunde et al. 2003; Fernandes et al. 2004; http://www.maizegdb.org/documentation/mgdp/index.php) has made several useful resources available. This project has sequenced over 180,000 maize cDNAs (expressed sequences tags; ESTs) and made this information publicly available. These ESTs have been characterized and unigene sets have been placed on microarrays that are available to researchers for analysis of gene expression. This project also makes use of an artificial Robertson's *Mutator* (Robertson 1978; Alleman and Freeling 1986; Chandler and Hardeman 1992; Bennetzen 1996) element that contains an internal pBluescript sequence (*RescueMu*; Raizada et al. 2001). When this element inserts into a locus, the gene can easily be isolated and sequenced. Through this process, this project has generated over 175,000 maize Genome Survey Sequences (GSS). As Robertson's *Mutator* has an extremely high propensity for inserting into genes (Hanley et al. 2000), the GSS generated by the Maize Gene Discovery project are providing very useful sequence information for many maize genes. The more than 40,000 stocks mutagenized by *RescueMu* have also been phenotypically screened for mutants (http://www.maizegdb.org/rescuemu-phenotype.php). For mutant adult plant traits, this project has organized an annual maize community mutant hunt in Urbana/Champaign, Illinois. The GSS generated from this project can also be searched by the Basic Local Alignment Search Tool (BLAST; http://www.maizegdb.org/blast.php) to find mutations in specific genes of interest in a reverse genetics approach. The mutagenized stocks generated by this and many other projects are available from the MGCSC.

The Maize Targeted Mutagenesis (MTM) project (May et al. 2003; http://mtm.cshl.org/) has generated more than 40,000 transposon-mutagenized stocks using Robertson's *Mutator*. As a service to maize researchers, the MTM project offers both forward and reverse genetics tools. Mutant phenotypes found in the MTM stocks are described in the MTM database (http://mtm.cshl.org/) and these mutant stocks are available to researchers. Most mutants in these stocks are due to a *Mutator* insertion, and the "tagged" gene thus can be readily isolated. Additionally, the MTM project offers researchers the ability to find mutations in a gene based solely on knowledge of that gene's DNA sequence. Using a "reverse genetics approach", a sequence can be submitted through the MTM website and mutations in this gene are screened at Cold Spring Harbor using the polymerase chain reaction (PCR). Those stocks found to carry a *Mutator* insertion in the target gene can then be used by researchers to analyze mutant phenotypes in a functional genomics approach. Researchers then report information about any observable phenotypes of mutants obtained by this process back to the MTM database, so links between genomic sequence and gene function can be made. In this way, researchers benefiting from the services provided by the MTM project contribute their findings back to it, so that other researchers will benefit from this information. The stocks from this project are being incorporated into the MGCSC. The public-sector MTM project is similar to the Trait Utility System for Corn (TUSC) developed by Pioneer Hi-Bred International, Inc. (Bensen et al. 1995; Meeley and Briggs 1995).

The maize Targeting Induced Local Lesions IN Genomes (TILLING) project (Till et al. 2004; http://genome.purdue.edu/maizetilling/) uses a reverse genetic strategy (McCallum et al. 2000) to identify mutations throughout the genome and a screening method that facilitates localization of these mutations. The mutations are induced in maize inbred lines by chemical mutagens, and particular regions can be screened for the presence of mutations by high-throughput PCR. The mutagenized material is also being phenotypically screened. Stocks from this project will be available from the MGCSC.

Other maize projects, funded by NSF through the Plant Genome Research Program, are also providing powerful mutant resources and tools for maize researchers, and stocks resulting from these projects are being distributed by the MGCSC (e.g., Brutnell 2002; Buckler et al. 2006; Carson et al. 2004; Chuck and Hake 2005; Cowperthwaite et al. 2002; Kolkman et al. 2005; Ma and Dooner 2004; McGinnis et al. 2005; Ostheimer et al. 2003; Settles et al. 2004; Springer et al. 2003; Stern et al. 2004; Zhang and Peterson 2004). Several other important maize projects that make public-sector research tools and resources available are listed at http://www.maizegdb.org/maizeprojects.php.

14.2.3 The Services Provided

The MGCSC is the main repository for maize mutants utilized in research by cooperators worldwide. It is an essential resource to maize scientists conducting basic and applied biological research. The Stock Center is designed primarily to provide a service to maize geneticists by assembling, perpetuating, and supplying seed samples for use in research. It is the goal of the MGCSC to acquire, maintain, and make available stocks containing all known allelic and cytological variation in maize and information about them. Any available stock(s) will be sent upon request without charge and without restrictions.

In order to allow more effective communication of maize genetics information and to allow for easier methods for scientists to request stocks, there was a need to develop a database for maize genetic stocks and to integrate the information into the Germplasm Resources Information Network (GRIN; the database of the NPGS; http://www.ars-grin.gov/).

In this regard, USDA/ARS supported the creation, in 1991, of the Maize Genetics/Genomics Database (currently MaizeGDB; http://www.maizegdb.org/; Lawrence et al. 2005) that provides a central repository for public maize information (including mutants and genetic stocks). MaizeGDB will also provide access to, and analysis tools for, the enormous amount of maize genomic sequence information that is becoming readily available. This database presently includes information about maize mutants and chromosomal aberrations as well as their use in biological research.

Data about stocks at the MGCSC are entered into MaizeGDB (and also into GRIN) to allow users access to the latest information about available maize genetic stocks. Presently, the list of available stocks is accessible on-line (http://www.maizegdb.org/cgi-bin/stockcatalog.cgi). This catalog serves as a basis for seed requests. During past few years, MGCSC has annually averaged approximately 3,000 seed samples supplied in response to 300 requests.

In addition to traditional methods for requesting stocks (mail, phone, fax, and more recently e-mail), a user can now find stocks of interest in the on-line database and directly request them using an on-line order-form (a 'shopping-cart' feature has been added to MGCSC's individual stock listings in MaizeGDB, and also to the on-line stock catalog). The request is then transmitted through the Internet. Seed requests can also be made through NPGS's GRIN database.

14.2.4 The Value of the Stocks

Maize occupies a pre-eminent position among higher plants with regard to its excellence as a test organism for cytogenetic investigations, studies that correlate gene transmission and expression with observable features of the physical chromosomes. A significant feature of the maize collection is the large number of chromosome aberrations that are included. Many of these were deliberately induced by various forms of radiation prior to the mid-1940s. The great majority of them, however, were induced in seed samples exposed to atomic bomb tests at Bikini Atoll in 1946 or Eniwetok Atoll in 1948 (Anderson 1948; Longley 1961). Over the years, several maize geneticists have made important contributions in assembling and maintaining stocks of chromosome aberrations.

While the vast majority of the mutants in the collection are too extreme for commercial use, and they are not usually evaluated and maintained with a view to their direct use in improving agricultural production or products, some of the mutants in the MGCSC collection clearly have had a major impact of commercial importance

(Coe et al. 1988; Neuffer et al. 1997; Sachs 2005). These include the white endosperm mutants, several of the mutants involved in starch biosynthesis (e.g., *su1* and *sh2* have been important in sweet corn production, *wx1* gives starch high in amylopectin, *ae1* gives starch high in amylose), and *ig1* for use in making doubled haploids enabling the rapid production of new inbred lines or placing a desired inbred genotype into a new cytoplasmic background (e.g., male sterile). MGCSC also has tools for mapping traits to chromosomal locations (e.g., recombinant inbred lines, translocations, and trisomics).

However, the vast majority of mutants at the MGCSC are chosen primarily to serve as tools for basic research; in this use it is important that the traits be clearly classifiable, that is, somewhat extreme. Mutant alleles are useful to maize scientists in many different areas of research. These mutations act as coordinates on genetic and physical maps of the maize genome. In addition, many define critical steps in metabolic, developmental, and other pathways of great interest to geneticists, physiologists, breeders, molecular biologists, chemists, and other plant scientists. Mutant traits may be studied directly to investigate metabolic blocks in biosynthetic pathways or, alternatively, they may be used as tools for such purposes as locating genes or controlling the inheritance of particular chromosome segments. These mutants give maize scientists a greater understanding of corn as a biological organism and thus can lead to applications that will improve corn agronomically. Genetic stocks represent the basic tools of maize researchers and MGCSC provides a service to geneticists and other biologists similar to that provided to chemists by chemical supply houses.

14.3 Other Maize Germplasm

In addition to the MGCSC, which focuses on genetic stocks, two major public-sector maize germplasm banks exist that specialize in providing researchers with inbred lines, synthetics, improved populations, landraces, and wild relatives of maize. These collections maintain the rich natural diversity that exists in maize and wild *Zea*.

14.3.1 The North Central Regional Plant Introduction Station (NCRPIS; NC7)

NCRPIS is administered by the USDA/ARS, and is located at Iowa State University, Ames, Iowa, USA (http://www.ars.usda.gov/main/site_main.htm?modecode=36-25-12-00). NCRPIS is also part of the National Plant Germplasm System (NPGS).

The introduction of potentially useful plant species into the USA dates back to the early nineteenth century when embassies were asked to collect and send these materials to the USA. A more organized effort was initiated when the Section of

Seed and Plant Introduction was formed in 1898 within the USDA. However, no provisions were made to store these materials adequately and most of these initial collections were lost over the years (Wilson et al. 1985).

Subsequently, the four original Regional Plant Introduction Stations were established under the Research and Marketing Act of 1946. The North Central Regional Plant Introduction Station (NCRPIS) at Ames, Iowa, which was the first station established, began operation in 1948 (White et al. 1989). NCRPIS was established as Regional Research Project NC7, entitled "Introduction, Multiplication, Evaluation, Preservation, Cataloging, and Utilization of Plant Germplasm" (Wilson et al. 1985). Maize represents the species with by far the largest number of accessions held at NCRPIS. Their maize and wild *Zea* accessions (http://www.ars-grin.gov/cgi-bin/npgs/html/tax_search.pl?Zea) represent most of the existing natural diversity (Liu et al. 2003). NCRPIS presently has approximately 20,000 accessions of maize and wild *Zea* germplasm and also maintains small seed collections of *Coix* and *Tripsacum*, which are closely related to *Zea*.

Proprietary maize inbred lines that have been protected by the US Patent and/or US Plant Variety Protection Act (PVPA), whose intellectual property protection has expired, are also available at NCRPIS (Mikel 2006).

The Germplasm Enhancement of Maize (GEM) project (http://www.ars.usda.gov/Main/docs.htm?docid=10647) is also housed at NCRPIS. GEM's objectives are to widen the germplasm base of commercial hybrid corn through the introduction and incorporation of novel and useful traits found in the germplasm of Latin American landraces (Pollak 2003).

Germplasm accessions held at NCRPIS are listed in the Germplasm Resources Information Network database (GRIN; http://www.ars-grin.gov/npgs/acc/acc_queries.html). Seed requests from NCRPIS can be made through GRIN.

14.3.2 Centro Internacional de Mejoramiento de Maiz y Trigo (CIMMYT)

The Maize Germplasm Bank of the International Maize and Wheat Improvement Center (CIMMYT, Mexico City, Mexico; http://www.cimmyt.org/) grew out of a pilot program in Mexico in 1943, sponsored by the Government of Mexico and the Rockefeller Foundation. The project developed into an innovative, sustained collaboration with Mexican and international researchers. The social and economic achievements of the Green Revolution were recognized worldwide when the Nobel Peace Prize was awarded to Norman Borlaug in 1970 (Khush 2001). The following year, a small cadre of development organizations, national sponsors, and private foundations organized the Consultative Group on International Agricultural Research (CGIAR) to spread the impact of research to more crops and nations. CIMMYT was one of the first international research centers to be supported through the CGIAR (Taba 1990; Taba et al. 2004).

CIMMYT's Maize Germplasm Bank maintains and distributes seed of an extensive germplasm collection of maize and of some of its wild relatives in the Western hemisphere. CIMMYT specializes in tropical maize germplasm, and it presently has more than 22,000 accessions. New introductions are constantly being added from the Cooperative Regeneration Project (Taba 1990; Warburton et al. 2002).

CIMMYT's accessions are organized into two collections, base and active. The base collection seed is kept in sealed containers at subzero temperatures and low humidity, allowing these accessions to remain viable for 50–100 years. Seed in the active collection is maintained at just above freezing (0–2°C) and constitutes the "working" bank from which seed requests are filled (Taba 1990; Taba et al. 2004).

Germplasm held at CIMMYT is listed in the System-wide Information Network for Genetic Resources (SINGER; http://singer.cgiar.org/) database. Passport data (descriptors) on CIMMYT maize germplasm bank accessions have been compiled and are available on CD-ROM (Taba et al. 2004). Seed requests from CIMMYT can be made on-line at http://www.cimmyt.cgiar.org/english/wps/obtain_seed/frmseedrequest.htm.

14.3.3 The National Center for Genetic Resources Preservation (NCGRP)

NCGRP is a USDA/ARS facility located in the Northern Plains Area. NCGRP is on the Colorado State University campus in Fort Collins, Colorado, USA.

Originally designated the National Seed Storage Laboratory (NSSL), it was built in 1958 to consolidate the plant collections in the NPGS into a single facility that uses state-of-the-art preservation practices. The use of liquid nitrogen (cryogenic storage) to store seeds at NSSL was introduced in 1977 and became a routine practice by 1990. Cryogenic storage also made it possible to preserve germplasm from vegetative cuttings and recalcitrant seeds. In 1992, the NSSL building was expanded and the capacity to store germplasm increased ten-fold. The greater security and access to liquid nitrogen made the facility an attractive place to establish the USDA's first animal genebank, the National Animal Germplasm Program (NAGP). The expanded mission of preserving germplasm of animals in the form of semen, plant genetic resources in the form of graftable buds or in vitro plantlets, and eventually insects and microbes prompted the designation of the facility as a Center in 2002 and a name change to the National Center for Genetic Resources Preservation (Walters et al. 2005).

The NCGRP facility contains backups of the active collections held at the MGCSC, NCRPIS, and CIMMYT. If certain accessions are lost at the active collection site, the curator can request seeds from the backup at NCGRP and rescue the lost germplasm.

14.4 Conclusions

Genetic stocks and germplasm accessions are important resources that will continue to be available to researchers, breeders, educators, and others due to the efforts of the gene banks at the Maize Genetics Cooperation Stock Center, the North Central Regional Plant Introduction Station, and the International Maize and Wheat Improvement Center. The genetic resources that these active centers categorize, maintain, and distribute are backed up and preserved at the National Center for Genetic Resources Preservation. These genetic resources will have profound significance in furthering the understanding of maize biology and in future breeding efforts to improve maize as a crop plant.

The true impact of the service is measured by its contribution to the sum total of knowledge. Applications flow from this knowledge. There is little doubt that the operation of these services vastly increases the efficiency and overall productivity of the maize research and breeding communities.

References

Alleman M, Freeling M (1986) The *Mu* transposable elements of maize: evidence for transposition and copy number regulation during development. Genetics 112:107–119

Anderson EG (1948) On the frequency and transmitted chromosome alterations and gene mutations induced by atomic bomb radiations in maize. Proc Natl Acad Sci USA 34:387–390

Anderson E, Cutler HC, (1942) Races of *Zea mays*: I. Their recognition and classification. Ann Missouri Bot Gard 29:69–88

Bennetzen JL (1996) The *Mutator* transposable element system of maize. Curr Top Microbiol Immunol 204:195–229

Bensen RJ, Johal GS, Crane VC, Tossberg JT, Schnable PS, Meeley RB, Briggs SP, (1995) Cloning and characterization of the maize *an1* gene. Plant Cell 7:75–84

Bird R McK (1982) Maize and teosinte germplasm banks. In: Sheridan WF (ed) Maize for biological research. Plant Molecular Biology Association, Charlottesville, pp 351–356

Bretting PK, Sachs MM (1995) Maize genetic resource management. Am J Bot 82:7–8

Bretting PK, Widrlechner MP (1995) Genetic markers and plant genetic resource management. Plant Breed Rev 13:11–86

Brown WL (1949) Numbers and distribution of chromosome knobs in United States maize. Genetics 34:524–536

Brown WL (1985) Maize variability of potential interest to plant molecular geneticists. Maydica 30:225–233

Brown WL, Goodman MM (1977) Races of corn. In: Sprague GF (ed) Corn and corn improvement. American Society of Agronomy, Madison, pp 49–88

Brutnell TP (2002) Transposon tagging in maize. Funct Integr Genomics 2:4–12

Buckler ES, Gaut BS, McMullen MD (2006) Molecular and functional diversity of maize. Curr Opin Plant Biol 9:172–176

Carlson WR (1988) The cytogenetics of corn. In: Sprague GF, Dudley JW (eds) Corn and corn improvement, 3rd edn. American Society of Agronomy, Madison, pp 259–344

Carson C, Robertson J, Coe E (2004) High-volume mapping of maize mutants with simple sequence repeat markers. Plant Missouri Biol Rep 22:131–143

Chandler VL, Hardeman KJ (1992) The *Mu* elements of *Zea mays*. Adv Genet 30:77–122

Chin E, Senior L, Shu H, Smith JSC (1996) Maize simple repetitive DNA sequences: abundance and allele variation. Genome 39:866–873

Chuck G, Hake S (2005) Regulation of developmental transitions. Curr Opin Plant Biol 8:67–70

Coe EH (2001) The origins of maize genetics. Nat Rev Genet 2:898–905

Coe EH, Neuffer MG, Hoisington DA (1988) The genetics of corn. In: Sprague GF, Dudley JW (eds) Corn and corn improvement, 3rd edn. American Society of Agronomy, Madison, pp 81–258

Cook RJ (1998) Toward a successful multinational crop plant genome initiative. Proc Natl Acad Sci USA 95:1993–1995

Cowperthwaite M, Park W, Xu Z, Yan X, Maurais S, Dooner HK (2002) Use of the transposon Ac as a gene-searching engine in the maize genome. Plant Cell 14:713–726

Crossa J, Taba S, Eberhart SA, Bretting P, Vencovsky R (1994) Practical considerations for maintaining germplasm in maize. Theor Appl Genet 89:98–95

De Vincente MC (ed) (2004) The evolving role of genebanks in the fast-developing field of molecular genetics. Issues in genetic resources no. 11. Genetic Resources Science and Technology, IPGRI, Rome

Dillmann C, BarHen A, Guerin D, Charcosset A, Murigneux A (1997) Comparison of RFLP and morphological distances between maize *Zea mays* L. inbred lines. Consequences for germplasm protection purposes. Theor Appl Genet 95:92–102

Dudley JW (1988) Evaluation of maize populations as sources of favorable alleles. Crop Sci 28:486–490

Duvick DN (1981) Genetic diversity in corn improvement. Proc 36th Annual Corn and Sorghum Research Conference, pp 48–60

Einset J (1943) Chromosome length in relation to transmission frequency of maize trisomics. Genetics 28:349–364

Fernandes J, Dong Q, Schneider B, Morrow DJ, Nan G-L, Brendel V, Walbot V (2004) Genome-wide mutagenesis of *Zea mays* L. using RescueMu transposons. Genome Biol 5:R82

Freeling M, Walbot V (1994) The maize handbook. Springer, New York

Galinat WC (1961) Corn's evolution and its significance for breeding. Econ Bot 15:320–325

Goodman MM (1968) The races of maize: I. Use of multivariate analysis of variance to measure morphological similarity. Crop Sci 8:693–698

Goodman MM (1990) Genetic and germ plasm stocks worth conserving. J Hered 81:11–16

Goodman MM, Stuber CW (1983) Maize. In: Tanksley SD, Orton TJ (eds) Isozymes in plant genetics and breeding, part B. Elsevier, Amsterdam, pp 1–33

Guo M, Rupe M, Zinselmeier C, Habben J, Bowen B, Smith OS (2004) Allelic variation of gene expression in maize hybrids. Plant Cell 16:1707–1716

Hanley S, Edwards D, Stevenson D, Haines S, Hegarty M, Schuch W, Edwards KJ (2000) Identification of transposon-tagged genes by the random sequencing of *Mutator*-tagged DNA fragments from *Zea mays*. Plant J 23:557–566

Ho JC, Kresovich S, Lamkey KR (2005) Extent and distribution of genetic variation in US maize: historically important lines and their open-pollinated Dent and Flint progenitors. Crop Sci 45:1891–1900

Hoisington D, Khairallah M, Reeves T, Ribaut J, Skovmand B, Taba S, Warburton M (1999) Plant genetic resources: what can they contribute toward increased crop productivity? Proc Natl Acad Sci USA 96:5937–5943

Iltis HH (1972) Taxonomy of *Zea mays*. Phytologia 23:248–249

Janick J (ed) (1989) The national plant germplasm system of the United States. Plant Breed Rev 7:1–230

Kass LB, Bonneuil C, Coe EH Jr (2005) Cornfests, cornfabs and cooperation: the origins and beginnings of the Maize Genetics Cooperation News Letter. Genetics 169:1787–1797

Khush GS (2001) Green revolution: the way forward. Nat Rev Genet 2:815–821

Kolkman JM, Conrad LJ, Farmer PR, Hardeman K, Ahern KR, Lewis PE, Sawers RJ, Lebejko S, Chomet P, Brutnell TP (2005) Distribution of *Activator* (*Ac*) throughout the maize genome for use in regional mutagenesis. Genetics 169:981–995

Lawrence CJ, Seigfried TE, Brendel V (2005) The maize genetics and genomics database. The community resource for access to diverse maize data. Plant Physiol 138:55–58

Lee M (1998) Genome projects and gene pools: new germplasm for plant breeding? Proc Natl Acad Sci USA 95:2001–2004

Lee M, Sharopova N, Beavis WD, Grant D, Katt M, Blair D, Hallauer A (2002) Expanding the genetic map of maize with the intermated B73 × Missouri17 (IBM) population. Plant Missouri Biol 48:453–461

Liu KJ, Goodman MM, Muse S, Smith JSC, Buckler E, Doebley JF (2003) Genetic structure and diversity among maize inbred lines as inferred from DNA microsatellites. Genetics 165:2117–2128

Longley AE (1961) Breakage points for four corn translocation series and other corn chromosome aberrations maintained at the California Institute of Technology. USDA Crops Research ARS 34:16

Lunde CF, Morrow DJ, Roy LM, Walbot V (2003) Progress in maize gene discovery: a project update. Funct Integr Genomics 3:25–32

Ma Z, Dooner HK (2004) Establishing an *Ac/Ds*-based enhancer detection and gene trap system in maize. Plant J 37:92–103

May BP, Liu H, Vollbrecht E, Senior L, Rabinowicz P, Roh D, Pan X, Stein L, Freeling M, Alexander DE, Martienssen RA (2003) Maize-targeted mutagenesis: a knockout resource for maize. Proc Natl Acad Sci USA 100:11541–11546

McCallum CM, Comai L, Greene EA, Henikoff S (2000). Targeted screening for induced mutations. Nat Biotech 18:455–457

McGinnis K, Chandler V, Cone KC, Kaeppler HF, Kaeppler SM, Kerschen A, Pikaard CS, Richards E, Sidorenko LV, Smith T, Springer N, Wulan T (2005) Transgene-induced RNA interference as a tool for plant functional genomics. Methods Enzymol 392:1–24

Meeley B, Briggs SP (1995) Reverse genetics for maize. Maize Genet Newsl 69:67–82

Mikel MA (2006) Availability and analysis of proprietary Dent corn inbred lines with expired U.S. plant variety protection. Crop Sci 46:2555–2560

National Research Council of the National Academies (2002) Committee on Objectives for the National Plant Genome Initiative: the National Plant Genome Initiative: Objectives for 2003–2008. National Academies Press, Washington, DC. http://www.nap.edu/books/0309085217/html/

Neuffer MG, Coe EH, Wessler SR (1997) The mutants of maize. Cold Spring Harbor Press, New York

Ostheimer G, Williams-Carrier R, Belcher S, Osborne E, Gierke J, Barkan A (2003) Group II intron splicing factors derived by diversification of an ancient RNA-binding domain. EMBO J 22:3919–3929

Peterson PA, Bianchi A (1999) Maize genetics and breeding in the 20th century. World Scientific Publishing Co., Hackensack

Pollak LM (2003) The history and success of the public–private project on germplasm enhancement of maize (GEM). Adv Agron 78:45–87

Raizada M, Nan G, Walbot V (2001) Somatic and germinal mobility of the *RescueMu* transposon in transgenic maize. Plant Cell 13:1587–1608

Randolph LF (1932) Some effects of high temperature on polyploidy and other variations in maize. Proc Natl Acad Sci USA 18:222–229

Rhoades MM (1949) Rollins Adams Emerson (1873–1947). Natl Acad Sci Biogr Mem 25:313–325

Rhoades MM (1984) The early years of maize genetics. Ann Rev Genet 18:1–29

Robertson DS (1978) Characterization of a mutator system in maize. Mutat Res 51:21–28

Sachs MM (2005) Maize mutants: resources for research. Maydica 50:305–309

Scholl R, Sachs M, Ware D (2003) Maintaining collections of mutants for plant functional genomics. Meth Molec Biol 236:311–326

Settles AM, Latshaw S, McCarty D (2004) Molecular analysis of high-copy insertion sites in maize. Nucl Acid Res 32(6):e54

Shands HL (1990) Plant genetic resources conservation: the role of the gene bank in delivering useful genetic materials to the research scientist. J Hered 81:7–10

Shands HL (1995) The U.S. National Plant Germplasm System. Can J Plant Sci 75:9–15

Shands HL, Fitzgerald PJ, Eberhart SA (1989) Program for plant germplasm preservation in the United States. In: Knutson L, Stoner AK (eds) Biotic diversity and germplasm preservation: global imperatives. Kluwer, Dordrecht, pp 97–116

Sheridan WF (1982) Maize for biological research. Plant Molecular Biology Association, Charlottesville

Springer N, Napoli CM, Selinger D, Pandey R, Cone KC, Chandler V, Kaeppler HF, Kaeppler S (2003) Identification of cis-acting regulatory sequences and trans-acting regulatory factors are the stepping stones to the elucidation of the mechanism of p1 paramutation. Plant Physiol 132:907–925

Stern DB, Hanson MR, Barkan A (2004) Genetics and genomics of chloroplast biogenesis: maize as a model system. Trends Plant Sci 9:293–301

Stuber CW, Moll RH, Goodman MM, Schaffer HE, Weir BS (1980) Allozyme frequency changes associated with selection for increased grain yield in maize (*Zea mays* L.). Genetics 95: 225–236

Taba S (1990) Conservation and use of maize genetic resources at CIMMYT. Maydica 35:183–186

Taba S, Eberhart SA, Pollak LM (2004) Germplasm resources. In: Smith CW, Betrán J, Runge ECA (eds) Corn: origin, history, technology and production. Wiley, Hoboken

Till BJ, Reynolds SH, Weil C, Springer N, Burtner C, Young K, Bowers E, Codomo CA, Enns LC, Odden AR, Greene EA, Comai L, Henikoff S (2004) Discovery of induced point mutations in maize genes by TILLING. BMC Plant Biol 4:28

Timothy DH, Goodman MM (1979) Germplasm preservation: the basis of future feast or famine – genetic resources of maize – an example. In: Rubenstein I, Phillips RL, Green CE, Gengenbach BG (eds) The plant seed: development, preservation, and germination. Academic, New York, pp 171–200

Troyer AF (1990) A retrospective view of corn genetic resources. J Hered 81:17–24

Vigouroux Y, McMullen M, Hittinger C, Houchins K, Schulz L, Kresovich S, Matsuoka Y, Doebley JF (2002) Identifying genes of agronomic importance in maize by screening microsatellites for evidence of selection during domestication. Proc Natl Acad Sci USA 99:9650–9655

Vigouroux Y, Mitchell S, Matsuoka Y, Hamblin M, Kresovich S, Smith JSC, Jaqueth J, Smith OS, Doebley J (2005) An analysis of genetic diversity across the maize genome using microsatellites. Genetics 169:1617–1630

Walters C, Wheeler LM, Grotenhuis JM (2005) Longevity of seeds stored in a genebank: species characteristics. Seed Sci Res 15:1–20

Warburton M, Xia XC, Crossa J, Franco J, Melchinger AE, Frisch M, Bohn M, Hoisington D (2002) Genetic characterization of CIMMYT inbred maize lines and open pollinated populations using large scale fingerprinting methods. Crop Sci 42:1832–1840

White GA, Shands HL, Lovell GR (1989) History and operation of the National Plant Germplasm System. Plant Breed Rev 7:5–56

Wilson RL, Clark RL, Widrlechner MP (1985) A brief history of the North Central Regional Plant Introduction Station and a list of the genera maintained. Proc Iowa Acad Sci 92:63–66

Xia XC, Reif JC, Melchinger AE, Frisch M, Hoisington DA, Beck D, Pixley K, Warburton ML (2005) Genetic diversity among CIMMYT maize inbred lines investigated with SSR markers: I. Subtropical, tropical midlatitude, and highland maize inbred lines and their relationships with elite US and European maize. Crop Sci 45:2573–2582

Yamasaki M, Tenaillon MI, Bi IV, Schroeder SG, Sanchez-Villeda H, Doebley JF, Gaut BS, McMullen MD (2005) A large-scale screen for artificial selection in maize identifies candidate agronomic loci for domestication and crop improvement. Plant Cell 17:2859–2872

Zhang J, Peterson T (2004) Transposition of reversed *Ac* element ends generates chromosome rearrangements in maize. Genetics 167:1929–1937

Part IV
The Corn Genome

Chapter 15
The Structure of the Maize Genome

Joachim Messing

15.1 Introduction

Agriculture is reaching a new era. While in the past, breeding was based on phenotypes, future breeding will be based on knowledge of the genotype. Therefore, the future of agriculture is written into the genome. What conventional breeding cannot achieve is to separate and combine one or a select few genes with the rest of the genes, because during crossing all genes of one parent are transmitted, even those that could neutralize the benefit of another one. But how would one identify single genes of interest and their regulatory components within the total gene pool? Recent estimates are that the maize genome contains between 42,000 and 56,000 genes (Haberer et al. 2005). To identify a gene of interest could be like finding a needle in a haystack. While it has been possible to clone genes based on their gene products or their function, they only represent a tiny portion of the entire gene set. To obtain knowledge about all genes in the genome requires first that we know their structures and position in the genome.

The first genome of a flowering plant that was sequenced was *Arabidopsis thaliana*, mainly because it has one of the smallest genomes (Arabidopsis Genome Initiative, 2000). Furthermore, it was assumed that the C-value paradox teaches that the complexity of a multicellular organism was not proportional to the size of its genome (Thomas 1971). In other words, the smaller genome could serve as a reference gene set for the larger ones. However, many of the most important crop plants on earth belong to the monocotyledons, and *Arabidopsis* belongs to the dicotyledons. Indeed, it became clear that the sequence of the *Arabidopsis* genome is too distant to serve as a reference to monocot crop species. On the other hand, unique genes of the Poaceae, a monocot family, also known as the grasses, are conserved across these species to a degree that they could be used as heterologous probes to

Joachim Messing
Waksman Institute of Microbiology, Rutgers University, 190 Frelinghuysen Road, Piscataway, NJ 08854-8020, USA
e-mail: messing@waksman.rutgers.edu

A.L. Kriz, B.A. Larkins (eds.), *Molecular Genetic Approaches to Maize Improvement*
Biotechnology in Agriculture and Forestry, Vol. 63

detect homologous gene sequences. Therefore, cross-hybridization of genetically mapped gene sequences made it possible to examine syntenic relationships among Poaceae (Hulbert et al. 1990; Whitkus et al. 1992; Ahn and Tanksley 1993). Because this family includes the cereals, it became possible to align entire chromosomal segments of the most important crops regardless of the sizes of their genomes (Moore et al. 1995; Gale and Devos 1998).

15.2 The Gold Standard of Genome Sequence

As a consequence it became attractive to sequence the rice genome as one of the smallest genomes among the cereals, still with a size of 390 Mb, two and a half times larger than *Arabidopsis* (International Rice Genome Sequencing Project 2005). Because of its larger size and its economic significance, the rice genome has been sequenced by different strategies to investigate speed, cost, and accuracy for future plant genome sequencing projects (Goff et al. 2002; Yu et al. 2002). Current DNA sequencing strategies are all based on the fragmentation of target DNA into uniform-sized clones into a sequencing vector. Cloning serves as a purification step of individual DNA fragments and sequencing with universal primers (Vieira and Messing 1982). Because fragmentation creates overlapping DNA fragments, sequence information itself can be used to concatenate the sequence of DNA fragments to contiguous sequences (contigs). Indeed, there is no apparent limitation to the length of such contigs and the process could yield the sequence of entire chromosomes. This highly parallel sequencing approach has been termed shotgun DNA sequencing (Messing et al. 1981; Larson and Messing 1982). Critical for the assembly is sequence links. Application of two universal primers yields two sequences whose distance is determined prior to cloning through the fragmentation process.

In the case of rice, we have comparative data where the same cultivar has been sequenced by overlapping BAC clones and by whole-genome shotgun (WGS) sequencing. The latter covered only 78.2% of the BAC-based sequence. As a consequence WGS sequencing missed 29.6% of the gene models discovered in the BAC-based sequence (International Rice Genome Sequencing Project 2005). The reason for this difference is that the C-value paradox is largely based on a relative increase in repetitive DNA. Fingerprinted BACs are concatenated because of restriction maps rather than sequence alignments and form fingerprinted contigs (FPCs) bridging even tandemly repeated sequences (Nelson et al. 2005). In fact, the centromeric regions of rice chromosome 4 and 8 have been completely sequenced (International Rice Genome Sequencing Project 2005). Clearly, the greater the size of the shotgun library, the more difficult the assembly of contigs and placement of contigs on the genetic map. Given these options, a map-based sequence is now regarded as the gold standard of genome sequence.

15.3 Fractionation Methods of the Maize Genome

Zea mays or corn has a much larger genome than *Arabidopsis* and rice; it is about six times larger than rice (Bennett and Leitch 2005). Therefore, it is not surprising that the larger size prompted the exploration of whether the genic content of maize could be separated from repetitive DNA, which was deemed of little information value. Based on sequencing a few contiguous regions smaller than 0.01% of the total size, the hypothesis was advanced that the structure of the genome was composed of genes that were concentrated in gene islands interrupted by large blocks of retrotransposons (Bennetzen et al. 1998). If this structure were correct, one could fractionate the maize genome into gene-enriched and repetitive DNA elements and sequence the genome for a fraction of the price of the entire genome. Because of the paucity of contiguous sequence information from random rather than selected regions of the genome, it became necessary to obtain a larger sample of maize genomic sequences from regions not selected by known gene sequences and to determine the coverage of those with genomic sequences derived from various fractionation methods.

One of the fractionation methods is based on the observation that repetitive DNA in contrast to genic sequences is frequently methylated. If one clones genomic DNA and passes the ligated recombinant DNA through *E. coli* strains that cleave methylated DNA, then only non-methylated DNA would be recovered (Rabinowicz 1999). The cloning step acts like a filter of methylated sequences (also called methylation filtering (MF)). Hence, the cloned MF DNA would contain pieces of genes, thereby serving as a gene-enrichment method. Another approach is to take advantage of the fact that DNA transposable elements (TEs) preferentially insert into genes. Using DNA amplification, sequences linked to TEs could be obtained. A collection of genic junction sequences from the insertion of a transgenic mutator element is referred to as the RescueMu data set (Lunde et al. 2003). Another has been obtained from a collection of mutator insertions of a uniform background, where parental insertions can be subtracted from new insertions (McCarty et al. 2005). The disadvantage of those TE junction sequences is the incomplete coverage of genes, but they would have a high gene-hit rate and can serve as a reference for the efficacy of gene enrichment methods. Furthermore, they provide critical resources for the functional analysis of genes (May et al. 2003). Yet a different approach to the ones mentioned is the use of reassociation kinetics to separate low copy sequences by chromatography of single-stranded from double-stranded DNA. If DNA is denatured at high temperature it becomes single-stranded. Reassociation of double-stranded DNA depends on the concentration of homologous sequences and the reannealing time, which is determined by C_0t curves (Yuan et al. 2003). The latter approach, also referred to as high C_0t or HC, has the advantage that it separates non-methylated repetitive DNA, representing a surprisingly large fraction of the maize genome, as will be discussed in more detail below. Sequences of genomic fragment libraries are categorized as genome survey sequences (GSSs).

15.4 Distribution of Methylated and Repetitive DNA in the Maize Genome

The evaluation of these fractionation methods also provides us with a survey of how sequences are organized within chromosomes. One can take two approaches to evaluate gene-enrichment methods. For instance, one can compare sequences obtained from fractionated genomic libraries with random genomic libraries. Such a comparison could measure the degree of enrichment. A different evaluation would be the alignment of filtered sequences with contiguous sequence information, thereby placing filtered sequences in respect to a chromosomal position. The latter could also indicate at which level of redundancy sequences would begin to cluster, a sign of diminished return of new sequence information. It also would signal that additional sequencing of clones from a fractionated genomic library is unlikely to fill remaining sequence gaps. Furthermore, it would provide us with topological organization of methylated DNA in contiguous sequences.

For instance, when genomic DNA is sheared and cloned, sequences from random or unfiltered ends (UF) would yield sequences that meet a certain threshold of copy number in the genome depending on the number of clones sequenced (Meyers et al. 2001). Sequences from MF and HC genomic libraries would then be expected to be free of high copy number sequences (Whitelaw et al. 2003). Indeed, both filtration methods have a substantial reduction, but are not void of TE-related DNA sequences (Table 15.1). If one increases the number of random end sequences to a threshold, where single copy sequences can be detected as well, then one can even determine what the level of enrichment of the filtered sequences is. A collection of such sequences has been obtained from 475,000 BAC end sequences (BESs), which yield a 650-bp sequence on average every 6.2 Kb along all chromosomes, sufficient to tag about 20% of all maize genes (Messing et al. 2004). If these BESs are compared to the filtered sequences, one could estimate a three- to six-fold reduction in non-genic sequences after enrichment.

The preliminary nature of these numbers is the lack of a comprehensive repeat database of the maize genome. Many repeat elements remain initially undetected because of their sequence divergence and the lack of sequences of larger chromosomal

Table 15.1 Percentage fractionation of repeat elements. *UF* Unfiltered sequences; *MF* methylation filtering; *HC* high C_0t; *MITES* miniature inverted-repeat transposable elements; *SSR* simple sequence repeats. (Composed from published data (Messing et al. 2004))

Type of sequence	UF	MF	HC	RescueMu
Class I elements	5773	15.70	6.55	4.91
Class II elements	0.92	1.14	1.58	4.84
Copia-like	17.50	8.37	2.54	1.98
Gypsy-like	31.25	4.32	2.22	1.71
Unclassified retros	8.94	2.95	1.62	1.09
MITES	0.19	0.39	0.62	1.08
SSRs	1.66	1.27	0.19	0.17

segments (Messing et al. 2004). One would expect the degree of enrichment to decrease with more known repeat elements. Still, if one breaks down repeat elements into different repeat families, one can observe certain biases in each non-random collection. The most surprising result is that half of the *copia*-like elements appear to be hypomethylated (Table 15.1). Although the majority of retroelements in maize are gypsy-like elements, one wonders why this class of elements is more heavily methylated. Another striking feature is the hypomethylation of simple sequence repeats (SSRs). Because SSRs are part of centromeric regions and these are also transcribed, the relative enrichment of these sequences might not be unexpected. The residual presence of TEs in the HC fraction, especially class I elements, could be due to the frequency of nested elements in the maize genome, which are chimeric in nature and could create double-stranded DNA with portions of single-stranded DNA covalently linked to it after reannealing junction sequences. Such structure would be retained in the single-stranded DNA fraction and therefore also cloned in the gene-enriched fraction. Analysis of a small number of fully sequenced BAC clones indicated that 27% of the retrotransposons were nested, but these regions were selected by known genetic markers (Du et al. 2006). On the other hand, when DNA is sheared before it is denatured for HC fractionation, the proportion of junction sequences should be rather short compared to the length of contiguous sequences.

Compared to the MF and HC methods, RescueMu is the most effective gene-enrichment method (Table 15.1). In addition, as discussed above, Mu insertions into genes can help in the functional analysis of genes. Therefore, Mu junction sequences can be used to locate knockout alleles of genes (McCarty et al. 2005). However, common to these gene enrichment methods is the cost of template preparation for sequencing and sequence contig assembly. For most clones one can sequence only one end. Insert sizes of MF and HC clones tend to be so small that they frequently yield only one sequence read per clone. The same is true for the junction sequence of a mutator insertion. Having a single sequence per clone is not only expensive for template purification, but also problematic for the reconstruction of contiguous sequence information; the latter is usually not the goal for junction sequences but is for filtered sequences. As discussed above, sequences that are linked over a predetermined distance are critical for the assembly algorithm.

15.5 One Hundred Random Regions of the Maize Genome

Even if one could overcome the limitation of insert length with filtered genomic sequences, what would be the probability of reconstructing gene sequences over the entire length and what sequence bias if any would occur? These questions could not be answered with other single sequence reads despite their deep coverage of the genome. One needs contiguous sequences containing intact genes. To obtain a set of genes from different regions of the genome, BAC clones from 100 random regions of the genome were selected for sequencing. The first obvious result from those 100 random regions is that previously sequenced BAC clones had on average a higher gene density (Haberer et al. 2005). As a consequence, repeat libraries of the genome were suboptimal and the masking of sequences prior to gene modeling was incomplete. For instance, based on BESs the repeat content appeared to be 58%, lower than the 63% from small insert libraries derived from sheared DNA (Messing et al. 2004). Clearly, restriction enzymes counter-select cleavage of repetitive DNA, in particular EcoRI (50%) compared to HindIII (60%). While the repeat content in 100 random BACs appeared to be 66%, it was only 53% in 117 non-random BACs using the same repeat library.

Given the improved repeat library built on the 100 random regions of the genome, it now became possible to assess the distribution of genes and repeat elements within contiguous sequences. Within these sequences there is no apparent support for gene islands. Moreover, spacing between genes appears to be quite variable. Furthermore, repeat elements invade introns, making gene sizes also variable. Because these random regions were obtained from the same inbred line as the gene-enriched MF and HC sequences, it became possible to superimpose MF and HC relative to contiguous sequence information by applying very high stringencies (98% identity over 90% length). There was an excellent hit rate of genes (93%), where at least one alignment of MF or HC corresponded to a predicted gene model (Haberer et al. 2005). However, only 29% of the predicted genes were covered in their entirety (90%). Moreover, annotated BAC clones revealed deep clusters of filtered sequence reads of genes, low-copy and repetitive intergenic sequences. Furthermore, promoter regions and untranslated coding sequences were underrepresented relative to exons. This bias of covering some sequences deeply and genes only partially indicates that even high redundant sequencing of gene-enriched libraries would probably fall short in identifying all genes, including regulatory sequence elements in the genome. In this respect it was interesting that genetic analysis of alleles of the *B*1 and *Tb1* genes revealed that enhancer elements were separated from the coding regions by large sets of retrotransposon insertions (Stam et al. 2002; Clark et al. 2004). Therefore, one can conclude that gene-enrichment methods provide a rapid and cheap access to gene tags at a low redundancy. However, to fully exploit the genetics of corn, in particular if traits could be associated with non-coding sequences, the gold standard is a critical goal for the maize genome sequence.

15.6 Physical Map of the Maize Genome

To achieve this goal, a physical map of a reference inbred needed to be constructed. Inbred B73 was selected because it has been the starting material for many breeding efforts. Heterozygosity has been avoided by successive selfing, which is very straightforward with maize. Genomic DNA from the same lot has been used to construct three large insert libraries using three different restriction enzymes, *Hin*dIII, *Eco*RI, and *Mbo*I. The average insert size for the three libraries was 154 Kb (Yim et al. 2002). Based on a set of unique probes it then became possible to calculate the redundancy of the libraries to be 30-fold and the size of the genome to be 2.3 Gb. Construction of a physical map from these libraries required fingerprinting and sequencing the ends of inserts (BES). Because of the size of the maize genome, it was necessary to introduce a high-throughput method to process the magnitude of clones in a reasonable time. A robotic pipeline comprised template preparation, DNA fingerprinting, and insert end sequencing in microtitre plates. To adopt fingerprinting to this streamline, DNA was cleaved with enzymes to produce fragment sizes under 1 Kb so that they could be resolved on capillary DNA sequencers (Nelson et al. 2005). Therefore, capillary sequencers could be used for fingerprinting and end sequencing. By masking BESs with the repeat library described above the remaining BESs served as sequence tags of contigs that were formed from fingerprints, also called FPCs. Masked BESs were further extended with EST resources, where possible, resulting in 9,129 anchored sequences. These anchored sequences provided four important structural features of the maize genome. Because many genes are duplicated on different contigs the maize genome appears to be the result of a whole-genome duplication (WGD) event. However, not all genes seem to be duplicated, which would indicate gene loss after WGD. A large proportion of genes appear to be tandemly duplicated because of their physical linkage, which is indicative of gene amplification and gene families. The fourth feature would be an estimate of total genes in the genome. Under the assumption that gene sizes on average would be similar between maize and rice, a gene space of 7.5% would amount to 59,000 genes. However, this estimate turns out to be incorrect. It became clear from recently annotated maize genes and a better knowledge of repeat elements that maize genes on average are larger than rice genes because of increased intron sizes. Maize introns expanded because of the invasion of TE-related sequences, which is not surprising given the overall TE activity in the maize genome. The increased gene size in the same gene space would lower the total gene number between 42,000 to 56,000 depending on the stringencies of gene calling methods. The lower numbers were somewhat surprising. If rice is diploid and maize is an ancient tetraploid, one could expect up to twice as many genes in maize than in rice. With an estimated 32,000 gene models in rice (Itoh et al. 2007), the gene counts in maize appear to be on the low side. However, the apparent gene loss after the WGD event would be consistent with this reduction (Messing et al. 2004).

High-density filters of the BAC libraries were also hybridized with large sets of gene sequences, mostly using overgos for rapid throughput. As a consequence, a total of 24,006 gene sequences could be placed on FPCs. Using these anchored sequences FPCs could be ordered along the rice genome sequence and further manually edited. The final map consisted of 721 FPCs representing about 93.5% of total genome length (Wei et al. 2007). Because of the hybridization of 1,902 genetic markers to high-density BAC filters, 421 FPCs could be anchored to about 86.1% of the genetic map, illustrating that mainly small FPCs are not assigned. This coverage indicates that probably some regions in the maize genome are difficult to assemble by restriction mapping because of their sizes and types of tandem repeats.

15.7 Evolution of Maize Chromosome Numbers

The physical map has two instant uses. One can relate genetic distances to physical distances, which are quite variable throughout the genome. The 1,902 genetic markers and the 24,006 sequence tags provide a total of 25,908 anchored gene sequences, about half of all estimated genes. The advantage of such a gene-dense map is that one can now use the rice genome as a reference to determine how many chromosome breakages it would take to reconstruct the maize genome from the rice genome. Such a hypothetical reconstruction would then suggest that maize arose by the hybridization of two progenitors that each had 10 chromosomes (Wei et al. 2007). Previously, it had been suggested that maize arose by allotetraploidization (Gaut and Doebley 1997). More recent studies of orthologous regions of the duplicated chromosomal segments of the maize genome with sorghum and rice permitted the clustering of genes derived from common ancestral chromosomes. If the ancestor of rice, sorghum, and maize separated 50 million years ago (mya), then the two progenitors of maize and the progenitor of sorghum split 11.9 mya (Swigonova et al. 2004). All three evolved independently, but two of them hybridized as recently as 4.8 mya to form maize. While sorghum remained diploid with 10 chromosomes, maize initially had 20 chromosomes. However, to prevent pairing of non-homoeologous chromosomes, the 20 chromosomes underwent chromosome breakages, fusions, and loss of 10 centromeres, resulting in 10 chromosomes nearly twice the size of sorghum chromosomes. In addition, waves of retrotranspositions nearly doubled the genome size again by expanding chromosomes in size. Analysis of the orthologous regions described above also shows that in about 50% of the cases only one copy of the genes that got duplicated by the whole-genome duplication (WGD) event was retained instead of the two. This analysis is consistent with the earlier observations obtained from BESs and the estimates of the lower gene count in maize. Based on the alignment with rice, one could estimate that as many as 62 major breakages and fusions occurred to form today's 10 chromosomes of maize (Wei et al. 2007). On the other hand, the large size of the maize chromosomes has been the mainstay of cytogenetic studies.

15.8 Diploidization of the Maize Genome

A major use of the physical map is to obtain contiguous sequences. To test the use of the map for this purpose, two homoeologous telomeric regions of the genome were selected for pilot sequencing, one in bin 1.00–1.01 and one in bin 9.07 (Bruggmann et al. 2006). The one on chromosome 1S spans about 17.4 cM of the genetic map and is about 7.8 Mb in length, the one on chromosome 9L is about 6.6 Mb (25.6 cM) in length, or both are about 0.6% of total genome length. It is immediately apparent that recombination frequencies differ significantly between these two regions. While the average recombination rate of the chromosome 1S region is about 450 Kb/cM, the one on chromosome 9L is about 256 Kb/cM. Both rates would be significantly lower than the average for the genome with 1,200 Kb/cM. These differences correlate to some degree with gene density, which is the highest for the chromosome 9L segment with 3.7 genes/100 Kb, 3.0 genes/100 Kb for the chromosome 1S segment, and 2.3 genes/100 Kb for the 100 random regions described earlier. This correlation is consistent with the observation that meiotic crossover preferentially occurs within genes (Thuriaux 1977); otherwise unequal crossover within retrotransposons would create large deletions and differences in genome sizes (see also below). The latter might still occur, but rather infrequently and also might be disadvantageous because of embedded genes.

If these two regions are aligned with the rice genome via their predicted genes, one can visualize how maize arose from a WGD event because there is only one orthologous region in rice and that is on rice chromosome 3S. Interestingly, it is also close to the telomeric end of the chromosome. Based on these alignments, several observations can be made in respect of maize genome structure. Gene collinearity can be divided into intervals, also defined as synteny blocks. Divisions can be marked by stretches of non-conserved sequences between both maize regions that can be several 100 Kb in size or one maize chromosome may contain an inversion of the entire synteny block relative to the homoeologous region. For instance, based on rice as a reference, maize chromosome 9L has a synteny block of about 2 Mb inverted relative to maize chromosome 1S. These types of inversions have been observed genome-wide in maize based on the 26K marker map described above (Wei et al. 2007). Inversions like the one observed here not only divide chromosomal intervals into synteny blocks, but also play a critical role in disrupting the pairing of chromosomes during meiosis. Therefore, one can envision that after the hybridization of the two progenitors of maize, small inversions in homoeologous chromosomes provided a selective advantage in converting maize to a diploid. However, one could view any structural change that makes highly homologous chromosomes dissimilar as a pathway of a tetraploid species to become diploid and call this process diploidization.

15.9 Retrotransposition

From these observations it appears that the tetraploid origin of maize constituted an unstable state of the genome and that stability was achieved by massive chromosome breakage and fusion events. Still, it is unclear whether these events were triggered by the tetraploid status or by other environmental cues. Nevertheless, the consequence is that today's maize genome behaves like a diploid and has a diploid set of chromosomes. Although the maize genome is still prone to small structural changes, it appears that its structure had already reached a reasonable stability before domestication and that most massive changes occurred in waves (Du et al. 2006). Probably the most visible changes are based on retrotranspositions. Retrotransposition occurs by a replicative mode. If an element is transcribed, it makes many RNA copies that can be reverse transcribed into DNA, which subsequently gets somewhat randomly integrated into the genome by illegitimate recombination. One can easily visualize how such a process can result in extensive amplification of a few sequences. Indeed, BES analysis indicated that more than half of the retroelements consists of five families, Ji (21.35%), Cinful (12.08%), Opie (11.12%), Zeon (9%), and Huck (5.01%). A smaller set of random sequences had Huck (10.7%), Ji (9.4%), Opie (7.1%), Zeon (4.8%), and Cinful (3.5%) as the top five (Meyers et al. 2001), which could reflect the bias of restriction sites in the end sequences. Another indication that the different families might have preferential target sites comes from the study of sequenced BAC clones that contain known genetic markers as described above. Here, Huck, Ji, and Opie are the front-runners with 24.8%, 21.5%, and 14.3%, respectively (Du et al. 2006). Because the latter case represents complete elements, the length of elements can also be compared. The average size of Huck elements is 14.7 Kb, Ji elements 9.4 Kb, and Opie elements 9.0 Kb. Therefore, the Huck family could dominate because of their length.

At the time of insertion the long terminal repeat (LTR) sequences of retrotransposons are 100% the same. The older they are the more nucleotide substitutions they have in their sequences. One can assume that the pace of substitutions is faster than in exons, but probably very similar among those elements (Ma and Bennetzen 2004). Therefore, the complete structures of elements are also useful to obtain estimates of when retrotranspositions occurred. Based on those studies, it appears that retrotransposition occurred after the hybridization of the two progenitors of maize 4.8 mya (Swigonova et al. 2005; Du et al. 2006). The oldest element in the BAC study inserted into the genome 4.6 mya. Although elements can be found to have inserted into the genome at different times after that, the majority of them inserted into the genome less than 1 mya. Because retrotransposition could occur in either of the two homoeologous regions of the two progenitor chromosomes, they represent the major drive to render homoeologous regions dissimilar. Therefore, even more than inversions retrotranspositions are advantageous from the point of diploidization. If that is the case, one wonders why retrotranspositions occurred in waves even long after the initial WGD event. One possibility is that it took only a few rearrangements initially after WGD to achieve a diploid status, but that additional changes accumulated over time to enhance the dissimilarities of homologous sequences reminiscent of entropic processes in nature described by the German physicist Rudolf Clausius (Clausius 1868).

15.10 Chromosome Expansion and Contraction

Interestingly, the process of making chromosomes more dissimilar also resulted in chromosome expansion. When we compare the two sequenced homoeologous regions on chromosome 1S and chromosome 9L described above, one can not only divide these regions in synteny blocks, but also observe that retrotranspositions led to a differential expansion of the maize genome (Bruggmann et al. 2006). If one assumes that the rice chromosome structure has experienced the least changes since the progenitors of rice and maize diverged, the B1 block stands out as an

Table 15.2 Chromosome expansion rates (in length in base pairs divided by length in base pairs)

Expansion	Size	Gypsy	Copia	Genes
Zm1S A	1.9	9	22	0.4
Zm9L A	1.1	5	11	0.4
Zm1S B1	4.2	380	833	1.4
Zm9L B1	0.8	54	79	0.5
Zm1S B2	4.2	14	142	1.2
Zm9L B2	0.8	4	27	0.4

example of change in the maize chromosomes. Blocks B1 and B2 of maize chromosome 1S have an expansion factor of 4.2 by size, which is above the average factor of 3.2 (Wei et al. 2007). However, block B1 of both maize chromosomal regions appears to be a hotspot of retrotransposon insertions if one considers the significantly smaller size of the B1 block on maize chromosome 9L (Table 15.2). In all synteny blocks, copia elements dominate between 1.5- and 10-fold, whereas the reverse is true genome-wide for both maize and rice with 0.62- and 0.22-fold, respectively. However, the greater increase of copia elements in maize relative to rice would suggest that copia elements played a greater role in chromosome expansion in gene-rich regions. Besides the role of copia elements in chromosome expansion, the synteny blocks indicate a very uneven rate of expansion. Some of the maize chromosome 9L blocks appear to have even undergone a contraction relative to rice or rice has expanded more than these maize regions (Table 15.2). The latter is less likely because maize has also lost genes that are present in chromosome 1S. If one compares the expansion of both maize regions by gene numbers, it appears that chromosome 1 has gained genes and chromosome 9 has lost genes relative to rice. As already discussed, unequal crossovers between LTRs could lead to chromosomal deletions with embedded genes. Because of the duplication on the other chromosome such deletions could be tolerated, but it would affect the total gene number. If these features were exemplary for the genome, then they would explain why despite the WGD event maize does not have twice as many genes as rice.

Having these contiguous chromosomal sequences also has the advantage that one can investigate features of the epigenome of the maize genome. As discussed above, the sequenced contigs are derived from the same inbred as the MF-enriched sequences. One can therefore align MF sequences with the contigs and place the distribution of hypomethylated sequences along the contigs. If multiple alignments are performed, one can compare the position of genes, repeat elements, and hypomethylated sequences. As a control for the MF fraction, one can also use the distribution of the HC fraction. Intuitively, one would expect a correlation between the distribution of gene and hypomethylated sequences. Such a correlation does not hold consistently throughout the regions and the fractionation methods seem to cover both maize regions equally well. Therefore, the expansion of chromosomal regions is not correlated with regional increase in methylated sequences (Bruggmann et al. 2006). These results are also consistent with the earlier observations that genome-wide half of the copia elements are hypomethylated (Table 15.1).

15.11 Orthologous and Paralogous Gene Copies

Probably the most striking feature of diploidization is the dynamics of the maize genes themselves. If the two aligned homoeologous regions had at the time of the WGD event each one copy of each gene, two thirds now have lost one of the two copies, depending on the interval that is compared (Table 15.3). Interestingly, loss is not random, but appears to be stronger in the chromosome 9L regions (Bruggmann et al. 2006). As explained above the reverse expansion could contribute to this preferential loss. Previous studies came to the same conclusion. However, they could not be as precise because they had either only sequence tags (BES) or only small contigs of the size of a single or a few (two or three) overlapping BAC clones.

Interestingly, this apparent gene loss is counteracted by mechanisms that add genes to each genome. Using rice as a reference for orthology, it appears that nearly a fifth of the genes in each maize region are non-collinear. Because these genes can be found somewhere else in the rice genome, it could be that these genes were copied or excised from different locations of the maize genome and possessed the ability to insert into the intergenic space of the two homoeologous regions. However, this process seems to be unlinked to the diploidization process because we can find it also in rice. The orthologous region in rice also has genes that are missing in both maize regions (Table 15.3). Taking advantage of the BES, MF, and HC GSSs, it appears that those are present somewhere else in the maize genome. Therefore, both grass genomes diverged by gene mobility. In addition, each genome appears to have a unique set of genes (6–10%) not found elsewhere in rice or maize. These features of paralogous gene copies intermittent with orthologous genes indicates an ability of these genomes to respond to whatever challenge to its structure with striking gene mobility, including a high percentage of rapidly evolving genes.

While we are still gaining a more global view of these dynamic features of these genomes, we can learn through the plasticity of storage protein gene copies more about the potential role of gene duplications in plants. When maize was domesticated some 10,000 years ago, one of the quantitative traits that gave rise to morphological changes was the *pbf* locus (Jaenicke-Despres et al. 2003). This locus controls the synthesis of alpha zein genes during kernel development (Ueda et al. 1994; Vicente-Carbajosa et al. 1997; Wang et al. 1998). Alpha zein genes are a multigene family. They arose after the progenitors of maize and sorghum separated from the

Table 15.3 Percentage gene collinearity after WGD

Aligned regions	Syntenic	Mobile	New
Zm1S/Os3S	73	21	6
Zm9L/Os3S	72	20	8
Os3S/Zm1S, Zm9L	48	N/a	10
	Syntenic	*Non-syntenic*	
Zm1S/Zm9L	36	64	
Zm9L/Zm1S	27	73	

progenitor of rice (Song et al. 2002). This can be established by aligning orthologous regions with collinear genes in chromosomal contigs of maize, sorghum, and rice. Because of their chromosomal positions, clustering of duplicated genes can be used to determine their origin in space *and* time. Based on such an analysis, the founding member of alpha zein genes seems to have arisen in a region on maize chromosome 1 before the time the progenitors of maize and sorghum split 11.8 mya. This gene gave rise to a second gene on chromosome 4 before the split of the lineages. However, after the two progenitors of maize hybridized, the copy in the homoeologous region was lost. On the other hand, additional copies were generated and inserted into another region of chromosome 4 after allotetraploidization and on one of the progenitors of maize on chromosome 7 before allotetraploidization. These movements did not occur in sorghum and are unique to maize (J.-H. Xu and J. Messing, in prep.). Insertion of these copies in other regions resulted also in tandem duplications of the inserted copies. Although the mechanism of this gene mobility is not known, it appears to be a copy and paste mechanism. Since zein genes do not have introns, one is reminded of retroposons. However, they would have poly-A tracts as zein mRNAs are polyadenylated. Furthermore, sequence alignments of the *z1C1* zein cluster indicated an average size of 4.4 Kb per amplicon, much larger than the 1-Kb transcribed region of zein genes, which is consistent with the retention of its function and tissue-specific expression (Llaca and Messing 1998). Based on these results there appears to be a greater degree of gene mobility and amplification in maize than in sorghum. Still, one might ask that if gene movement and subsequent tandem amplification occurred long before the domestication of maize, what role would gene duplications play in domestication of maize some 10,000 years ago? Agronomically, the major change in seed structure that occurred in the transition from teosinte to maize is grain filling and kernel size. If one considers that both aspects are dictated by starch and to a lesser degree by protein accumulation, the easiest way to change protein accumulation could be through the change of regulation of a regulatory factor that is common to the entire gene family. Therefore, duplicated genes could amplify any effect on one regulatory factor by a greater degree the more copies of the gene family are available and protected from inactivation by various mechanisms. Would such a concept be unique for storage proteins? In this respect it is interesting to note that although the rice genome does not have these storage protein genes, it has a large percentage (29%) of its genes in tandem gene clusters (International Rice Genome Sequencing Project 2005). Based on BESs, the number of gene clusters could be even higher in maize (35%) than in rice, but the annotation of intact genes from total genome sequence might lower these estimates.

15.12 Haplotype Variation

We have seen that different mechanisms have operated in the maize genome to make homoeologous chromosomes that were formed from two progenitors as a

WGD event dissimilar over time. They include chromosome breakage and fusion, the loss of centromeric regions, massive retrotranspositions largely from as few as five founding members, uneven expansion, inversion, or contraction of chromosome segments forming synteny blocks, loss of duplicated genes, and significant formation of paralogous gene copies. One can speculate that some of these dynamic features of the genome are triggered by the WGD event, but others appear to be more general and more likely triggered by other stimuli as well.

With this insight, earlier genetic mapping data of zein genes appear in a different light. Using the physical differences of zein genes in protein gels, it has been possible to map individual paralogous copies (Wilson 1989). While previously one could have assumed that this is due to sequence polymorphisms between different inbreds, another explanation appeared to be simply absence or presence of paralogous copies. This became clear when contigs of two haplotypes of the same locus containing zein genes were sequenced (Song and Messing 2003). There was a dramatic difference between the *z1C1* locus of BSSS53 and B73, when the sequenced regions were aligned. Now expansion was different not only between homoeologous regions but also even between allelic regions. The 360 Kb region in BSSS53 corresponded to only 263 Kb in B73, a 37% increase that was 50% due to transposition events and 50% due to genic and intergenic space. Taking advantage of such divergence, one can also investigate how non-allelic copies are expressed in hybrids. Intuitively, one would expect expression levels to simply follow a dosage pattern. Since these genes are expressed in the triploid endosperm, one would expect differences between the reciprocal crosses. If the non-allelic gene copy passes through the female flower, one would expect higher levels of expression than in the reciprocal cross. However, non-allelic copies of zein genes did not follow a simple dosage pattern in gene expression measurements of reciprocal hybrid crosses, suggesting variability even in genes encoding trans-acting factors (Song and Messing 2003).

Besides zein genes, the same region contained a copy of a cytidine deaminase (CDA) gene present in BSSS53, but not B73. Subsequent analysis showed that this copy was indeed copied about 4.5 mya, soon after the WGD event, from one of the orthologous copies (Xu and Messing 2006). In this case, both orthologous copies are retained in the maize genome, diverged at the time when the two progenitors of maize split, and are present in all inbred lines tested. In contrast, the paralogous copy is present in some inbreds but absent in others. Interestingly, when the paralogous copy was compared to the empty site in B73, sequence features became apparent that it arose by helitron movement (Kapitonov and Jurka 2001). Helitrons are abundant sequences that when clustered appear to have homologies to helicase coding regions, hence the name. It has been proposed that these helicases could have the same function as the replication protein of single-stranded DNA phage-like M13. M13 replicates via a rolling circle mechanism, which goes through a single-stranded intermediate. It is therefore conceivable that circular single-stranded DNA produced as a copy from an orthologous gene recombines by illegitimate recombination elsewhere in the genome. Hijacking gene copies by such a mechanism appears to be widespread in the maize genome, but usually fails to capture entire genes, except for the CDA example in maize chromosome 4S so far (Lai et al. 2005; Morgante et al. 2005; Xu and Messing 2006). Nevertheless, it represents one of several mechanisms to generate paralogous gene copies, which are manifested in haplotypes that differ by the presence and absence of genes. In general, the finding of non-allelic copies of genes in the maize genome is consistent with the entropy concept discussed above. The interesting additional aspect is now that new combinations of the linear arrangement of sequences in the maize genome, retrotransposons or genes, can be achieved by the crossing of two different inbreds. A unique array of haplotypes that differ in the absence and presence of sequences can then be maintained by selfing, generating inbreds with unique genotypes (Messing and Dooner 2006).

Acknowledgements Work reported here was supported in part by National Science Foundation Plant Genome grants #9975618 and #0211851, and Department of Energy grant # DE-FG05-95ER20194.

References

Ahn S, Tanksley SD (1993) Comparative linkage maps of the rice and maize genomes. Proc Natl Acad Sci USA 90:7980–7984

Arabidopsis Genome Initiative (2000) Analysis of the genome sequence of the flowering plant *Arabidopsis thaliana*. Nature 408:796–815

Bennett MD, Leitch IJ (2005) Nuclear DNA amounts in angiosperms: progress, problems and prospects. Ann Bot (Lond) 95:45–90

Bennetzen JL, SanMiguel P, Chen M, Tikhonov A, Francki M, Avramova Z (1998) Grass genomes. Proc Natl Acad Sci USA 95:1975–1978

Bruggmann R, Bharti AK, Gundlach H, Lai J, Young S, Pontaroli AC, Wei F, Haberer G, Fuks G, Du C, Raymond C, Estep MC, Liu R, Bennetzen JL, Chan AP, Rabinowicz PD, Quackenbush J, Barbazuk WB, Wing RA, Birren B, Nusbaum C, Rounsley S, Mayer KF, Messing J (2006) Uneven chromosome contraction and expansion in the maize genome. Genome Res 16:1241–1251

Clark RM, Linton E, Messing J, Doebley JF (2004) Pattern of diversity in the genomic region near the maize domestication gene tb1. Proc Natl Acad Sci USA 101:700–707

Clausius R (1868) On the mechanical theory of heat. Philos Mag 40:122

Du C, Swigonova Z, Messing J (2006) Retrotranspositions in orthologous regions of closely related grass species. BMC Evol Biol 6:62

Gale MD, Devos KM (1998) Comparative genetics in the grasses. Proc Natl Acad Sci USA 95:1971–1974

Gaut BS, Doebley JF (1997) DNA sequence evidence for the segmental allotetraploid origin of maize. Proc Natl Acad Sci USA 94:6809–6814

Goff SA, Ricke D, Lan TH, Presting G, Wang R, Dunn M, Glazebrook J, Sessions A, Oeller P, Varma H, Hadley D, Hutchison D, Martin C, Katagiri F, Lange BM, Moughamer T, Xia Y, Budworth P, Zhong J, Miguel T, Paszkowski U, Zhang S, Colbert M, Sun WL, Chen L, Cooper B, Park S, Wood TC, Mao L, Quail P, Wing R, Dean R, Yu Y, Zharkikh A, Shen R, Sahasrabudhe S, Thomas A, Cannings R, Gutin A, Pruss D, Reid J, Tavtigian S, Mitchell J, Eldredge G, Scholl T, Miller RM, Bhatnagar S, Adey N, Rubano T, Tusneem N, Robinson R, Feldhaus J, Macalma T, Oliphant A, Briggs S (2002) A draft sequence of the rice genome (*Oryza sativa* L. ssp. *japonica*). Science 296:92–100

Haberer G, Young S, Bharti AK, Gundlach H, Raymond C, Fuks G, Butler E, Wing RA, Rounsley S, Birren B, Nusbaum C, Mayer KF, Messing J (2005) Structure and architecture of the maize genome. Plant Physiol 139:1612–1624

Hulbert SH, Richter TE, Axtell JD, Bennetzen JL (1990) Genetic mapping and characterization of sorghum and related crops by means of maize DNA probes. Proc Natl Acad Sci USA 87:4251–4255

International Rice Genome Sequencing Project (2005) The map-based sequence of the rice genome. Nature 436:793–800

Itoh T, Tanaka T, Barrero RA, Yamasaki C, Fujii Y, Hilton PB, Antonio BA, Aono H, Apweiler R, Bruskiewich R, Bureau T, Burr F, Costa de Oliveira A, Fuks G, Habara T, Haberer G, Han B, Harada E, Hiraki AT, Hirochika H, Hoen D, Hokari H, Hosokawa S, Hsing Y, Ikawa H, Ikeo K, Imanishi T, Ito Y, Jaiswal P, Kanno M, Kawahara Y, Kawamura T, Kawashima H, Khurana JP, Kikuchi S, Komatsu S, Koyanagi KO, Kubooka H, Lieberherr D, Lin YC, Lonsdale D, Matsumoto T, Matsuya A, McCombie WR, Messing J, Miyao A, Mulder N, Nagamura Y, Nam J, Namiki N, Numa H, Nurimoto S, O'Donovan C, Ohyanagi H, Okido T, Oota S, Osato N, Palmer LE, Quetier F, Raghuvanshi S, Saichi N, Sakai H, Sakai Y, Sakata K, Sakurai T, Sato F, Sato Y, Schoof H, Seki M, Shibata M, Shimizu Y, Shinozaki K, Shinso Y, Singh NK, Smith-White B, Takeda J, Tanino M, Tatusova T, Thongjuea S, Todokoro F, Tsugane M, Tyagi AK, Vanavichit A, Wang A, Wing RA, Yamaguchi K, Yamamoto M, Yamamoto N, Yu Y, Zhang H, Zhao Q, Higo K, Burr B, Gojobori T, Sasaki T (2007) Curated genome annotation of *Oryza sativa* ssp. *japonica* and comparative genome analysis with *Arabidopsis thaliana*. Genome Res 17:175–183

Jaenicke-Despres V, Buckler ES, Smith BD, Gilbert MT, Cooper A, Doebley J, Paabo S (2003) Early allelic selection in maize as revealed by ancient DNA. Science 302:1206–1208

Kapitonov VV, Jurka J (2001) Rolling-circle transposons in eukaryotes. Proc Natl Acad Sci USA 98:8714–8719

Lai J, Li Y, Messing J, Dooner HK (2005) Gene movement by helitron transposons contributes to the haplotype variability of maize. Proc Natl Acad Sci USA 102:9068–9073

Larson R, Messing J (1982) Apple II software for M13 shotgun DNA sequencing. Nucleic Acids Res 10:39–49

Llaca V, Messing J (1998) Amplicons of maize zein genes are conserved within genic but expanded and constricted in intergenic regions. Plant J 15:211–220

Lunde CF, Morrow DJ, Roy LM, Walbot V (2003) Progress in maize gene discovery: a project update. Funct Integr Genomics 3:25–32

Ma J, Bennetzen JL (2004) Rapid recent growth and divergence of rice nuclear genomes. Proc Natl Acad Sci USA 101:12404–12410

May BP, Liu H, Vollbrecht E, Senior L, Rabinowicz PD, Roh D, Pan X, Stein L, Freeling M, Alexander D, Martienssen R (2003) Maize-targeted mutagenesis: a knockout resource for maize. Proc Natl Acad Sci USA 100:11541–11546

McCarty DR, Settles AM, Suzuki M, Tan BC, Latshaw S, Porch T, Robin K, Baier J, Avigne W, Lai J, Messing J, Koch KE, Hannah LC (2005) Steady-state transposon mutagenesis in inbred maize. Plant J 44:52–61

Messing J, Dooner HK (2006) Organization and variability of the maize genome. Curr Opin Plant Biol 9:157–163

Messing J, Crea R, Seeburg PH (1981) A system for shotgun DNA sequencing. Nucleic Acids Res 9:309–321

Messing J, Bharti AK, Karlowski WM, Gundlach H, Kim HR, Yu Y, Wei F, Fuks G, Soderlund CA, Mayer KF, Wing RA (2004) Sequence composition and genome organization of maize. Proc Natl Acad Sci USA 101:14349–14354

Meyers BC, Tingey SV, Morgante M (2001) Abundance, distribution, and transcriptional activity of repetitive elements in the maize genome. Genome Res 11:1660–1676

Moore G, Devos K, Wang Z, Gale MD (1995) Grasses, line up and form a circle. Curr Biol 5:737–739

Morgante M, Brunner S, Pea G, Fengler K, Zuccolo A, Rafalski A (2005) Gene duplication and exon shuffling by helitron-like transposons generate intraspecies diversity in maize. Nat Genet 37:997–1002

Nelson WM, Bharti AK, Butler E, Wei F, Fuks G, Kim H-R, Wing RA, Messing J, Soderlund C (2005) Whole-genome validation of high-information-content fingerprinting. Plant Physiol 139:27–38

Rabinowicz PD, Schutz K, Dedhia N, Yordan C, Parnell LD, Stein L, McCombie WR, Martienssen RA (1999) Differential methylation of genes and retrotransposons facilitates shotgun sequencing of the maize genome. Nat Genet 23:305–308

Song R, Messing J (2003) Gene expression of a gene family in maize based on noncollinear haplotypes. Proc Natl Acad Sci USA 100:9055–9060

Song R, Llaca V, Messing J (2002) Mosaic organization of orthologous sequences in grass genomes. Genome Res 12:1549–1555

Stam M, Belele C, Ramakrishna W, Dorweiler JE, Bennetzen JL, Chandler VL (2002) The regulatory regions required for B' paramutation and expression are located far upstream of the maize b1 transcribed sequences. Genetics 162:917–930

Swigonova Z, Lai J, Ma J, Ramakrishna W, Llaca V, Bennetzen JL, Messing J (2004) Close split of sorghum and maize genome progenitors. Genome Res 14:1916–1923

Swigonova Z, Bennetzen JL, Messing J (2005) Structure and evolution of the r/b chromosomal regions in rice, maize and sorghum. Genetics 169:891–906

Thomas CA, Jr (1971) The genetic organization of chromosomes. Annu Rev Genet 5:237–256

Thuriaux P (1977) Is recombination confined to structural genes on the eukaryotic genome? Nature 268:460–462

Ueda T, Wang Z, Pham N, Messing J (1994) Identification of a transcriptional activator-binding element in the 27-kilodalton zein promoter, the –300 element. Mol Cell Biol 14:4350–4359

Vicente-Carbajosa J, Moose SP, Parsons RL, Schmidt RJ (1997) A maize zinc-finger protein binds the prolamin box in zein gene promoters and interacts with the basic leucine zipper transcriptional activator Opaque2. Proc Natl Acad Sci USA 94:7685–7690

Vieira J, Messing J (1982) The pUC plasmids, an M13mp7-derived system for insertion mutagenesis and sequencing with synthetic universal primers. Gene 19:259–268

Wang Z, Ueda T, Messing J (1998) Characterization of the maize prolamin box-binding factor-1 (PBF-1) and its role in the developmental regulation of the zein multigene family. Gene 223:321–332

Wei F, Coe E, Nelson W, Bharti AK, Engler F, Butler E, Kim H, Goicoechea JL, Chen M, Lee S, Fuks G, Sanchez-Villeda H, Schroeder S, Fang Z, McMullen M, Davis G, Bowers JE, Paterson AH, Schaeffer M, Gardiner J, Cone K, Messing J, Soderlund C, Wing RA (2007) Physical and genetic structure of the maize genome reflects its complex evolutionary history. PLoS Genet 3:e123

Whitelaw CA, Barbazuk WB, Pertea G, Chan AP, Cheung F, Lee Y, Zheng L, van Heeringen S, Karamycheva S, Bennetzen JL, SanMiguel P, Lakey N, Bedell J, Yuan Y, Budiman MA, Resnick A, Van Aken S, Utterback T, Riedmuller S, Williams M, Feldblyum T, Schubert K, Beachy R, Fraser CM, Quackenbush J (2003) Enrichment of gene-coding sequences in maize by genome filtration. Science 302:2118–2120

Whitkus R, Doebley J, Lee M (1992) Comparative genome mapping of sorghum and maize. Genetics 132:1119–1130

Wilson CM, Spraque GF, Nelsen TC (1989) Linkages among zein genes determined by isoelectric focusing. Theor Appl Genet 77:217–226

Xu JH, Messing J (2006) Maize haplotype with a helitron-amplified cytidine deaminase gene copy. BMC Genet 7:52

Yim YS, Davis GL, Duru NA, Musket TA, Linton EW, Messing JW, McMullen MD, Soderlund CA, Polacco ML, Gardiner JM, Coe EH, Jr (2002) Characterization of three maize bacterial artificial chromosome libraries toward anchoring of the physical map to the genetic map using high-density bacterial artificial chromosome filter hybridization. Plant Physiol 130: 1686–1696

Yu J, Hu S, Wang J, Wong GK, Li S, Liu B, Deng Y, Dai L, Zhou Y, Zhang X, Cao M, Liu J, Sun J, Tang J, Chen Y, Huang X, Lin W, Ye C, Tong W, Cong L, Geng J, Han Y, Li L, Li W, Hu G, Li J, Liu Z, Qi Q, Li T, Wang X, Lu H, Wu T, Zhu M, Ni P, Han H, Dong W, Ren X, Feng X, Cui P, Li X, Wang H, Xu X, Zhai W, Xu Z, Zhang J, He S, Xu J, Zhang K, Zheng X, Dong J, Zeng W, Tao L, Ye J, Tan J, Chen X, He J, Liu D, Tian W, Tian C, Xia H, Bao Q, Li G, Gao H, Cao T, Zhao W, Li P, Chen W, Zhang Y, Hu J, Liu S, Yang J, Zhang G, Xiong Y, Li Z, Mao L, Zhou C, Zhu Z, Chen R, Hao B, Zheng W, Chen S, Guo W, Tao M, Zhu L, Yuan L, Yang H (2002) A draft sequence of the rice genome (*Oryza sativa* L. ssp. *indica*). Science 296:79–92

Yuan Y, SanMiguel PJ, Bennetzen JL (2003) High-Cot sequence analysis of the maize genome. Plant J 34:249–255

Chapter 16
Molecular Markers

Patrick S. Schnable, An-Ping Hsia, Ling Guo, and W. Brad Barbazuk

16.1 Utility of Molecular Markers

Molecular markers are valuable tools for basic and applied research. Any detectable type of polymorphism in proteins or DNA can potentially be used as a molecular marker. After genotyping multiple members of a population of genetically related individuals with multiple markers, it is possible to generate a genetic map. Molecular markers and genetic maps can be used to associate specific chromosomal intervals (and depending on marker density, even specific genes) with phenotypes and traits. The genetic mapping of mutants or qualitative variation facilitates marker assisted selection (MAS) and is the first step in cloning the affected genes via chromosome walking. Sequence-based genetic markers can be used to cross-link genetic, physical, and cytological maps of maize and reveal patterns of microsynteny among homeologous and orthologous chromosomal segments within the genome and syntenic relationships with other species. Genetic maps can enhance our understanding of the organization and evolution of the maize genome. They also reveal

Patrick S. Schnable
Center for Plant Genomics, Iowa State University, Ames, IA 50011, USA
e-mail: schnable@iastate.edu

An-Ping Hsia
Department of Agronomy, Iowa State University, Ames, IA 50011, USA
e-mail: hsia@iastate.edu

Ling Guo
Center for Plant Genomics, Iowa State University, Ames, IA 50011, USA
Current address: Pioneer Hi-Bred Inc., 7100 NW 62nd Avenue, P O Box 1014, Johnston, IA 50131-1014, USA
e-mail: lingguo1@gmail.com

W. Brad Barbazuk
Donald Danforth Plant Science Center, 975 N Warson Road, St. Louis, MO 63132, USA
e-mail: bbarbazuk@danforthcenter.org

A.L. Kriz, B.A. Larkins (eds.), *Molecular Genetic Approaches to Maize Improvement*
Biotechnology in Agriculture and Forestry, Vol. 63

genome-wide patterns of chromosome structure, gene distribution, and meiotic recombination.

16.2 Molecular Markers

16.2.1 Detection of Polymorphisms

A variety of techniques can be used to detect the many types of DNA polymorphisms (Kristensen et al. 2001). Each marker type has its own advantages, and the choice of a specific marker type depends on both biological and technological factors. Biological factors associated with a given marker type include the numbers of alleles per locus, mutation rates, and the distribution of polymorphisms across the genome. Technological factors include the repeatability, ease of use, throughput, and labor and cost investments required for polymorphism detection. The polymorphism detection technique of choice also depends on whether the DNA sequences associated with the polymorphism are known in advance of detection. For example, for some detection technologies it is feasible to simply survey parental lines for polymorphisms and then use the polymorphic markers for mapping. This approach is not practical for other detection technologies.

This chapter focuses on those types of DNA-based molecular markers that are currently widely used in maize, namely simple sequence repeats (SSRs), indel polymorphisms (IDPs), and single-nucleotide polymorphisms (SNPs). Only markers, maps, and populations that are freely available are discussed.

16.2.2 SSRs (Simple Sequence Repeats)

The first molecular markers used to construct maize genetic maps were restriction fragment length polymorphisms (RFLPs), which are highly polymorphic in maize (Helentjaris and Gesteland 1983). However, RFLPs were quickly supplanted by SSRs (also termed microsatellites or simple sequence length polymorphisms, SSLPs), which consist of short repeats (e.g., di- nucleotide repeats such as $(AG)_n$, where the value of n varies among alleles). Like RFLPs, SSRs are highly polymorphic and usually co-dominant. However, unlike RFLP, SSRs do not require a costly, low-throughput, and technically challenging hybridization step. Initially SSRs were often discovered by preparing and sequencing SSR-enriched libraries but are now more typically discovered by screening EST databases (e.g., Sharopova et al. 2002; Jayashree et al. 2006). Primers flanking discovered repeats are used to screen inbreds for length/size polymorphisms, which are detected via high-resolution gel electrophoresis. As of December 2007, the maize genetics database, MaizeGDB (http://www.maizegdb.org/), contained ∼2,000 mapped SSRs. In maize, SSRs have been widely used to study diversity, linkage disequilibrium, and evolution (e.g., Vigouroux et al. 2005; Stich et al. 2005; Reif et al. 2005).

16.2.3 IDPs (InDel Polymorphisms)

IDPs are co-dominant (i.e., size) and dominant (i.e., presence/absence) PCR-based polymorphisms that can be detected via agarose gel-electrophoresis. Hence, like SSRs, IDP markers can be used in virtually any molecular laboratory without the need for substantial investment in equipment or training. In addition, like SSRs, IDP markers can be developed in the absence of existing DNA sequence polymorphism data.

To do this in maize, the over 1 million maize genomic sequences of B73 maize that had been deposited in GenBank, including gene-enriched genomic survey sequences (GSSs) (Palmer et al. 2003; Whitelaw et al. 2003), BAC shotgun reads generated by the Consortium for Maize Genomics, and random whole genome shotgun (WGS) sequences generated by the Joint Genome Institute (JGI), were first assembled into maize assembled genomic islands (MAGIs) (Emrich et al. 2004; Fu et al. 2005). Within the MAGIs, gene structures were predicted based on EST alignments and using FGENESH software. Pairs of primers designed to amplify 3′ UTRs and intronic regions from the MAGIs and 3′ UTRs from the over half million public maize expressed sequence tags (ESTs) from diverse genotypes were used to screen B73 and Mo17 genomic DNA for polymorphisms that can be detected via low-resolution agarose gel-electrophoresis. Almost 4,800 IDPs have been identified in this fashion and placed on the IBM genetic map (http://magi.plantgenomics.iastate.edu/).

A significant advantage of IDPs is that they are substantially more common than SSRs. Approximately 1.5% of maize ESTs contain an SSR (Kantety et al. 2002). In contrast, ∼10% of tested primer pairs exhibit IDPs that can be detected via low-resolution agarose gel-electrophoresis (L. Guo and P.S. Schnable, unpublished results).

16.2.4 SNPs (Single Nucleotide Polymorphisms)

Although SSRs and IDPs offer significant advantages over the early types of molecular markers (e.g., RFLPs, RAPD, and AFLPs), like the early markers, their detection requires electrophoresis (for review see Schlotterer 2004). Hence, they are not readily amenable to the very high-throughput analyses required for large-scale genetic studies. In contrast, SNPs can be detected in a very high-throughput manner using hybridization to short-oligo chips or mass spectrometry (e.g., Leushner and Chiu 2000).

Compared to SSRs, SNPs exhibit increased marker data quality and quantity (Jones et al. 2007). In contrast, a set of ~90 SSRs clustered lines into populations better than did a set of ~850 SNPs (Hamblin et al. 2007). As the number of available SNPs increases the advantage is expected to shift to the SNPs.

SNPs also have the advantage of being much more plentiful than SSRs and IDPs. On average, two random maize lines exhibit one single-nucleotide polymorphism (SNP)/100 bp (Zhao et al. 2006). Even so, until recently, the significant challenge to using SNP markers in maize was SNPs discovery. It has been possible to identify limited numbers of SNPs via comparisons of ESTs from various inbreds (Batley et al. 2003), but this approach was quickly exhausted because most maize sequences in GenBank are derived from a single inbred (B73).

16.2.4.1 Temperature Gradient Capillary Electrophoresis (TGCE)-Based SNP Detection

It is possible to detect SNPs even in the absence of comparative sequence data. This is because PCR products, which are amplified from inbreds that carry alleles that differ by an SNP, form heteroduplex molecules when disassociated and allowed to re-anneal. Heteroduplexes can be detected via a variety of techniques, including an endonuclease (CELI) that cleaves heteroduplexes (McCallum et al. 2000; Colbert et al. 2001) and denaturing HPLC (dHPLC; Kota et al. 2001). Neither of these techniques is, however, well suited to genetic mapping. In contrast, temperature gradient capillary electrophoresis (TGCE; Li et al. 2002; Spectrumedix, http://www.spectrumedix.com/), which can detect heteroduplexes due to the presence of SNPs (and IDPs), has been adapted for use in genetic mapping experiments (Hsia et al. 2005; Maher et al. 2006). Unlike CELI-based heteroduplex detection systems, it is not necessary to purify samples after amplification and less PCR product is required for detection, thus minimizing reagent cost. Unlike dHPLC, individual primer pairs do not require optimization prior to TGCE. TGCE can detect a single SNP in amplicons as large as 800 bp (Hsia et al. 2005). It has a throughput of 12–24 96-well plates per day. Over 1,800 co-dominant TGCE-detected markers have been discovered and placed on the IBM genetic map (http://magi.plantgenomics.iastate.edu/).

16.2.4.2 SNP Discovery

The NSF Plant Genome project led by John Doebley (University of Wisconsin) has discovered SNPs by amplifying genes from multiple genotypes and sequencing the resulting amplicons using Sanger technology. Data are available at the project website: http://www.panzea.org/ (Zhao et al. 2006; Canaran et al. 2007). More recently, algorithmic and technological advances have made it possible to efficiently identify many thousands of SNPs.

The algorithmic advances that make it possible to identify SNPs are based on the observation that the presence of SNPs affects hybridization results obtained using short oligo microarrays such as those developed by Affymetrix. Several algorithms have been developed that extract putative SNPs from the hybridization signals obtained using different inbred lines (Rostoks et al. 2005, Cui et al. 2005; Greenhall et al. 2007; Kumari et al. 2007). For example, signals obtained by hybridizing B73 and Mo17 cDNAs to Affymetrix chips (Stupar and Springer 2006) have allowed for the identification of thousands of SNPs (H. Wu and P.S. Schnable, unpublished results). It is also possible to hybridize genomic DNA to Affymetrix chips and thereby genetically map SNPs (Kirst et al. 2006), as has been done in Arabidopsis (Borevitz et al. 2007; Kim et al. 2007). However, the repetitive content of the maize genome is likely to present challenges for such analyses (Gore et al. 2007).

Because the next generation sequencing technology from 454 Life Science (Margulies et al. 2005) makes it possible to quickly and cheaply generate vast amounts of sequence data, it is now straightforward to experimentally discover thousands of SNPs (Barbazuk et al. 2007). Maize shoot apical meristems (SAMs) from B73 and Mo17 were collected via laser capture microdissection (LCM; Ohtsu et al. 2007). Transcripts were reverse transcribed to cDNA and sequenced using 454 technology (Emrich et al. 2007). A computational pipeline that uses the PolyBayes polymorphism detection system (Marth et al. 1999) was adapted for 454 ESTs and used to detect SNPs between the two inbred lines. The deep coverage produced by this technology compensates for the somewhat higher error rate of 454 sequencing (1.5%, Emrich et al. 2007) compared to Sanger sequencing. Over 36,000 putative SNPs were detected within ~10,000 unique B73 genomic anchor sequences (MAGIs; Fu et al. 2005). Stringent post-processing reduced this number to $> 7,000$ putative SNPs. Over 85% (94/110) of a sample of these putative SNPs were validated by Sanger resequencing. Subsequently, 1,045 SNPs were validated and genetically mapped on ~300 IBM recombinant inbred lines (RILs) using mass spectrometry-based technology (H.D. Chen and P.S. Schnable, unpublished results), demonstrating that 454-based transcriptome sequencing is an excellent method for the high-throughput acquisition of gene-associated SNPs. Sequence analysis of methylation-filtered libraries (Rabinowicz et al. 1999) should make it possible to discover SNPs located outside of spliced transcripts. It is likely that this approach could also be applied to the Solexa and SOLiD systems from Illumina and ABI, respectively. The ready availability of large numbers of SNPs is expected to enable genetic association studies (Yu and Buckler 2006).

16.3 Maize Mapping Populations

The creation of high-resolution genetic maps requires access to high quality mapping populations. A variety of mapping populations have been developed for maize, including F1BC populations, F2 populations, immortalized F2 populations, RILs, and, most recently, intermated RILs. At the beginning of the molecular marker era, the most widely used public mapping populations were RILs derived from two crosses (Tx303 × CO159 and T232 × CM37; Burr and Burr 1991) and "immortalized F2s" from the Tx303×CO159 cross (Gardiner et al. 1993). Although these populations served as useful resources for the development of the first genetic maps to be populated with molecular markers, they consist of few individuals, which limits the resolution of these maps.

To address the issue of mapping resolution the intermated B73 × Mo17 (IBM) population of RILs ($n = 350$) was developed by intermating an F2 population derived from the single cross of the inbreds B73 and Mo17 for five generations prior to the extraction of recombinant inbred lines (Lee et al. 2002). As a consequence of the additional opportunities for recombination provided during the intermating generations (Beavis et al. 1992), the IBM RILs provide 17 times more mapping resolution than do previously available mapping populations (Coe et al. 2002).

16.4 Genetic Maps of Maize

Using 302 IBM IRILs, the Missouri Mapping Project (MMP) constructed a linkage map (IBM2) that contains ~2,000 markers of various types (Davis et al. 1999; Coe et al. 2002; Cone et al. 2002). About 57% of these markers are sequence based (Fu et al. 2006). Falque et al. (2005) placed ~1,500 mostly RFLP markers on an IBM-based map they termed IBM_Gnp2004. Subsequently, the MAGI team at Iowa State University produced a genetic map based on 91 IBM RILs (ISU-IBM Map4) which contains 2,029 of the MMP markers plus 1,329 IDPs (Fu et al. 2006). MaizeGDB (www.maizegdb.org) generated a map, IBM2 2005 Neighbors, that integrates IBM mapping data from IBM2, IBM_Gnp2004, and ISU-IBM Map4, as well as data from several non-IBM populations and data from the physical map. It is likely that future versions of this map will be released that incorporate mapping data that were not available at the time IBM2 2005 Neighbors was generated.

In February 2007, the MAGI team released ISU-IBM Map7, which was constructed by integrating existing markers from IBM2, IBM_Gnp2004, and ISU-IBM Map4 with 4,700 new IDPs derived from genes and predicted genes. In total, ISU-IBM Map7 contains ~9,660 molecular markers. Because many of these markers are sequence based, they can be used to integrate genetic and physical maps using sequence similarity rather than via hybridization-based approaches such as overgos. A graphical view of the ISU-IBM Map7 is available at the MAGI website (http://magi.plantgenomics.iastate.edu/). This website permits users to browse and search Map7 markers by chromosome or polymorphism type and to view each IDP marker's PCR primer and design details. Users can also use the Blast algorithm to determine whether a query sequence (or related sequence) has been mapped.

16.5 Future Perspectives

The genetic map has been an important resource for anchoring to chromosomes the BAC contigs that are being sequenced as part of the maize genome sequencing project. With the completion of the draft maize genome sequence scheduled for 2008, the genome sequence will become an important resource for analyzing recombination events. For example, it will soon be possible to explore the relationships between genetic and physical distances across thousands of intervals throughout the maize genome. In addition, during the post-genomic era, genetic markers will continue to play a critical role in helping to link phenotypes to specific genes.

References

Barbazuk WB, Emrich SJ, Chen HD, Li L, Schnable PS (2007) SNP discovery via 454 transcriptome sequencing. Plant J 51(5):910–918

Batley J, Barker G, O'Sullivan H, Edwards KJ, Edwards D (2003) Mining for single nucleotide polymorphisms and insertions/deletions in maize expressed sequence tag data. Plant Physiol 132(1):84–91

Beavis WD, Lee M, Grant D, Hallauer AR, Owens T, Katt M, Blair D (1992) The influence of random mating on recombination among RFLP loci. Maize Genet Coop Newsl (66): 52–53

Borevitz JO, Hazen SP, Michael TP, Morris GP, Baxter IR, Hu TT, Chen H, Werner JD, Nordborg M, Salt DE, Kay SA, Chory J, Weigel D, Jones JD, Ecker JR (2007) Genome-wide patterns of single-feature polymorphism in *Arabidopsis thaliana*. Proc Natl Acad Sci USA 104(29): 12057–12062

Burr B, Burr FA (1991) Recombinant inbreds for molecular mapping in maize: theoretical and practical considerations. Trends Genet 7(2):55–60

Canaran P, Buckler ES, Glaubitz JC, Stein L, Sun Q, Zhao W, Ware D (2007) Panzea: an update on new content and features. Nucleic Acids Res Nov 19 (Epub ahead of print)

Coe E, Cone K, McMullen M, Chen SS, Davis G, Gardiner J, Liscum E, Polacco M, Paterson A, Sanchez-Villeda H, Soderlund C, Wing R (2002) Access to the maize genome: an integrated physical and genetic map. Plant Physiol 128(1):9–12

Colbert T, Till BJ, Tompa R, Reynolds S, Steine MN, Yeung AT, McCallum CM, Comai L, Henikoff S (2001) High-throughput screening for induced point mutations. Plant Physiol 126(2):480–484

Cone KC, McMullen MD, Bi IV, Davis GL, Yim YS, Gardiner JM, Polacco ML, Sanchez-Villeda H, Fang Z, Schroeder SG, Havermann SA, Bowers JE, Paterson AH, Soderlund CA, Engler FW, Wing RA, Coe EH, Jr (2002) Genetic, physical, and informatics resources for maize. On the road to an integrated map. Plant Physiol 130(4):1598–1605

Cui X, Xu J, Asghar R, Condamine P, Svensson JT, Wanamaker S, Stein N, Roose M, Close TJ (2005) Detecting single-feature polymorphisms using oligonucleotide arrays and robustified projection pursuit. Bioinformatics 21(20):3852–3858

Davis GL, McMullen MD, Baysdorfer C, Musket T, Grant D, Staebell M, Xu G, Polacco M, Koster L, Melia-Hancock S, Houchins K, Chao S, Coe EH, Jr (1999) A maize map standard with sequenced core markers, grass genome reference points and 932 expressed sequence tagged sites (ESTs) in a 1736-locus map. Genetics 152(3):1137–1172

Emrich SJ, Aluru S, Fu Y, Wen TJ, Narayanan M, Guo L, Ashlock DA, Schnable PS (2004) A strategy for assembling the maize (*Zea mays* L.) genome. Bioinformatics 20(2):140–147

Emrich SJ, Barbazuk WB, Li L, Schnable PS (2007) Gene discovery and annotation using LCM-454 transcriptome sequencing. Genome Res 17(1):69–73

Falque M, Decousset L, Dervins D, Jacob AM, Joets J, Martinant JP, Raffoux X, Ribiere N, Ridel C, Samson D, Charcosset A, Murigneux A (2005) Linkage mapping of 1454 new maize candidate gene loci. Genetics 170(4):1957–1966

Fu Y, Emrich SJ, Guo L, Wen TJ, Ashlock DA, Aluru S, Schnable PS (2005) Quality assessment of maize assembled genomic islands (MAGIs) and large-scale experimental verification of predicted genes. Proc Natl Acad Sci USA 102(34):12282–12287

Fu Y, Wen TJ, Ronin YI, Chen HD, Guo L, Mester DI, Yang Y, Lee M, Korol AB, Ashlock DA, Schnable PS (2006) Genetic dissection of intermated recombinant inbred lines using a new genetic map of maize. Genetics 174(3):1671–1683

Gardiner JM, Coe EH, Melia-Hancock S, Hoisington DA, Chao S (1993) Development of a core RFLP map in maize using an immortalized F2 population. Genetics 134(3):917–930

Gore M, Bradbury P, Hogers R, Kirst M, Verstege E, Van Oeveren J, Peleman J, Buckler E, Van Eijk M (2007) Evaluation of target preparation methods for single-feature polymorphism detection in large complex plant genomes. Crop Sci 47(2):S135–S148

Greenhall JA, Zapala MA, Caceres M, Libiger O, Barlow C, Schork NJ, Lockhart DJ (2007) Detecting genetic variation in microarray expression data. Genome Res 17(8):1228–1235

Hamblin MT, Warburton ML, Buckler ES (2007) Empirical comparison of simple sequence repeats and single nucleotide polymorphisms in assessment of maize diversity and relatedness. PLoS ONE 2(12):e1367

Helentjaris T, Gesteland R (1983) Evaluation of random cDNA clones as probes for human restriction fragment polymorphisms. J Mol Appl Genet 2(2):237–247

Hsia AP, Wen TJ, Chen HD, Liu Z, Yandeau-Nelson MD, Wei Y, Guo L, Schnable PS (2005) Temperature gradient capillary electrophoresis (TGCE) – a tool for the high-throughput discovery and mapping of SNPs and IDPs. Theor Appl Genet 111(2):218–225

Jayashree B, Punna R, Prasad P, Bantte K, Hash CT, Chandra S, Hoisington DA, Varshney RK (2006) A database of simple sequence repeats from cereal and legume expressed sequence tags mined in silico: survey and evaluation. In Silico Biol 6(6):607–620

Jones ES, Sullivan H, Bhattramakki D, Smith JS (2007) A comparison of simple sequence repeat and single nucleotide polymorphism marker technologies for the genotypic analysis of maize (*Zea mays* L.). Theor Appl Genet 115(3):361–371

Kantety RV, La Rota M, Matthews DE, Sorrells ME (2002) Data mining for simple sequence repeats in expressed sequence tags from barley, maize, rice, sorghum and wheat. Plant Mol Biol 48(5–6):501–510

Kim S, Plagnol V, Hu TT, Toomajian C, Clark RM, Ossowski S, Ecker Jr, Weigel D, Nordborg M (2007) Recombination and linkage disequilibrium in *Arabidopsis thaliana*. Nat Genet 39(9):1151–1155

Kirst M, Caldo R, Casati P, Tanimoto G, Walbot V, Wise RP, Buckler ES (2006) Genetic diversity contribution to errors in short oligonucleotide microarray analysis. Plant Biotechnol J 4(5):489–498

Kota R, Wolf M, Michalek W, Graner A (2001) Application of denaturing high-performance liquid chromatography for mapping of single nucleotide polymorphisms in barley (*Hordeum vulgare* L.). Genome 44(4):523–528

Kristensen VN, Kelefiotis D, Kristensen T, Borresen-Dale AL (2001) High-throughput methods for detection of genetic variation. Biotechniques 30(2):318–322, 324, 326–332

Kumari S, Verma LK, Weller JW (2007) AffyMAPSDetector: a software tool to characterize Affymetrix GeneChip expression arrays with respect to SNPs. BMC Bioinformatics 8:276

Lee M, Sharopova N, Beavis WD, Grant D, Katt M, Blair D, Hallauer A (2002) Expanding the genetic map of maize with the intermated B73 $\times$ Mo17 (IBM) population. Plant Mol Biol 48(5–6):453–461

Leushner J, Chiu NH (2000) Automated mass spectrometry: a revolutionary technology for clinical diagnostics. Mol Diagn 5(4):341–348

Li Q, Liu Z, Monroe H, Culiat CT (2002) Integrated platform for detection of DNA sequence variants using capillary array electrophoresis. Electrophoresis 23(10):1499–1511

Maher PM, Chou HH, Hahn E, Wen TJ, Schnable PS (2006) GRAMA: genetic mapping analysis of temperature gradient capillary electrophoresis data. Theor Appl Genet 113(1): 156–162

Margulies M, Egholm M, Altman WE, Attiya S, Bader JS, Bemben LA, Berka J, Braverman MS, Chen YJ, Chen Z, Dewell SB, Du L, Fierro JM, Gomes XV, Godwin BC, He W, Helgesen S, Ho CH, Irzyk GP, Jando SC, Alenquer ML, Jarvie TP, Jirage KB, Kim JB, Knight JR, Lanza JR, Leamon JH, Lefkowitz SM, Lei M, Li J, Lohman KL, Lu H, Makhijani VB, Mcdade KE, Mckenna MP, Myers EW, Nickerson E, Nobile JR, Plant R, Puc BP, Ronan MT, Roth GT, Sarkis GJ, Simons JF, Simpson JW, Srinivasan M, Tartaro KR, Tomasz A, Vogt KA, Volkmer GA, Wang SH, Wang Y, Weiner MP, Yu P, Begley RF, Rothberg JM (2005) Genome sequencing in microfabricated high-density picolitre reactors. Nature 437(7057): 376–380

Marth GT, Korf I, Yandell MD, Yeh RT, Gu Z, Zakeri H, Stitziel NO, Hillier L, Kwok PY, Gish WR (1999) A general approach to single-nucleotide polymorphism discovery. Nat Genet 23(4): 452–456

McCallum CM, Comai L, Greene EA, Henikoff S (2000) Targeted screening for induced mutations. Nat Biotechnol 18(4):455–457

Ohtsu K, Smith MB, Emrich SJ, Borsuk LA, Zhou R, Chen T, Zhang X, Timmermans MC, Beck J, Buckner B, Janick-Buckner D, Nettleton D, Scanlon MJ, Schnable PS (2007) Global gene expression analysis of the shoot apical meristem of maize (*Zea mays* L.). Plant J 52(3):391–404

Palmer LE, Rabinowicz PD, O'Shaughnessy AL, Balija VS, Nascimento LU, Dike S, de la Bastide M, Martienssen RA, McCombie WR (2003) Maize genome sequencing by methylation filtration. Science 302(5653):2115–2117

Rabinowicz PD, Schutz K, Dedhia N, Yordan C, Parnell LD, Stein L, McCombie WR, Martienssen RA (1999) Differential methylation of genes and retrotransposons facilitates shotgun sequencing of the maize genome. Nat Genet 23(3):305–308

Reif JC, Hamrit S, Heckenberger M, Schipprack W, Peter Maurer H, Bohn M, Melchinger AE (2005) Genetic structure and diversity of European flint maize populations determined with SSR analyses of individuals and bulks. Theor Appl Genet 111(5):906–913

Rostoks N, Borevitz JO, Hedley PE, Russell J, Mudie S, Morris J, Cardle L, Marshall DF, Waugh R (2005) Single-feature polymorphism discovery in the barley transcriptome. Genome Biol 6(6):R54

Schlotterer C (2004) The evolution of molecular markers – just a matter of fashion? Nat Rev Genet 5(1):63–69

Sharopova N, McMullen MD, Schultz L, Schroeder S, Sanchez-Villeda H, Gardiner J, Bergstrom D, Houchins K, Melia-Hancock S, Musket T, Duru N, Polacco M, Edwards K, Ruff T, Register JC, Brouwer C, Thompson R, Velasco R, Chin E, Lee M, Woodman-Clikeman W, Long MJ, Liscum E, Cone K, Davis G, Coe EH, Jr (2002) Development and mapping of SSR markers for maize. Plant Mol Biol 48(5–6):463–481

Stich B, Melchinger AE, Frisch M, Maurer HP, Heckenberger M, Reif JC (2005) Linkage disequilibrium in European elite maize germplasm investigated with SSRs. Theor Appl Genet 111(4):723–730

Stupar RM, Springer NM (2006) Cis-transcriptional variation in maize inbred lines B73 and Mo17 leads to additive expression patterns in the F1 hybrid. Genetics 173(4):2199–2210

Vigouroux Y, Mitchell S, Matsuoka Y, Hamblin M, Kresovich S, Smith JS, Jaqueth J, Smith OS, Doebley J (2005) An analysis of genetic diversity across the maize genome using microsatellites. Genetics 169(3):1617–1630

Whitelaw CA, Barbazuk WB, Pertea G, Chan AP, Cheung F, Lee Y, Zheng L, Van Heeringen S, Karamycheva S, Bennetzen JL, SanMiguel P, Lakey N, Bedell J, Yuan Y, Budiman MA, Resnick A, Van Aken S, Utterback T, Riedmuller S, Williams M, Feldblyum T, Schubert K, Beachy R,

Fraser CM, Quackenbush J (2003) Enrichment of gene-coding sequences in maize by genome filtration. Science 302(5653):2118–2120

Yu J, Buckler ES (2006) Genetic association mapping and genome organization of maize. Curr Opin Biotechnol 17(2):155–160

Zhao W, Canaran P, Jurkuta R, Fulton T, Glaubitz J, Buckler E, Doebley J, Gaut B, Goodman M, Holland J, Kresovich S, McMullen M, Stein L, Ware D (2006) Panzea: a database and resource for molecular and functional diversity in the maize genome. Nucleic Acids Res 34 (Database issue):D752–D757

Chapter 17
Applied Cytogenetics

R. Kelly Dawe

17.1 Chromosome Analysis on Mitotic Chromosome by FISH

Historically, most cytogenetics was carried out on (meiotic) pachytene cells where individual chromosomes can be readily identified (e.g. Anderson et al. 2004). A weakness of pachytene analysis is that whole plants must be grown to near maturity to collect samples. Root tip chromosomes offer a simpler way to collect chromosome information, but they have been viewed as too small to accurately identify chromosome variants and cytological features.

The power of mitotic chromosome analysis changed dramatically with the discovery of new FISH methods to label and identify root tip chromosomes. Birchler and colleagues showed that by mixing a collection of repetitive probes labeled with differently colored tags (fluorophores) it was possible to rapidly identify all ten maize chromosomes (Kato et al. 2004). They also introduced an important nitrous oxide method for increasing the number of condensed chromosomes from a single root tip. Subsequently, the same group went on to show that the sensitivity of FISH could be increased dramatically by increasing the amount of a key enzyme (DNA polymerase) in the labeling protocol (Kato et al. 2006).

Several important advances have been made with this new technology. Perhaps most importantly, two groups have convincingly accomplished single-gene localization for targets as small as 3 kb. These efforts have considerably improved cytological maps (Wang et al. 2006; Amarillo FI and Bass 2007; Lamb et al. 2007b). Others have used the ease of chromosome painting to show that retroelement abundance and content differs significantly between *Zea mays* and *Zea diploperennis* and between both *Zea* species and *Tripsacum* (Lamb and Birchler 2006), and to show that the *Robertson's Mutator* element moves by a cut and paste mechanism (Yu et al. 2007a). The technology has also been heavily used in the process of

R. Kelly Dawe
Department of Plant Biology, Miller Plant Sciences Bldg, University of Georgia, Athens, GA 30602, USA
e-mail: kelly@plantbio.uga.edu

A.L. Kriz, B.A. Larkins (eds.), *Molecular Genetic Approaches to Maize Improvement* 241
Biotechnology in Agriculture and Forestry, Vol. 63

producing an engineerable minichromosome (discussed below; Yu et al. 2006; Han et al. 2007).

17.2 Histones, ChIP, Genes and Histones

It has been suspected for many years that nucleosomes and chromatin packaging underlie many of the visible characteristics of chromosomes. However, the tools were not available to understand the relationship. This changed with discoveries in yeast and animals that tied gene expression to core histones, and gradually expanded into a broad framework that relates simple histone modifications to complex events underlying gene inactivation and heterochromatin formation (Strahl and Allis 2000; Loidl 2004; Matzke and Matzke 2004).

In the simplest cases, repression of transcription can be tied to methylation of lysine 9 and 27 on histone H3, and gene activation can be tied to methylation of lysine 4 and acetylation at several residues (Loidl 2004). The basic concept is remarkably well conserved, but there are important variations among species, primarily with respect to how many methyl groups are present on the lysines (Strahl and Allis 2000; Loidl 2004; Ebert et al. 2006). There are also clear differences between small and large genome plants (Houben et al. 2003; Shi and Dawe 2006; Zhang et al. 2007). Maize transposable elements (Tes) are distributed throughout the genome, with different classes targeting particular 'niches', such as intergenic spaces, promoters, or introns (Feschotte et al. 2002). Correspondingly, the intergenic spaces of maize are condensed (as chromomeres) and rich in 'off' marks (Shi and Dawe 2006), while the gene space contains a mixture of 'on' and 'off' marks (Shi and Dawe 2006; Haring et al. 2007).

The technique of chromatin immunoprecipitation (ChIP) is widely used to interpret chromatin structure around known sequences. In this method, chromatin is purified and incubated with anti-histone antisera (Zhong et al. 2002; Haring et al. 2007). The nucleosome–antibody complexes are precipitated and the associated DNA characterized by PCR. In other species, ChIP samples have been hybridized to arrays to reveal the profiles of histone modification within genes and promoters (Mito et al. 2007; Zhang et al. 2007). However, ChIP methods have only recently been worked out for single copy genes in maize (Luce et al. 2006; Haring et al. 2007; Hernandez et al. 2007). The data reveal that methylation of lysine 4 and acetylation of other lysines mark gene activity, whereas a mixture of presumed 'on' and 'off' marks can be present on inactive genes and repetitive domains (Alvarez-Venegas and Avramova 2005; Earley et al. 2007; Danker et al. 2008). This variation probably reflects the fact that TEs may or may not be close to a gene, and the fact that some TEs are actively transcribed (Kashkush and Khasdan 2007). Whether the histone profile around single genes and TEs is consistent within inbreds and tissues is not yet known. More detailed analyses are sure to arrive once the complete genome sequence is available.

17.3 Centromere Cytogenetics

While the vast majority of the genome contains the canonical histone H3 (H3.1), there are also important histone variants known as H3.3 and CENH3. Histone H3.3 is incorporated into actively transcribed genes as a replacement histone (since nucleosomes are disassembled during transcription; Mito et al. 2005). In contrast, the histone variant CENH3 is only found at active centromeres (Zhong et al. 2002; Yan et al. 2006). The N-terminal tail on H3.1 that is heavily modified is entirely different on CENH3 – in fact, it appears to have rapidly evolved away from any relationship to H3.1 (Malik and Henikoff 2001; Zhong et al. 2002). The CENH3 tails among different species have no homology, and at least in some cases entirely different CENH3s can be substituted for each other (Wieland et al. 2004). CENH3 appears to serve as a 'blank slate' that rejects H3-associated binding proteins and, perhaps by default, recruits kinetochore proteins.

Centromeres in most species can be identified by the absence of genes and the presence of tandem repeats. In maize the primary repeat is CentC, which is present in long arrays. In maize and other cereals the centromeres are also rich in a novel class of transposons known as centromeric retroelements (CRM). CRMs of several subclasses can be extremely abundant and densely nested into each other. Indeed, on some chromosomes it may be more accurate to say that maize centromeres are composed of CRMs interspersed with occasional CentC arrays. Although the same basic repeats are present at all maize centromeres, the overall size and arrangement of repeats is highly variable. Centromere 7 was measured in excess of 3 Mb, but the average is probably closer to 700 kb. On centromeres like 7, the occupied kinetochore domain must be much smaller than the foundation of repeats, and, conversely, the pericentromeric domains contain vast arrays of repeats that would normally be referred to as centromeric.

Under such conditions where the type and size of repeats does not dictate kinetochore size, the centromeres must be defined by epigenetic mechanisms. This has been widely confirmed in many species (Dawe 2005). ChIP analysis and extended fiber studies have established that while all kinetochores form over CentC and CRM arrays, neither repeat is always associated with CENH3 (Zhong et al. 2002; Jin et al. 2004). Similarly in rice, the kinetochores are invariably found over arrays of the CRM homologue (CRR), but there are many CRR elements that do not associate with CENH3 (Yan et al. 2005, 2006). Further, it appears that centromeres can be entirely removed and replaced with new sequences. In barley, an ancient centromere was removed by deletion and a new centromere formed in flanking DNA that had no obvious similarities to the other centromeres (Nasuda et al. 2005). A similar phenomenon was described for maize centromere 8. During a routine inspection by FISH, the authors found a maize line that appeared to have two centromere domains (Lamb et al. 2007a). This event was caused or associated with an inversion on the long arm. Although only a small part of the large centromeric-repeat domain was split off, the inversion moved the entire functional centromere including the primary constriction. When antibodies to CENH3 were used, the authors discovered that the new centromere was interacting with the spindle while the original centromere had been inactivated. These and other data support the view that centromeres are not defined by DNA sequences, but by the proteins that bind to them (Karpen et al. 1997; Choo 2001).

17.4 Minichromosomes – Using Cytogenetics to Produce a Better Vector

17.4.1 Background

The facts that centromeres dictate chromosome movement and are easily visible under a microscope make them a natural focus for basic chromosome research. However, several recent studies have focused on a more applied rationale – using centromeres as a cornerstone for developing new plant transformation vectors.

These efforts have closely followed similar studies in animals (Harrington et al. 1997; Basu and Willard 2005; Suzuki et al. 2006). There are two basic strategies in use. The 'top down' approach involves cloning (or otherwise recreating) centromeric arrays and transforming them into cells. It is based on the idea that unique repeats recruit key inner kinetochores by base-pair-level DNA specificity, akin in concept to the binding of transcription factors to promoters. The technique and the assumptions appear to be at least partly true in animals, but only for a limited number of cell lines (Irvine et al. 2005).

The 'top down' strategy assumes that centromeres are difficult to assemble and it is more effective to use an existing centromere. This idea is based on the view that centromere specification is largely sequence independent (Choo 2001). From a practical perspective, the top down approach involves first modifying an existing chromosome so that a native centromere is separated from the bulk of its chromosome arm(s). Generally, the engineered chromosome is supernumerary or functionally trisomic. The arm-trimming step can be accomplished by inserting telomere repeats, which when present on a chromosome arm tend to break the chromosomes at the integration site (Farr et al. 1991). After trimming, the minichromosome is modified by some form of site-directed recombination (Lim and Farr 2004). An important strength of using a known exchange site is that much of the expression variation associated with transformation can (in principle) be avoided.

17.4.2 Efforts in Plants

Efforts towards 'bottom up' and 'top down' maize artificial chromosomes have been recently published (Carlson et al. 2007; Houben and Schubert 2007; Yu et al. 2007b). The 'bottom up' effort roughly followed the method used in humans (Carlson et al. 2007). Basically, the authors cloned a variety of repetitive regions into BACs and then transformed and screened for plants with independently segregating artificial chromosomes. The selectable markers were transmitted through meiosis at remarkably high frequencies, in most cases at Mendelian levels. Sequences as diverse as knob repeats showed the same properties as sequences from centromeric regions. Small, dot-like chromosomes were routinely observed in metaphase spreads, and for one BAC the authors concluded that a single ∼30-kb plasmid was segregating autonomously and carrying the selectable marker with it. However, these data are difficult to reconcile, with extensive evidence showing that plant centromeres are established and maintained by epigenetic processes (which would be erased following passage through bacteria) and with a very large literature showing that small chromosomes are unstable in meiosis (discussed below). The FISH methodology used in the Carlson study has also been questioned (Houben et al. 2008). Further studies will be necessary to show that the minichromosomes contain CENH3 and to confirm that they segregate independently of other chromosomes.

A top down method based on the B chromosome may be more appropriate for maize (Yu et al. 2007b). The B chromosome is ideal for use as a minichromosome since it evolved to segregate effectively as a single (univalent) chromosome. It is also dispensable and has few if any functional genes (Carlson 1986). The current strategy began with the demonstration that interstitial telomere repeats are remarkably efficient at inducing chromosome breakage in maize (Yu et al. 2006). In a subsequent paper, several tiny B chromosomes containing almost no detectable sign of the long arm were described (Yu et al. 2007b). Segregation data showed that the minichromosomes were heritable, though at reduced levels (12–39%). Yu and coworkers also demonstrated that a site-specific (Lox) exchange site can mediate gene insertion at the truncation construct. By crossing a second transgenic line containing a Lox-flanked marker gene and the necessary Cre recombinase, they showed that the marker was transferred to the minichromosome. We can expect next-generation truncation constructs to contain more robust insertion technology. For instance, it should be possible to add a series of different exchange sites that can be engineered to add multiple useful genes.

Since B chromosomes are not present in other major crop plants (soybean, wheat), strategies for engineering A chromosomes will be required if the technology is to extend beyond maize. To this end, the same group showed that a minichromosome from chromosome 7 could be created using telomere truncation. The tiny chromosome initially arose in a spontaneous polyploid. Indeed, a polyploid of some form is a prerequisite for this approach, since an arm-trimmed chromosome will not be transmitted unless a second homologous chromosome complements the missing sections. It is also important that the bulk of the existing genetic information on the engineered chromosome be deleted; genes that are not removed will be included on the vector molecule and affect the final product. In most cases this will be an arduous process. However, once a minichromosome is created it can be used for a variety of downstream applications.

17.4.3 Limitations and Outlook for Engineered Chromosomes

A major concern with minichromosomes is that they generally do not segregate in a Mendelian manner, either in the single (monosomic, univalent) or homozygous (disomic) state (Dawe 1998). Indeed, the same is true for any otherwise normal chromosome that does not pair and recombine with a homologue.

At issue is the mechanism of chromosome alignment, which relies on chiasmata. Recombination and chiasmata provide the only attachments between chromosomes in the first division – tethers that allow homologues to faithfully swing away from each other and find opposite poles. Once stable connections are made, the chiasmata are broken and the homologues separate. If there are no chiasmata, the univalents generally align along their sister chromatids and separate precociously. Although single chromatids will often make it to a pole in the first division, in the second division there is nothing to divide, and the chromatid will often stay at the spindle midzone. The problem will not normally be remedied by producing disomic individuals. Even when a small chromosome has a pairing partner, recombination usually fails since the small chromosome arms cannot mediate recombination. This produces not one but two unpaired chromosomes that cannot orient properly. Some will segregate to pollen mother cells, but many will end up in the cytoplasm and degrade (Dawe 1998).

These expectations were born out in a recent study of the pairing, segregation, and transmission behavior of 22 small B chromosome derivatives. The authors observed severely sub-Mendelian segregation for all chromosomes in both the monosomic and disomic states (Han et al. 2007). Since size and pairing behavior did not correlate well with recovery of the chromosome, the authors ascribed the reduced transmission to qualitative differences in pericentromeric sequences. The reported transmission levels of 10–30% are viewed as workable, though are no doubt a challenge. It is likely that modest increases in transmission can be achieved by selecting for the best lines. A production-level product would presumably include some form of selection that ensures only the minichromosome pollen survives (Han et al. 2007). As a monoecious plant that produces vast excesses of pollen, maize certainly allows for further engineering at this level.

Although there are many obstacles yet to overcome in chromosome engineering, we should not underestimate the importance of these new developments, both for their potential in crop improvement and as yet another reminder of the power of cytogenetics.

References

Alvarez-Venegas R, Avramova Z (2005) Methylation patterns of histone H3 Lys 4, Lys 9 and Lys 27 in transcriptionally active and inactive Arabidopsis genes and in atx1 mutants. Nucleic Acids Res 33:5199–5207

Amarillo FI, Bass HW (2007) A transgenomic cytogenetic sorghum (*Sorghum propinquum*) bacterial artificial chromosome fluorescence in situ hybridization map of maize (*Zea mays* L.) pachytene chromosome 9, evidence for regions of genome hyperexpansion. Genetics 177:1509–1526

Anderson LK, Salameh N, Bass HW, Harper LC, Cande WZ, Weber G, Stack SM (2004) Integrating genetic linkage maps with pachytene chromosome structure in maize. Genetics 166:1923–1933

Basu J, Willard HF (2005) Artificial and engineered chromosomes: non-integrating vectors for gene therapy. Trends Mol Med 11:251–258

Carlson SR, Rudgers GW, Zieler H, Mach JM, Luo S, Grunden E, Krol C, Copenhaver GP, Preuss D (2007) Meiotic transmission of an in vitro-assembled autonomous maize minichromosome. PLOS Genet 3:1965–1974

Carlson WR (1986) The B-chromosome of maize. Crit Rev Plant Sci 3:201–226

Choo KH (2001) Domain organization at the centromere and neocentromere. Dev Cell 1: 165–177

Danker T, Dreesen B, Offermann S, Horst I, Peterhansel C (2008) Developmental information but not promoter activity controls the methylation state of histone H3 lysine 4 on two photosynthetic genes in maize. Plant J 53(3):456–474

Dawe RK (1998) Meiotic chromosome organization and segregation in plants. Ann Rev Plant Phys Plant Mol Biol 49:371–395

Dawe RK (2005) Centromere renewal and replacement in the plant kingdom. Proc Natl Acad Sci USA 102:11573–11574

Earley KW, Shook MS, Brower-Toland B, Hicks L, Pikaard CS (2007) In vitro specificities of Arabidopsis co-activator histone acetyltransferases: implications for histone hyperacetylation in gene activation. Plant J 52:615–626

Ebert A, Lein S, Schotta G, Reuter G (2006) Histone modification and the control of heterochromatic gene silencing in *Drosophila*. Chromosome Res 14:377–392

Farr C, Fantes J, Goodfellow P, Cooke H (1991) Functional reintroduction of human telomeres into mammalian cells. Proc Natl Acad Sci USA 88:7006–7010

Feschotte C, Jiang N, Wessler SR (2002) Plant transposable elements: where genetics meets genomics. Nat Rev Genet 3:329–341

Han F, Gao Z, Yu W, Birchler JA (2007) Minichromosome analysis of chromosome pairing, disjunction, and sister chromatid cohesion in maize. Plant Cell 19:3853–3863

Haring M, Offermann S, Danker T, Horst I, Peterhansel C, Stam M (2007) Chromatin immunoprecipitation: optimization, quantitative analysis and data normalization. Plant Methods 3:11

Harrington JJ, Bokkelen GV, Mays RW, Gustashaw K, Willard HF (1997) Formation of de novo centromeres and construction of first-generation human artificial chromosomes. Nature Genetics 15:345–355

Hernandez JM, Feller A, Morohashi K, Frame K, Grotewold E (2007) The basic helix-loop-helix domain of maize R links transcriptional regulation and histone modifications by recruitment of an EMSY-related factor. Proc Natl Acad Sci USA 104:17222–17227

Houben A, Schubert I (2007) Engineered plant minichromosomes: a resurrection of B chromosomes? Plant Cell 19:2323–2327

Houben A, Demidov D, Gernand D, Meister A, Leach CR, Schubert I (2003) Methylation of histone H3 in euchromatin of plant chromosomes depends on basic nuclear DNA content. Plant J 33:967–973

Houben A, Dawe R, Jiang J, Schubert I (2008) Engineered plant minichromosomes – a bottom-up success? Plant Cell 20:8–10

Irvine DV, Shaw ML, Choo KH, Saffery R (2005) Engineering chromosomes for delivery of therapeutic genes. Trends Biotechnol 23:575–583

Jin W, Melo JR, Nagaki K, Talbert PB, Henikoff S, Dawe RK, Jiang J (2004) Maize centromeres: organization and functional adaptation in the genetic background of oat. Plant Cell 16:571–581

Karpen GH, Allshire RC (1997) The case of epigenetic effects on centromere identity and function. Trends Genet 13:489–496

Kashkush K, Khasdan V (2007) Large-scale survey of cytosine methylation of retrotransposons and the impact of readout transcription from long terminal repeats on expression of adjacent rice genes. Genetics 177:1975–1985

Kato A, Lamb JC, Birchler JA (2004) Chromosome painting using repetitive DNA sequences as probes for somatic chromosome identification in maize. Proc Natl Acad Sci USA 101:13554–13559

Kato A, Albert PS, Vega JM, Birchler JA (2006) Sensitive fluorescence in situ hybridization signal detection in maize using directly labeled probes produced by high concentration DNA polymerase nick translation. Biotech Histochem 81:71–78

Lamb JC, Birchler JA (2006) Retroelement genome painting: cytological visualization of retroelement expansions in the genera *Zea* and *Tripsacum*. Genetics 173:1007–1021

Lamb JC, Meyer JM, Birchler JA (2007a) A hemicentric inversion in the maize line knobless Tama flint created two sites of centromeric elements and moved the kinetochore-forming region. Chromosoma 116:237–247

Lamb JC, Danilova T, Bauer MJ, Meyer JM, Holland JJ, Jensen MD, Birchler JA (2007b) Single-gene detection and karyotyping using small-target fluorescence in situ hybridization on maize somatic chromosomes. Genetics 175:1047–1058

Lim HN, Farr CJ (2004) Chromosome-based vectors for Mammalian cells: an overview. Methods Mol Biol 240:167–186

Loidl P (2004) A plant dialect of the histone language. Trends Plant Sci 9:84–90

Luce AC, Sharma A, Mollere OS, Wolfgruber TK, Nagaki K, Jiang J, Presting GG, Dawe RK (2006) Precise centromere mapping using a combination of repeat junction markers and chromatin immunoprecipitation-polymerase chain reaction. Genetics 174:1057–1061

Malik HS, Henikoff S (2001) Adaptive evolution of Cid, a centromere-specific histone in *Drosophila*. Genetics 157:1293–1298

Matzke MA, Matzke AJM (2004) Planting the seeds of a new paradigm. PLOS Biol 2:582–586

Mito Y, Henikoff JG, Henikoff S (2005) Genome-scale profiling of histone H3.3 replacement patterns. Nat Genetics 37:1090–1097

Mito Y, Henikoff JG, Henikoff S (2007) Histone replacement marks the boundaries of cis-regulatory domains. Science 315:1408–1411

Nasuda S, Hudakova S, Schubert I, Houben A, Endo TR (2005) Stable barley chromosomes without centromeric repeats. Proc Natl Acad Sci USA 102:9842–9847

Shi J, Dawe K (2006) Partitioning of the maize epigenome by the number of methyl groups on histone H3 lysines 9 and 27. Genetics 173:1571–1583

Strahl BD, Allis CD (2000) The language of covalent histone modifications. Nature 403: 41–45

Suzuki N, Nishii K, Okazaki T, Ikeno M (2006) Human artificial chromosomes constructed using the bottom-up strategy are stably maintained in mitosis and efficiently transmissible to progeny mice. J Biol Chem 281:26615–26623

Wang CJ, Harper L, Cande WZ (2006) High-resolution single-copy gene fluorescence in situ hybridization and its use in the construction of a cytogenetic map of maize chromosome 9. Plant Cell 18:529–544

Wieland G, Orthaus S, Ohndorf S, Diekmann S, Hemmerich P (2004) Functional complementation of human centromere protein A (CENP-A) by Cse4p from *Saccharomyces cerevisiae*. Mol Cell Biol 24:6620–6630

Yan H, Jin W, Nagaki K, Tian S, Ouyang S, Buell CR, Talbert PB, Henikoff S, Jiang J (2005) Transcription and histone modifications in the recombination-free region spanning a rice centromere. Plant Cell 17:3227–3238

Yan HH, Ito H, Nobuta K, Ouyang S, Jin WW, Tian SL, Lu C, Venu RC, Wang GL, Green PJ, Wing RA, Buell CR, Meyers BC, Jiang JM (2006) Genomic and genetic characterization of rice Cen3 reveals extensive transcription and evolutionary implications of a complex centromere. Plant Cell 18:2123–2133

Yu W, Lamb JC, Han F, Birchler JA (2006) Telomere-mediated chromosomal truncation in maize. Proc Natl Acad Sci USA 103:17331–17336

Yu W, Lamb JC, Han F, Birchler JA (2007a) Cytological visualization of DNA transposons and their transposition pattern in somatic cells of maize. Genetics 175:31–39

Yu W, Han F, Gao Z, Vega JM, Birchler JA (2007b) Construction and behavior of engineered minichromosomes in maize. Proc Natl Acad Sci USA 104:8924–8929

Zhang XY, Clarenz O, Cokus S, Bernatavichute YV, Pellegrini M, Goodrich J, Jacobsen SE (2007) Whole-genome analysis of histone H3 lysine 27 trimethylation in Arabidopsis. PLOS Biol 5:1026–1035

Zhong CX, Marshall JB, Topp C, Mroczek R, Kato A, Nagaki K, Birchler JA, Jiang J, Dawe RK (2002) Centromeric retroelements and satellites interact with maize kinetochore protein CENH3. Plant Cell 14:2825–2836

Chapter 18
The Wonderland of Global Expression Profiling

David W. Galbraith

18.1 Introduction

18.1.1 Some Definitions

"When I use a word," Humpty Dumpty said in rather a scornful tone, "it means just what I choose it to mean – neither more nor less."

One wonders what Humpty might have said about compound words, such as "expression profiling", for example. Only a few years ago, life was simple: rapid progress was being made towards defining the identities of specific genes, and the development of high-throughput methods for simultaneously analyzing genes in parallel led to the development of the concept of expression profiles which could uniquely represent the individual cell types and cellular states found within tissues and organs (Hughes et al. 2000). However, more recently, the primary concepts underpinning our understanding of the meaning of gene expression have blurred somewhat. Defining the physical bounds for a gene, for example, is no longer a question simply of determining mileposts up and downstream from the point of initiation of transcription, since these may be modulated and influenced by chromosomal context and chromatin state. Again, defining expression no longer solely reflects the concept of translation of the encoded information content of an mRNA into its cognate protein; we must include in our understanding the concepts of RNA processing, transport, sequestration, and degradation, and the roles of genes producing transcripts that are non-coding. Finally, it is becoming clear that specific sequences of DNA can interact via a number of mechanisms that influence expression, including antisense transcription, the occurrence of alternative splicing, and the production of small RNA molecules. The complexities can be maddening, as Alice in Wonderland

David W. Galbraith
BIO5 Institute for Collaborative Bioresearch and Department of Plant Sciences, University of Arizona, Thomas W. Keating Bioresearch Building, 1657 E. Helen St., Tucson, AZ 85721, USA
e-mail: galbraith@arizona.edu

A.L. Kriz, B.A. Larkins (eds.), *Molecular Genetic Approaches to Maize Improvement*
Biotechnology in Agriculture and Forestry, Vol. 63

observed: "But I don't want to go among mad people!" "Oh, you can't help that," said the cat. "We're all mad here!"

Fortunately, clarity can be effected by combining Humpty Dumpty's axiom with strictly empirical considerations. This means defining gene expression exclusively in terms of the platforms available for obtaining quantitative data, and recognizing that newly emerging methods then allow a progressively more nuanced understanding of the mechanisms that link, most generally, the stored information of the genotype to the defined and observed phenotype. So let us start with the platforms.

18.1.2 Available Platforms for Global Expression Profiling

To be useful, platforms for expression profiling must provide a high density of information content per individual assay, but at a reasonable cost. The term 'global' implies comprehensive coverage of the genome, to the extent possible. Platforms roughly divide into two general classes: those based on hybridization and those based on DNA sequencing.

18.1.2.1 Hybridization-Based Expression Profiling

DNA microarrays comprise by far the greatest representation in these types of methods. Microarrays are solid surfaces, typically glass, pre-coated in various ways, on which individual DNA elements (termed "probes") representing specific gene sequences are immobilized. Microarrays are produced either by mechanical spotting or by in situ synthesis, and are queried by hybridization, most commonly using fluorescent "targets" produced from RNA samples of the tissues or cells of interest. Spotted microarrays initially employed amplicons produced from cDNA libraries, and thence unigene collections. More recently, spotted microarrays have almost exclusively been produced using single-stranded presynthesized oligonucleotides, typically 50–70 bases in length. Specificity of hybridization regulates the information quantity and quality that can be obtained.

Whereas spotted arrays are produced predominantly by academic users, microarrays having elements synthesized in situ are entirely the products of the commercial domain, with Affymetrix, NimbleGen, and Agilent being the major players. Affymetrix and NimbleGen employ light-mediated deprotection to spatially regulate sequential addition of DNA bases (Lipshutz et al. 1999; Singh-Gasson et al. 1999), whereas Agilent uses guided droplet deposition for this purpose (Hughes et al. 2001). As a general observation, the dynamic ranges reported by microarrays span three linear orders of magnitude, which can result in compression of the actual magnitudes of effects measured by these platforms. Affymetrix GeneChips® differ from Agilent and NimbleGen microarrays in utilizing sets of much shorter array elements (25-mers) to report transcript levels of specific genes, and in implementing a Perfect-Match (PM)/Mismatch (MM) element design strategy to improve selectivity (Dalma-Weiszhausz et al. 2006). In brief, this involves designing 10–30 pairs of 25-mer probes (termed a "probe-set"), in which the MM probe differs from its cognate PM probe at a single central position. Specific algorithms are provided by Affymetrix for the purposes of converting probe set intensity values to transcript amounts. There is considerable debate as to the value of using the PM/MM approach compared to alternative methods for GeneChip data extraction (see, for example, Irizarry et al. 2003), and this debate is particularly significant for highly complex and variable genomes such as maize (see below).

Critical to the use of microarray platforms is the appropriate employment of statistics and statisticians in experimental design and data analysis. When searching for alterations in transcript abundances, it is important to establish the accuracy with which the measurements are made, the relationship between treatment and observed changes, and the nature and type of unsuspected variation that might confound the observations. For a recent discussion of statistical issues in microarrays, see Allison et al. (2006).

For spotted and synthesized array platforms, as the maize genome moves toward a state of relatively complete genome sequence information, the sequence composition of the individual probes and their numbers can be globally adjusted to provide more comprehensive and precise information concerning the transcriptional activities of the genome. An alternative approach is also possible, involving the production of "tiling" arrays (Yamada et al. 2003), in which the specific oligonucleotide sequences are designed such that they are sequentially indexed across the entire genomic sequence. To provide the highest resolution, both DNA strands are represented with a 1-bp indexing offset (see Clark et al. 2007 for a recent example of this approach using Affymetrix 25-mer probes for resequencing 20 different Arabidopsis accessions for single nucleotide polymorphism (SNP) discovery). Complete tiling of the 120-Mbp Arabidopsis genome required five large microarray wafers, and hybridization correspondingly involved volumes ($\sim$14 mL) of target solution that were very large by conventional microarray standards.

For all of the synthesized array platforms, individual element sizes are being engineered to progressively smaller dimensions, with the aim of providing extremely high element densities per array. Costs can also be decreased through use of tiling designs having lesser degrees of element sequence overlap. It remains to be seen whether tiling arrays can be economically realistic for use with maize, given the large size of its genome.

An alternative microarray platform has been recently described (Kris et al. 2007). Based on a multiplexed nuclease protection assay, and providing a luminescence read-out of exceptional dynamic range, this platform appears particularly suited to high throughput screens for effector molecules that modulate the transcript levels of predefined groups of genes. It also, uniquely, provides an absolute measure of individual transcripts, allowing comparisons of these levels across genes.

A final alternative platform for gene expression measurements based on hybridization involves quantitative RT-PCR (Czechowski et al. 2004). Again providing exceptional dynamic range, coupled to the potential of allele-specificity, this approach is limited only by cost and the complex manipulation pipeline that must be implemented.

18.1.2.2 Digital Read-out

Expression profiling based on digital read-out involves methods and platforms that rely on DNA sequencing to enumerate unambiguous identifiers of individual transcripts. This approach started with the "digital Northerns" produced as a consequence of large-scale Sanger-based sequencing of cDNA libraries. The key factor underlying the accuracy of this approach is the depth of sequencing achieved (see below); thus whereas this approach is a natural requisite for the characterization of unsequenced genomes (i.e. gene discovery), it does not represent a reasonable approach for characterization of transcript levels across multiple samples, due to cost considerations.

The digital read-out approach greatly expanded in popularity for global expression profiling following the development of lower cost methods for serial sequencing of concatenated cDNA-derived tags, particularly serial analysis of gene expression (SAGE; Velculescu et al. 1995) and massively parallel signature sequencing (MPSS; Brenner et al. 2000). These latter two platforms rely on the observation that a transcript can be uniquely identified through sequencing of a short (15- to 25-bp) tag derived from that transcript. Both approaches have the effect of reducing the cost per digital tag sequenced, as contrasted to the cost per sample processed. The general power of the MPSS technique, as developed by Lynx, Inc., has been recently exemplified in rice (Nobuta et al. 2007). This approach has gained additional impetus, and MPSS incidentally rendered obsolete, from the recent development of Next-Generation DNA sequencing instruments (for example, Solexa, 454 Life Sciences, Applied Biosystems, and polony-based methods (Kim et al. 2007)), which produce highly parallel short reads ideally suited for expression tag identification and enumeration. They also provide the potential to monitor allele-specific expression, as well as tags derived from the 5′-ends of transcripts (Gowda et al. 2006).

The criterion for comparative quality of expression profiling information between microarrays and sequencing fundamentally involves the granularity of the measurement: sufficient tags must be enumerated to provide a digital data set that avoids errors due to binning over a dynamic range comparable to microarrays. It also, of course, involves cost considerations on a per sample basis: ultimately, expression profiling methods must account for biological variation, which requires appropriate experimental design, the use of statistics, and adequate replication to provide significant contrasts. Finally, overall sample throughput, namely the time required for sample preparation through data production, must be taken into consideration. Currently, none of the sequencing-based methods can match microarrays based on these considerations.

18.1.3 Fractionation and Prepurification Procedures for Complex Systems

Global analysis of gene expression, to be fully useful, needs to accommodate the structural complexities associated with the presence of different cell types in complex tissue and organs and with defining the process of gene expression itself. Dealing with different cell types can be done in one of several ways: the first involves prepurification of the different cell types, utilizing specific cell separation and enrichment methods such as fluorescence-activated sorting (FAS) and laser-capture microdissection (LCM). The cell types are then employed for gene expression analysis.

FAS methods, which have been developed using Arabidopsis as the model (Birnbaum et al. 2003, 2005; Nawy et al. 2005; Galbraith 2006; Galbraith and Birnbaum 2006), involve production of protoplasts from plants having specific cell types that are highlighted through transgenic expression of fluorescent proteins. The fluorescent protoplasts are separated from their non-fluorescent counterparts by flow sorting. RNA is then extracted from the sorted protoplasts and employed for microarray hybridization. In principle, other downstream methods of global measurement of gene expression, such as sequencing, can also be employed. The applicability of this approach is restricted to tissues that are amenable to protoplast production and that lack confounding levels of endogenous autofluorescence. In this respect, Arabidopsis roots are nearly ideal. However, the general applicability of this method to crops such as maize will also be restricted by difficulties associated with producing transgenics.

LCM methods have been extensively explored and implemented using maize (Kerk et al. 2003; Nakazono et al. 2003; Woll et al. 2005; Nelson et al. 2006; Day et al. 2007; Ohtsu et al. 2007). These methods exclusively employ fixed or frozen, embedded, and sectioned tissues. Lasers are then used to selectively capture (Arcturus, Molecular Devices), cut (Leica), or catapult (PALM Microlaser Technologies) the cells of interest. Due to the limited amounts of these cells that can be recovered routinely, global transcript profiling following LCM requires RNA amplification, typically based on linear in vitro methods (Van Gelder et al. 1990). These amplification methods appear reproducible (Day et al. 2007; Ohtsu et al. 2007), but it is also critical they amplify equally over the entire transcript space, if differential gene expression is to be accurately identified and measured.

Deconvolution of the contributions to gene expression of different cell types within complex tissues can also be achieved using a strategy alternative to prepurifying the different cell types. This involves tagging of specific cellular components that are specifically involved in the process of gene expression, and the use of associated methods for their selective enrichment (for example, flow sorting of labeled organelles, or immunoprecipitation of polyribosomes (Zanetti et al. 2005)) of these components from general tissue or organ homogenates. The latter approach lends itself to flexibility as more cellular components are discovered that are involved in specific steps in gene expression, such as P-bodies (Parker and Sheth 2007) and RNA-binding proteins.

18.1.4 The Information Content of the Maize Genome

Maize comprises a genome size of approximately 2,300 Mbp (Wei et al. 2007). The genome of the important inbred line B73 is currently being sequenced using a clone-by-clone approach, requiring the availability of a high resolution physical map; see http://www.maizegdb.org/sequencing_project.php for details of progress. Results to date indicate that two rounds of genome duplication have accompanied the evolution of the maize genome, from a common cereal ancestor approximately 50 million years ago (Wei et al. 2007). About two-thirds of the genome comprises transposable elements, of which most are retrotransposons and recombinationally inert.

Maize appears to have the most genetic diversity of any domesticated grass (Gore et al. 2007), with the genomic sequences of any two different inbreds differing on average by one SNP every 70 bp (Tenaillon et al. 2001). Diversity is also a function of the occurrence of insertion-deletion (indel) polymorphisms which are often uncovered in SNP discovery projects (Bhattramakki et al. 2002), and of the presence of tandem gene arrays (Messing et al. 2004; Jander and Barth 2007), and is complicated by the observation that different maize lines differ in the copy numbers of specific genes, and in their degree of relatedness at the sequence level.

The high degree of overall polymorphism for maize has serious implications for array-based analysis of transcript concentrations, since the degree of hybridization reported by microarrays is evidently also sensitive to sequence similarity. Such sensitivity increases as probe length decreases, implying that expression platforms based on multiple short oligomer probe sets (i.e. Affymetrix GeneChips) will potentially be capable of extracting greater information content than those based on fewer but longer sequences of oligonucleotides (NimbleGen, Agilent), noting that this information content may be more confusing to interpret with respect to transcript profiling. A further difficulty is that, given the large size and complexity of the maize genome, one cannot simply employ whole genome hybridization for detection of polymorphisms, via incorporating fluorochromes into fragmented genomic DNA to produce targets, since most of the label would be incorporated into repetitive DNA. Methods to reduce target complexity and/or enrich for genic sequences therefore become essential.

Kirst et al. (2006) explored the use of mRNA-derived cRNA for this purpose: custom Affymetrix GeneChips containing up to 30 probes per probe set were hybridized with cRNA produced from mRNA isolated from four different maize lines. They found considerable evidence for probe-by-line interactions, with on average 5–8 of the 30 probes within each probe set exhibiting hybridization differences relating to the presence of DNA polymorphisms. Excluding probes from consideration that exhibit probe-by-line interactions was insufficient to correct for polymorphisms. Evidently biases in measurement of transcript abundances will be ameliorated through using longer oligomers as probes, particularly those due to the presence of SNPs, since empirical measurements indicate that sequence divergence of around 30% is needed before decreases in hybridization signal intensities are observed for 70-mer oligonucleotide probes (Xu et al. 2001). In further work, Gore et al. (2007) compared the mRNA-based approach with additional methods for reducing genomic complexity, including methyl-filtration, C_ot-filtration, and amplified fragment length polymorphism based methods. They found these methods afforded only modest ability to detect known single-feature polymorphism with the maize GeneChip, and suggest that additional replication may be needed combined with methods for more aggressively reducing genomic complexity.

Interest in the interaction of genotype with global expression profiling springs not only from the early theoretical observations of Jansen and Nap (2001), but also from the historical and archetypal observation of heterosis in maize. Uncovering the mechanistic underpinning of heterosis currently represents a primary goal of basic and applied research in plant biology, and platforms for global expression profiling are likely to be central to achieving this goal.

18.2 Global Transcript Analysis

18.2.1 Affymetrix GeneChips

The first reports of global expression profiling in maize using Affymetrix GeneChips were based on a design produced at Pioneer Hi-Bred (Hunter et al. 2002; Zinselmeier (2002)) representing around 1,500 ESTs (with 20 match and 20 mismatch probe sets per gene). Hunter et al. (2002) explored the effects of various opaque endosperm mutations on gene expression as measured by these GeneChips. They noted highly pleiotropic effects of the mutations, particularly given an over-representation of zein genes within the GeneChip, resulting in difficulties in global normalization of the data. They also reported the phenomenon of a proportion of probe sets for which lower signals were obtained from match than mismatch probes; this was thought likely to be a consequence of the presence, within the maize genome, of nearly identical alleles producing transcripts at very different levels. Zinselmeier (2002) examined the effects of abiotic stresses on gene expression, also comparing GeneChips with an Amersham spotted microarray platform. The limited numbers of genes represented on the arrays evidently restricted the scope of the conclusions of this work.

Reports of use of the Affymetrix platform within the public sector have only just started to emerge, this perhaps reflecting the problems associated with the extensive genomic polymorphism of maize (Kirst et al. 2006), and the fact that the commercially available GeneChip, since its probe sets are designed to avoid polymorphisms wherever possible, represents only around 13,000 genes. Stupar and Springer (2006) explored the patterns of gene expression detected in parental inbreds and their heterotic F1 hybrids. They found considerable evidence of differential hybridization signals across the B73 and Missouri 17 (Mo17) parents at the probe-set level, with analysis of the patterns of hybridization of the individual probes implying that, in the majority of cases, the differences were due to differences in transcript levels rather than to polymorphisms. Dividing the genes into groups according to whether interparental differences in expression were not (group 1) or were (group 2) observed, novel hybrid expression states were restricted to group 2 genes, with most displaying additive expression patterns. Finally, intraspecific expression seemed predominately subject to *cis*-regulatory variation. This group went on to study the role of epigenetic regulation, through comparative analysis of gene expression patterns in near isogenic derivatives of B73 and Mo17 containing a loss-of-function allele of the chromomethylase *ZMET2* gene (Makarevitch et al. 2007). They found that many targets of this gene represent epigenetic states displaying natural variation across maize inbreds and suggest these may contribute to phenotypic variation that can be, and has been, employed for selection.

Ma et al. (2007) used the same platform in part for studies in which they compared hybridization of maize targets to GeneChips and to Agilent microarrays comprising 60-mer synthetic single-stranded elements (see details below). They concluded that long-mer probes were superior in performance for expression profiling in complex genomes such as maize, particularly in the accurate quantification of low-level transcripts.

Given the low proportion of the genome represented on maize GeneChips, Jiang et al. (2006) employed the strategy of cross-species hybridization to increase potential gene coverage. They applied maize targets, prepared from microdissected tissues encompassing the root tip, the proximal meristem, the quiescent center, and the root cap, to GeneChips designed by Syngenta representing approximately 50% of the genes of Nipponbare rice (Zhu et al. 2003). They were able to identify a number of genes that were differentially regulated within the root cap, and confirmed these results using RT-PCR. The validity of the approach of employing heterologous platforms from expression profiling clearly derives from the close evolutionary relationships between the grasses.

18.2.2 Microarrays Employing PCR Amplicons as Probes

Other early analyses of global transcript profiling in maize employed PCR amplicons derived from cDNA libraries and unigene sets as microarray probes. Studies included broad comparisons of expression patterns in different tissues and across different platforms (Fernandes et al. 2002), and analysis of the impact of abiotic stresses (Casati and Walbot 2003, 2004; Wang et al. 2003; Yu and Setter 2003), using first-generation cDNA amplicon microarrays developed at the University of Arizona from ESTs produced, sequenced, and assembled at Stanford and Iowa State University. Nakazono et al. (2003) employed laser capture microdissection in combination with amplicon-based microarrays for the analysis of differential gene expression in epidermal and vascular cells isolated from fixed and sectioned coleoptiles. Woll et al. (2005) combined LCM with comparative analysis of a specific mutant defective in seminal and lateral root initiation, to define genes active in the pericycle prior to root emergence. Swanson-Wagner et al., (2006) covered ground similar to that described by Stupar and Springer (2006) but using amplicon-based microarrays. In concordance with those results, they found that most of the ESTs displayed additive patterns of expression. Of the minority displaying non-additive modes, all possible forms were observed, including high- and low-parent dominance and under- and over-dominance, consistent with multiple molecular mechanisms contributing to heterosis. Finally, Shi et al. (2007) recently reported an analysis of eQTLs associated with cell wall digestibility, utilizing amplicon-based arrays. Interestingly, of the 102 ESTs identified as being differentially expressed between high and low quality groups, only a minority detected a single eQTL and none mapped to the same location. This suggests that trans-acting effects predominate in the regulation of inheritance of this complex trait.

18.2.3 Microarrays Employing Long Oligonucleotides as Probes

One of the first examples of the use of long oligo microarrays in maize was provided by Schadt et al. (2003) in a general cross-kingdom study (mouse, human, maize) of the interactions between global expression profiles and genotype. This work, they employed custom microarrays produced by Rosetta using the in situ droplet deposition and synthesis process now owned by Agilent. For the maize studies, the microarrays comprised 24,437 elements chosen from maize and rice sequence information, but this detail was not provided in the publication or the associated supplementary materials on-line. The results indicated that most genes had single eQTLs, the majority of which were localized to the gene exhibiting the eQTL. Further interpretation was limited by a complete absence of detailed information in this publication.

Based on their relative degree of insensitivity to SNPs but improved specificity of hybridization in comparison to amplicon-based microarrays, platforms based on long oligonucleotide probes have become increasingly popular. The Arizona maize microarray consortium (www.maizearrays.org) has produced a spotted array employing 70-mer elements designed around sequence information primarily contained in the TIGR Maize Gene Index Release 13.0 (Gardiner et al. 2005). This information came from assembled expressed sequence tags, sequences emerging from analysis of gene-rich regions of the genome, and predicted genes. Organellar genes were included, as well as a number of controls. In some situations, gene index sequences were oriented via transcript hybridization using NimbleGen microarrays containing elements in both possible orientations. Version 1 microarrays comprised 57,442 probes and were printed on two microarray slide surfaces. Version 2 microarrays comprise 46,110 probes printed on a single slide surface. Probes were dropped based on a lack of detectable hybridization using targets across numerous tissues (Gardiner et al., unpublished data).

Sawers et al. (2007) employed the Version 1 platform for analysis of differential gene expression between bundle sheath and mesophyll cells, paying particular attention to the effects on false discovery of the stress imposed during separation of the different cell types. The lists of genes identified as differentially regulated using the microarrays confirmed many examples of genes known to exhibit differential regulation between bundle-sheath and mesophyll cells. They concluded that as much as 18% of the features represented on the microarrays might be differentially expressed.

An Agilent platform based on 60-mer oligonucleotide has been developed by the Walbot group at Stanford University (Ma et al. 2006, 2007). Version 1 comprised approximately 21,000 probes designed around the December 2003 maize EST assembly in MaizeGDB. Ma et al. (2006) described the genesis of these arrays, their use to survey transcript levels across four tissues and across different genotypes (including one hybrid) and to explore the extent of sense and antisense transcription. Broad concordance was observed between the performance of the Stanford and Arizona long oligonucleotide microarray platforms when identical or nearly identical array elements were considered. Agilent Version 2 maize microarrays comprise a similar number (21,000) of gene features, representing ~13,000 unique sense and ~5,000 antisense transcripts, 17,026 probes being shared with Version 1. Ma et al. (2007) employed these arrays for characterization of gene expression within maize anthers and during development of normal and male-sterile mutants. A further version of these arrays has become commercially available from Agilent, comprising 42,035 60-mer probes.

One unexpected result from microarray experiments was that of apparently high levels of antisense expression (Ma et al. 2006; also first reported in other species using tiling arrays (Yamada et al. 2003)). A recent report suggests at least a portion of this may be due to artefactual second-strand priming, which can be eliminated using actinomycin D (Perocchi et al. 2007). Since such artefactual priming might occur as a prelude to most if not all other forms of profiling, this observation deserves considerable attention.

18.2.4 Non-Microarray-Based Profiling

Examples in maize of EST enumeration through conventional Sanger sequencing include the work of Fernandes et al. (2002), and the approach has recently been generalized through bar-coding (Poroyko et al. 2007) for an examination of changes in root transcripts as a function of tissue location and consequent water stress. An example of the application of SAGE in maize has also been provided by this group, leading to the identification of novel genes expressed in the root (Poroyko et al. 2005).

The application of MPSS technologies to maize seems to have been largely restricted to work contracted to Lynx by Pioneer/DuPont and collaborators (Gao et al. 2004; Zhang et al. 2005; Ching et al. 2006; Woll et al. 2006). The approach was to produce cDNA libraries from the tissue or cellular state of interest, which were subjected to MPSS at Lynx. The resultant primary data sets are not available in the public sector, which reduces their utility. Nevertheless, summary results and conclusions have emerged, including mining for genes involved in alternative slicing (Gao et al. 2004), and description of tissue patterns of expression of genes encoding 12-oxo-phytodienoic acid reductases (Zhang et al. 2005) and of a putative glycosylphosphatidylinositol-anchored protein affecting cell wall mechanical strength (Ching et al. 2006), and of a root tip-specific gene (Woll et al. 2006).

Emrich et al. (2007) have described the combination of 454 Life Sciences sequencing with LCM. They captured shoot apical meristems from 14-day-old seedlings, amplified cDNA from this material, and subjected it to 454-sequencing. This resulted in 288,992 EST sequences, 70% of which were not previously captured in cDNA libraries from hand-dissected apices, and 30% of which were not found in the total extant maize EST collections. It appears that deep-sequencing, using Next-Generation sequencing, coupled with a means of isolation of specific tissue- or cell-types, will provide an efficient means of discovery of genes that produce rare transcripts.

18.3 MicroRNA Profiling

Discussion of global expression profiling cannot be complete without some comments concerning the analysis of the contributions of short RNA transcripts to cellular regulatory processes. Given their established importance in this arena (Rana 2007), there is considerable interest in developing high-throughput methods for characterizing their abundances within tissues and cells and in response to changes in state. MicroRNAs (miRNAs), being only ∼22 nucleotides in length, provide considerable challenges to conventional microarray-based methods (Baskerville and Bartell 2005; Davison et al. 2006) in terms of selectivity, specificity, and dynamic range. Wang et al. (2003) have demonstrated the importance of subtle features of conventional probe design and target production in addressing these issues. Other groups have adopted radically different approaches: Nelson et al. (2004) described the use of in situ RNA-primed microarray probe for miRNA detection and quantification. Castoldi et al. (2006) employed a microarray platform with probes produced from locked nuclei acids. Beuvink et al. (2007) described a combination of oligoribonucleotide probes with detection based on evanescence resonance. Finally, Liang et al. (2005) described a method of miRNA detection based on quantum dot and nanogold technologies. Given the level of activity in this area, it appears likely that robust microarray platforms will be established for miRNA analysis, and their application to maize is inevitable.

Comprehensive identification and enumeration of short RNA transcripts through MPSS has already been described for Arabidopsis and rice (Lu et al. 2005; Nobuta et al. 2007), and similar work involving next generation sequencing is underway for maize (B.C. Meyers, pers. comm.).

18.4 Conclusions and Future Prospects

Surveying the development of the field of expression profiling over the last decade or so, certain observations seem axiomatic. First, novel technologies will continue to emerge. Second, the capabilities of these technologies (and refinements of extant technologies) are likely to comply with Moore's law concerning output of data per unit cost over time. Third, our abilities to integrate and thereby benefit from the massive amounts of data flux will likely lag behind our abilities to generate this data; this is because any combinatorial analysis of such data must, by definition, occupy more computational time than required for data production. Finally, one must mention, at least in passing, the problems of curating and archiving such data and accessing these archives.

Given the complex nature of gene expression and the complicating factors that appear almost daily in the published literature, how confident can we be of achieving overall understanding in this field? Humpty Dumpty is associated with hubris culminating, of course, in a "great fall". Despite this precedent, given the numbers of investigators, the levels of funding, and our intrinsic inventiveness, I am confident an integrated picture of global gene expression will emerge from assembly of the pieces that are separately being discovered using the increasingly powerful techniques at our disposal, without getting too much egg on our faces!

Acknowledgements This work was supported by the National Science Foundation Plant Genome Research Program (grant number DBI-0321663).

References

Allison DB, Cui XQ, Page GP, Sabripour M (2006) Microarray data analysis: from disarray to consolidation and consensus. Nat Rev Genet 7:55–65

Baskerville S, Bartell DP (2005) Microarray profiling of microRNAs reveals frequent coexpression with neighboring miRNAs and host genes. RNA 11:241–247

Beuvink I, Kolb FA, Budach W, Garnier A, Lange J, Natt F, Dengler U, Hall J, Filipowicz W, Weiler J (2007) A novel microarray approach reveals new tissue-specific signatures of known and predicted mammalian microRNAs. Nucl Acids Res 35(e52)

Bhattramakki D, Dolan M, Hanafey M, Wineland R, Vaske D, Register JC, Tingey SV, Rafalski A (2002) Insertion-deletion polymorphisms in 3′ regions of maize genes occur frequently and can be used as highly informative genetic markers. Plant Mol Biol 48:539–547

Birnbaum K, Shasha DE, Wang JY, Jung JW, Lambert GM, Galbraith DW, Benfey PN (2003) A gene expression map of the *Arabidopsis* root. Science 302:1956–1960

Birnbaum K, Jung JW, Wang JY, Lambert GM, Hirst JA, Galbraith DW, Benfey PN (2005) Cell-type specific expression profiling in plants using fluorescent reporter lines, protoplasting, and cell sorting. Nat Meth 2:1–5

Brenner S, Johnson M, Bridgham J, Golda G, Lloyd DH, Johnson D, Luo S, McCurdy S, Foy M, Ewan M, Roth R, George D, Eletr S, Albrecht G, Vermaas E, Williams SR, Moon K, Burcham T, Pallas M, DuBridge RB, Kirchner J, Fearon K, Mao J-I, Corcoran K (2000) Gene expression analysis by massively parallel signature sequencing (MPSS) on microbead arrays. Nat Biotech 18:630–634

Casati P, Walbot V (2003) Gene expression profiling in response to ultraviolet radiation in maize genotypes with varying flavonoid content. Plant Physiol 132:1739–1754

Casati P, Walbot V (2004) Rapid transcriptome responses of maize (*Zea mays*) to UV-B in irradiated and shielded tissues. Genome Biol 5(R16)

Castoldi M, Schmidt S, Benes V, Noerholm M, Kulozik AE, Hentze MW, Muckenthaler MU (2006) A sensitive array for microRNA expression profiling (miChip) based on locked nucleic acids (LNA). RNA 12:913–920

Ching A, Dhugga KS, Appenzeller L, Meeley R, Bourett TM, Howard RJ, Rafalski A (2006) Brittle stalk 2 encodes a putative glycosylphosphatidylinositol-anchored protein that affects mechanical strength of maize tissues by altering the composition and structure of secondary cell walls. Planta 224:1174–1184

Clark RM, Schweikert G, Toomajian C, Ossowski S, Zeller G, Shinn P, Warthmann N, Hu TT, Fu G, Hinds DA, Chen H, Frazer KA, Huson DH, Schölkopf B, Nordborg M, Rätsch G, Ecker JR, Weigel D (2007) Common sequence polymorphisms shaping genetic diversity in *Arabidopsis thaliana*. Science 317:338–342

Czechowski T, Bari RP, Stitt M, Scheible WR, Udvardi MK (2004) Real-time RT-PCR profiling of over 1400 Arabidopsis transcription factors: unprecedented sensitivity reveals novel root- and shoot-specific genes. Plant J 38:366–379

Dalma-Weiszhausz DD, Warrington J, Tanimoto EY, Miyada G (2006) The Affymetrix GeneChip® platform: an overview. Methods Enzymol 410:3–28

Davison TS, Johnson CD, Andruss BF (2006) Analyzing micro-RNA expression using microarrays. Methods Enzymol 411:14–34

Day RC, McNoe LA, Macknight RC (2007) Transcript analysis of laser microdissected plant cells. Physiol Plant 129:267–282

Emrich SJ, Barbazuk WB, Li L, Schnable PS (2007) Gene discovery and annotation using LCM-454 transcriptome sequencing. Genome Res 17:69–73

Fernandes J, Brendel V, Gai X, Lal S, Chandler VL, Elumalai R, Galbraith DW, Pierson E, Walbot V (2002) Comparison of RNA expression profiles based on maize EST frequency analysis and microarray hybridization. Plant Physiol 128:896–910

Galbraith DW (2006) Microarray analyses in higher plants. OMICS J Integr Biol 10:455–473

Galbraith DW, Birnbaum K (2006) Global studies of cell type-specific gene expression in plants. Annu Rev Plant Biol 57:451–475

Gao HR, Gordon-Kamm WJ, Lyznik LA (2004) ASF/SF2-like maize pre-mRNA splicing factors affect splice site utilization and their transcripts are alternatively spliced. Gene 339: 25–37

Gardiner JM, Buell CR, Elumalai R, Galbraith DW, Henderson DA, Iniguez AL, Kaeppler SM, Kim JJ, Liu J, Zheng L, Chandler VL (2005) Design, production, and utilization of long oligonucleotide microarrays for expression analysis in maize. Maydica 50:425–435

Gore M, Bradbury P, Hogers R, Kirst M, Verstege E, van Oeveren J, Peleman J, Buckler E, van Eijk M (2007) Evaluation of target preparation methods for single-feature polymorphism detection in large complex plant genomes. Crop Sci 47: S135–S148

Gowda M, Li HM, Alessi J, Chen F, Pratt R, Wang GL (2006) Robust analysis of 5'-transcript ends (5'-RATE): a novel technique for transcriptome analysis and genome annotation. Nucleic Acids Res 34. doi:10.1093/nar/gkl522

Hughes TR, Marton MJ, Jones AR, Roberts CJ, Stoughton R, Armour CD, Bennett HA, Coffey E, Dai HY, He YDD, Kidd MJ, King AM, Meyer MR, Slade D, Lum PY, Stepaniants SB,

Shoemaker DD, Gachotte D, Chakraburtty K, Simon J, Bard M, Friend SH (2000) Functional discovery via a compendium of expression profiles. Cell 102:109–126

Hughes TR, Mao M, Jones AR, Burchard J, Marton MJ, Shannon KW, Lefkowitz SM, Ziman M, Schelter JM, Meyer MR, Kobayashi S, Davis C, Dai HY, He YDD, Stephaniants SB, Cavet G, Walker WL, West A, Coffey E, Shoemaker DD, Stoughton R, Blanchard AP, Friend SH, Linsley PS (2001) Expression profiling using microarrays fabricated by an ink-jet oligonucleotide synthesizer. Nat Biotechnol 19:342–347

Hunter BG, Beatty MK, Singletary GW, Hamaker BR, Dilkes BP, Larkins BA, Jung R (2002) Maize opaque endosperm mutations create extensive changes in patterns of gene expression. Plant Cell 14:2591–2612

Irizarry RA, Bolstad BM, Collin F, Cope LM, Hobbs B, Speed TP (2003) Summaries of Affymetrix GeneChip probe level data. Nucleic Acids Res 31(e15)

Jander G, Barth C (2007) Tandem gene arrays: a challenge for functional genomics. Trends Plant Sci 12:203–210

Jansen RC, Nap JP (2001) Genetical genomics: the added value from segregation. Trends Genet 17:388–391

Jiang K, Zhang SB, Lee S, Tsai G, Kim K, Huang HY, Chilcott C, Zhu T, Feldman LJ (2006) Transcription profile analyses identify genes and pathways central to root cap functions in maize. Plant Mol Biol 60:343–363

Kerk NM, Ceserani T, Tausta SL, Sussex IM, Nelson TM (2003) Laser capture microdissection of cells from plant tissues. Plant Physiol 132:27–35

Kim JB, Porreca GJ, Song L, Greenway SC, Gorham JM, Church GM, Seidman CE, Seidman JG (2007) Polony multiplex analysis of gene expression (PMAGE) in mouse hypertrophic cardiomyopathy. Science 316:1481–1484

Kirst M, Caldo R, Casati P, Tanimoto G, Walbot V, Wise RP, Buckler ES (2006) Genetic diversity contribution to errors in short oligonucleotide microarray analysis. Plant Biotechnol J 4:489–498

Kris RM, Felder S, Deyholos M, Lambert GM, Botros I, Martel R, Seligmann B, Galbraith DW (2007) High-throughput, high-sensitivity analysis of gene expression in *Arabidopsis thaliana*. Plant Physiol 144:1256–1266

Liang RQ, Li W, Li Y, Tan CY, Li JX, Jin YX, Ruan KC (2005) An oligonucleotide microarray for microRNA expression analysis based on labeling RNA with quantum dot and nanogold probe. Nucl Acids Res 33(e17). doi: 10.1093/nar/gni019

Lipshutz RJ, Fodor SPA, Gingeras TR, Lockhart DJ (1999) High density synthetic oligonucleotide arrays. Nat Genetics 21(Suppl S):20–24

Lu C, Tej SS, Luo SJ, Haudenschild CD, Meyers BC, Green PJ (2005) Elucidation of the small RNA component of the transcriptome. Science 309:1567–1569

Ma J, Morrow DJ, Fernandes J, Walbot V (2006) Comparative profiling of the sense and antisense transcriptome of maize lines. Genome Biol 7(R22)

Ma J, Duncan D, Morrow DJ, Fernandes J, Walbot V (2007) Transcriptome profiling of maize anthers using genetic ablation to analyze pre-meiotic and tapetal cell types. Plant J 50:637–648

Makarevitch I, Stupar RM, Iniguez AL, Haun WJ, Barbazuk WB, Kaeppler SM, Springer NM (2007) Natural variation for alleles under epigenetic control by the maize chromomethylase *Zmet2*. Genetics 177:749–760

Messing J, Bharti AK, Karlowski WM, Gundlach H, Kim HR, Yu Y, Wei FS, Fuks G, Soderlund CA, Mayer KFX, Wing RA (2004) Sequence composition and genome organization of maize. Proc Natl Acad Sci USA 101:14349–14354

Nakazono M, Qiu F, Borsuk LA, Schnable PS (2003) Laser-capture microdissection, a tool for the global analysis of gene expression in specific plant cell types: identification of genes expressed differentially in epidermal cells or vascular tissues of maize. Plant Cell 15:583–596

Nawy T, Lee JY, Colinas J, Wang JY, Thongrod SC, Malamy JE, Birnbaum K, Benfey PN (2005) Transcriptional profile of the Arabidopsis root quiescent center. Plant Cell 17:1908–1925

Nelson PT, Baldwin DA, Scearce LM, Oberholtzer JC, Tobias JW, Mourelatos Z (2004) Microarray-based, high-throughput gene expression profiling of microRNAs. Nat Methods 1:155–161

Nelson T, Tausta SL, Gandotra N, Liu T (2006) Laser microdissection of plant tissue: what you see is what you get. Annu Rev Plant Biol 57:181–201

Nobuta K, Venu RC, Lu C, Belo A, Vemaraju K, Kulkarni K, Wang WZ, Pillay M, Green PJ, Wang GL, Meyers BC (2007) An expression atlas of rice mRNAs and small RNAs. Nat Biotech 25:473–477

Ohtsu K, Takahashi H, Schnable PS, Nakazono M (2007) Cell type-specific gene expression profiling in plants by using a combination of laser microdissection and high-throughput technologies. Plant Cell Physiol 48:3–7

Parker R, Sheth U (2007) P bodies and the control of mRNA translation and degradation. Mol Cell 25:635–646

Perocchi F, Xu Z, Clauder-Münster S, Steinmetz LM (2007) Antisense artifacts in transcriptome microarray experiments are resolved by actinomycin D. Nucleic Acids Res doi: 10.1093/nar/gkm683

Poroyko V, Hejlek LG, Spollen WG, Springer GK, Nguyen HT, Sharp RE, Bohnert HJ (2005) The maize root transcriptome by serial analysis of gene expression. Plant Physiol 138:1700–1710

Poroyko V, Spollen WG, Hejlek LG, Hernandez AG, LeNoble ME, Davis G, Nguyen HT, Springer GK, Sharp RE, Bohnert HJ (2007) Comparing regional transcript profiles from maize primary roots under well-watered and low water potential conditions. J Exp Bot 58:279–289

Rana TM (2007) Illuminating the silence: understanding the structure and function of small RNAs. Nat Rev Mol Cell Biol 8:23–36

Sawers RJH, Liu P, Anufrikova K, Hwang JTG, Brutnell TP (2007) A multi-treatment experimental system to examine photosynthetic differentiation in the maize leaf. BMC Genomics 8(12)

Schadt EE, Monks SA, Drake TA, Lusis AJ, Che N, Colinayo V, Ruff TG, Milligan SB, Lamb JR, Cavet G, Linsley PS, Mao M, Stoughton RB, Friend SH (2003) Genetics of gene expression surveyed in maize, mouse and man. Nature 422:297–302.

Shi C, Uzarowska A, Ouzunova M, Landbeck M, Wenzel G, Lubberstedt T (2007) Identification of candidate genes associated with cell wall digestibility and eQTL (expression quantitative trait loci) analysis in a Flint × Flint maize recombinant inbred line population. BMC Genomics 8(22)

Singh-Gasson S, Green RD, Yue YJ, Nelson C, Blattner F, Sussman MR, Cerrina F (1999) Maskless fabrication of light-directed oligonucleotide microarrays using a digital micromirror array. Nat Biotechnol 17:974–978

Stupar RM, Springer NM (2006) *Cis*-transcriptional variation in maize inbred lines B73 and Mo17 leads to additive expression patterns in the F-1 hybrid. Genetics 173:2199–2210

Swanson-Wagner RA, Jia Y, DeCook R, Borsuk LA, Nettleton D, Schnable PS (2006) All possible modes of gene action are observed in a global comparison of gene expression in a maize F-1 hybrid and its inbred parents. Proc Natl Acad Sci USA 103:6805–6810

Tenaillon MI, Sawkins MC, Long AD, Gaut RL, Doebley JF, Gaut BS (2001) Patterns of DNA sequence polymorphism along chromosome 1 of maize (*Zea mays* ssp. *mays* L.). Proc Natl Acad Sci USA 98:9161–9166

Van Gelder RN, von Zastrow ME, Yool A, Dement WC, Barchas JD, Eberwine JH (1990) Amplified RNA synthesized from limited quantities of heterogeneous cDNA. Proc Natl Acad Sci USA 87:1663–1667

Velculescu VE, Zhang L, Vogelstein B, Kinzler KW (1995) Serial analysis of gene expression. Science 270:484–487

Wang H, Kawai K, Deyholos M, Galbraith DW, Bohnert HJ (2003) Temporal progression of gene expression responses to salt shock in maize roots. Plant Mol Biol 52:873–891

Wei F, Coe E, Nelson W, Bharti AK, Engler F, Butler E, Kim H, Goicoechea JL, Chen M, Lee S, Fuks G, Sanchez-Villeda H, Schroeder S, Fang Z, McMullen M, Davis G, Bowers JE, Paterson

AH, Schaeffer M, Gardiner J, Cone K, Messing J, Soderlund C, Wing RA (2007) Physical and genetic structure of the maize genome reflects its complex evolutionary history. PLOS Genetics 3:1254–1263

Woll K, Borsuk LA, Stransky H, Nettleton D, Schnable PS, Hochholdinger F (2005) Isolation, characterization, and pericycle-specific transcriptome analyses of the novel maize lateral and seminal root initiation mutant *rum1*. Plant Physiol 139:1255–1267

Woll K, Dressel A, Sakai H, Piepho HP, Hochholdinger F (2006) ZmGrp3: identification of a novel marker for root initiation in maize and development of a robust assay to quantify allele-specific contribution to gene expression in hybrids. Theor Appl Genet 113:1305–1315

Xu W, Bak S, Decker A, Paquette SM, Feyereisen R, Galbraith DW (2001) Microarray-based analysis of gene expression in very large gene families: the cytochrome P450 gene superfamily of *Arabidopsis thaliana*. Gene 272:61–74

Yamada K, Lim J, Dale JM, Chen HM, Shinn P, Palm CJ, Southwick AM, Wu HC, Kim C, Nguyen M, Pham P, Cheuk R, Karlin-Newmann G, Liu SX, Lam B, Sakano H, Wu T, Yu GX, Miranda M, Quach HL, Tripp M, Chang CH, Lee JM, Toriumi M, Chan MMH, Tang CC, Onodera CS, Deng JM, Akiyama K, Ansari Y, Arakawa T, Banh J, Banno F, Bowser L, Brooks S, Carninci P, Chao QM, Choy N, Enju A, Goldsmith AD, Gurjal M, Hansen NF, Hayashizaki Y, Johnson-Hopson C, Hsuan VW, Iida K, Karnes M, Khan S, Koesema E, Ishida J, Jiang PX, Jones T, Kawai J, Kamiya A, Meyers C, Nakajima M, Narusaka M, Seki M, Sakurai T, Satou M, Tamse R, Vaysberg M, Wallender EK, Wong C, Yamamura Y, Yuan SL, Shinozaki K, Davis RW, Theologis A, Ecker JR (2003) Empirical analysis of transcriptional activity in the Arabidopsis genome. Science 302:842–846

Yu LX, Setter TL (2003) Comparative transcriptional profiling of placenta and endosperm in developing maize kernels in response to water deficit. Plant Physiol 131:568–582

Zanetti ME, Chang I-F, Gong FC, Galbraith DW, Bailey-Serres J (2005) Immunopurification of polyribosomal complexes of Arabidopsis for global analysis of gene expression. Plant Physiol 138:624–635

Zhang JL, Simmons C, Yalpani N, Crane V, Wilkinson H, Kolomiets M (2005) Genomic analysis of the 12-oxo-phytodienoic acid reductase gene family of *Zea mays*. Plant Mol Biol 59:323–343

Zhu T, Budworth P, Chen WQ, Provart N, Chang HS, Guimil S, Su WP, Estes B, Zou GZ, Wang X (2003) Transcriptional control of nutrient partitioning during rice grain filling. Plant Biotech J 1:59–70

Zinselmeier C, Sun YJ, Helentjaris T, Beatty M, Yang S, Smith H, Habben J (2002) The use of gene expression profiling to dissect the stress sensitivity of reproductive development in maize. Field Crops Res 75:111–121

Part V
Molecular Biology and Physiological Studies

Chapter 19
Zein Storage Proteins

David R. Holding and Brian A. Larkins

19.1 Introduction

The seeds of angiosperms accumulate large amounts of nitrogen in the form of storage proteins that are hydrolyzed and utilized by the embryo during germination. Besides their importance for establishing the seedling, storage proteins are also a valuable source of amino acids for the animals that consume them. Storage proteins are the most abundant proteins in the seed, accounting for 50–70% of the total protein. As a result, their amino acid composition has important nutritional consequences, especially for monogastric animals such as humans and certain livestock. Usually one or more of the essential amino acids is deficient in seeds. As a result of this, a great deal of research has been devoted to finding ways of increasing the level of these amino acids. Aside from their nutritional value, the structure and solubility properties of storage proteins are also important, because they confer important functional characteristics to flours made from seeds. These properties affect food processing and manufacturing, and they also make these proteins useful for manufacturing a variety of industrial products. Therefore, it is not surprising that the structure and synthesis of seed storage proteins has been of interest for many years.

In angiosperms, seed storage proteins are found in the cotyledons of the embryo, where they form accretions, or protein bodies, within specialized protein storage vacuoles (Hermann and Larkins 1999). The most common forms of embryo storage proteins are saline-soluble globulins of three (7S or vicillin-type) or six (11S or legumin-type) subunits (Casey 1999). These proteins have been crystallized and the molecular details of their structure are being increasingly understood

David R. Holding
Department of Plant Sciences, University of Arizona, 303 Forbes Building, P.O. Box 210036, Tucson, AZ 85721, USA

Brian A. Larkins
Department of Plant Sciences, University of Arizona, Marley 822C, P.O. Box 210036, Tucson, AZ 85721-0036, USA
e-mail: larkins@ag.arizona.edu

A.L. Kriz, B.A. Larkins (eds.), *Molecular Genetic Approaches to Maize Improvement*
Biotechnology in Agriculture and Forestry, Vol. 63

(Adachi et al. 2003). Embryos also contain significant amounts of other proteins, including protease inhibitors (Domoney 1999; Shewry 1999), α-amylase inhibitors (Carbonero and García-Olmedo 1999), lectins (Peumans and Van Damme 1999) and ribosome inactivating proteins (RIPs) (Nielsen and Boston 2001), which presumably are synthesized as anti-nutritional factors that protect the seed from pathogens, insects and bird predators. However, these proteins also affect the food and feed value of certain types of seeds.

In cereal seeds, the endosperm is the principal site for storage protein accumulation. A novel type of storage protein known as "prolamin" evolved in cereals. Many of the unique features of the food products made from cereal grains are related to the types of prolamins they contain (Hamaker and Larkins 2000; Shewry and Halford 2002). Prolamins are structurally and biochemically quite different from the 7S and 11S globulins. They feature repeats of short peptides rich in proline and glutamine (hence the name), and appear to have evolved from protease and α-amylase inhibitors (Shewry and Tatham 1999). They contain few charged amino acids, and consequently are insoluble in aqueous solutions. In maize, the prolamin storage proteins, generically known as zeins, account for 50–70% of the protein in the endosperm. Besides storing nitrogen and sulfur for the embryo, zeins also appear to influence the texture and hardness of the grain. The prolamin proteins therefore influence many of the valuable features of cereal grains.

The structure and biochemical properties of seed storage proteins have been widely investigated over the past 30 years, and a great deal is now known about how these proteins are made and stored in the seed, as well as how they are hydrolyzed and absorbed by the embryo. For broader and more detailed reviews describing the nature and biochemistry of seed storage proteins, we refer the reader to the Larkins and Vasil (1997) and Shewry and Tatham (1999). However, in this chapter we will describe specifically the storage proteins of maize seeds, and the prolamin proteins of the endosperm in particular.

19.2 Storage Proteins in the Maize Kernel

19.2.1 Embryo Proteins

Two of the most abundant proteins in the maize embryo are the globulins encoded by the *Glb1* and *Glb2* genes (Belanger and Kriz 1989; Kriz and Wallace 1991). These water-insoluble, saline-soluble proteins share a low level of similarity and have molecular weights of 63 and 45 kDa respectively (Kriz 1989). The GLB1 amino acid sequence has homology with the 7S-type storage proteins of wheat and legumes (Belanger and Kriz 1989). It also has a predicted N-terminal signal peptide and like other storage globulins is likely stored in protein bodies derived from specialized storage vacuoles (Herman and Larkins 1999). GLB1 and GLB2 have no known enzymatic function, and since they are rich in amino acids such as glutamate and glutamine, they serve as a nitrogen sink for embryo growth during germination (Kriz 1989).

19.2.2 Endosperm Proteins

19.2.2.1 Characterization of Zein Proteins and Their Structure

The primary storage proteins in the maize kernel are prolamins called "zeins". The zein fraction is composed of several structurally different types of prolamin proteins that are soluble in aqueous-alcohol solutions, e.g. 60% isopropanol, 70% ethanol, or 95% methanol. Because zeins are deposited as insoluble accretions and surrounded by proteins cross-linked by disulfide bonds, they dissolve slowly in alcoholic solutions. Their extraction is accelerated by increasing the temperature of the alcohol and by including a disulfide reducing agent, but it is also affected by the particle size of the endosperm flour. These factors influence the recovery of the different types of zein proteins, and they are responsible for the complex and often confusing nomenclature that evolved to describe these proteins.

One of the first reports describing "zeine" was published in 1821 by J. Gorham, and the unique properties of this protein fraction attracted the interest of a number of other investigators (Lawton 2002). T.B. Osborne, the father of seed storage protein research (Osborne 1908), developed methods for extracting zein and other prolamin proteins from cereal grains based on their hydrophobic properties. The solvent he used to extract zein contained 95% ethyl alcohol, but other types of alcohol, such as methanol and isopropanol, proved to be more cost effective. The classical Osborne extraction procedure, which is still widely used today, involves extracting proteins from endosperm with water, 5% saline, 70% alcohol and 5% NaOH, and this sequentially removes the albumin, globulin, prolamin and glutelin fractions, respectively (Osborne 1897). In the older literature, two types of zeins were distinguished: "α-zein", which is soluble in 95% aqueous alcohol or 85% isopropanol, and "β-zein", which is soluble in 60% aqueous ethanol (McKinney 1958). Later, more efficient methods for extracting proteins from maize flour and separating zein from the other solubility classes of proteins were described by Paulis et al. (1969) and Landry and Moreaux (1970). With these procedures, following the removal of albumins and globulins, zein was recovered in several steps: the first used aqueous alcohol alone and the second used aqueous alcohol plus a reducing agent, such as β-mercaptoethanol. This produced what was classified as zein, as well as a second fraction identified as gluteliln-1 (Landry and Moreaux 1970), alcohol-soluble reduced glutelin (Paulis and Wall 1971), zein-2 (Sodek and Wilson 1971) and "zein-like" protein (Misra et al. 1975). Proteins in these two solubility groups had similar amino acid compositions, but there were some distinctive differences, suggesting they contained unique polypeptides. The resolution of their relationship was eventually made clear with application of SDS-PAGE, which showed differences in polypeptide composition.

In 1986, Asim Esen developed a new technique for extracting and identifying maize prolamin proteins. This procedure was based on the differential solubility of zeins in the presence of reducing agent. The total zein fraction, which could be efficiently recovered after extraction with 60% isopropanol containing 2% β-mercaptoethanol, was divided into three sub-fractions (SF): SF1, which was soluble in 60% isopropanol alone, principally contained polypeptides of 20 and

24 kDa and appeared to correspond to the α-zein described by McKinney (1958); SF2, containing polypeptides of 17–18 kDa that required 60% isopropanol containing 2% β-mercaptoethanol for solubility; and SF3, predominantly containing a 27-kDa polypeptide that was soluble in aqueous buffer, providing reducing agent was present. Esen proposed that SF1, SF2 and SF3 be designated as α-, β- and λ-zeins (Esen 1987). Subsequently, Kirihara et al. (1988) reported a methionine-rich 10-kDa protein in the zein fraction, and Chui and Falco (1995) described a related protein of 18 kDa. Larkins et al. (1989) suggested that these proteins be designated δ-zeins as an extension of Esen's nomenclature. The designations of α-, β-, γ- and δ-zeins not only distinguished four structurally distinctive types of proteins in alcoholic extracts of maize endosperm, but also fit with the order in which the genes encoding these proteins were isolated.

With the advent of large-scale sequencing of maize endosperm cDNA libraries, a global picture of the complexity and expression patterns of zein genes became more evident. The predominant proteins in the zein fraction, the α-zeins, have apparent molecular weights of 22 and 19 kDa and correspond to a group of proteins with predicted molecular weights between 20,000 and 24,000 (Fig. 19.1). These proteins are encoded by large multigene families (Woo et al. 2001; Song et al. 2001; Song and Messing 2002) and contain a high proportion of glutamine and proline and the hydrophobic amino acids alanine and leucine (Table 19.1). They are essentially devoid

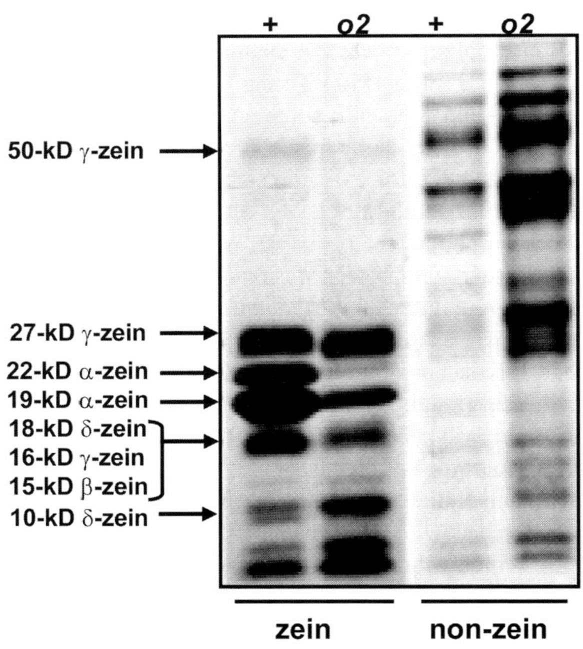

Fig. 19.1 SDS page analysis of zein and non-zein endosperm proteins from W64A wild type (+) and W64A*o2*. Zein and non-zein fractions were isolated according to the method of Wallace et al. (1990) from equal amounts of mature endosperm flour

Table 19.1 Amino acid composition of the predominant zein storage proteins (in mol %)

	19-kDa α-zein	22-kDa α-zein	27-kDa γ-zein	16-kDa γ-zein	10-kDa δ-zein
Asn	4.5	5.3	0.0	0.6	2.0
Asp	0.5	0.0	0.0	0.0	0.7
Thr	3.2	2.8	4.4	3.7	4.0
Ser	7.3	6.9	3.9	5.5	6.7
Gln	19.5	20.7	14.7	18.9	10.0
Glu	0.5	0.8	1.0	1.8	0.0
Pro	10.0	8.9	25.0	15.2	13.3
Gly	0.9	0.8	6.4	9.1	2.7
Ala	13.6	13.8	4.9	7.9	10.0
Cys	0.9	0.4	7.4	7.3	4.0
Val	2.7	6.9	7.4	4.9	3.3
Met	0.5	2.0	0.5	1.8	20.7
Ile	4.5	4.5	2.0	0.6	2.0
Leu	19.5	17.1	9.3	8.5	13.3
Tyr	3.6	2.8	2.0	4.9	0.7
Phe	5.5	3.3	1.0	4.3	4.0
His	0.5	1.2	7.8	2.4	2.0
Lys	0.0	0.0	0.0	0.0	0.7
Arg	1.4	1.6	2.5	1.8	0.0
Trp	0.0	0.0	0.0	0.6	0.0

of lysine and tryptophan, both of which are essential amino acids. The structure of α-zeins is largely defined by a series of tandemly repeated peptides of 20 amino acids: there are nine repeats in the 19-kDa α-zeins and ten in the 22-kDa α-zeins. Each repeat is flanked by clusters of glutamine residues and appears to have an α-helical conformation, based on circular dichroism and a solution conformational analysis (Argos et al. 1982; Tatham et al. 1993). The position of polar amino acids in the repeats is conserved, and it was proposed that they are important for establishing the conformation of the protein, or intermolecular interactions that influence its aggregation in the protein body (Garrat et al. 1993).

The γ-zeins comprise the second most abundant type of maize prolamin. γ-Zeins are cysteine-rich, disulfide-linked proteins. The 27-kDa γ-zein (a.k.a. glutelin I and alcohol-soluble reduced glutelin; see above) is the most abundant protein in this group, followed by the 16-kDa γ-zein (Fig. 19.1). Like the α-zeins, both of these proteins are essentially devoid of lysine and tryptophan (Table 19.1). A larger version of these proteins, the 50-kDa γ-zein (Fig. 19.1), features a longer N terminus, but shares the conserved cysteine-rich core. Surprisingly, this protein contains seven lysine residues (2.7%), making it unique among zeins. The protein originally designated as β-zein (Pedersen et al. 1986) was shown to have a high degree of homology with the γ-zeins and is now termed 15-kDa γ-zein (Woo et al. 2001). All the γ-zeins have six highly conserved cysteine-rich domains. The 27-kDa and 50-kDa γ-zeins contain a block of tandem amino acid repeats at the N terminus that are missing in the 15- and 16-kDa γ-zeins. This region of the 27-kDa γ-zein contains the hexapeptide PPPVHL, while the extension of the 50-kDa γ-zein N terminus has polyglutamine repeats.

Although the two sulfur-rich δ-zeins differ markedly in size (Fig. 19.1), their structures are highly conserved. The 18-kDa protein differs from the 10-kDa protein by a highly repetitive, methionine-rich, 53 amino acid insertion in the center of the protein. Little is known about the structure of these proteins, although they are clearly hydrophobic and occur in the center of the core of the protein body, along the α-zeins (Esen and Stetler 1992).

19.2.2.2 Zein Proteins Associate to Form Protein Bodies

Zein synthesis begins in starchy endosperm cells at around 9 days after pollination (DAP), and it continues throughout endosperm development, although the process becomes spatially restricted as the cells mature and undergo programmed cell death (Young and Gallie 2000). Zein proteins are synthesized as precursors containing a signal peptide that directs their secretion into the lumen of the rough endoplasmic reticulum (RER). Here they become organized into accretions called protein bodies which surround starch granules (Fig. 19.2A). The smallest protein bodies, which are found in cells beneath the aleurone layer, contain the cysteine-rich β- and γ-zeins proteins cross-linked by disulfide bonds (Fig. 19.2B; Lending and Larkins 1989; Lopes and Larkins 1991). During synthesis, the α- and δ-zeins penetrate the matrix of β- and γ-zeins and expand the protein body into a larger spherical structure that reaches a diameter of 1–2 μm. Our model describing this process (Fig. 19.2B) is supported by past studies where we examined the immunolocalization of zeins within developing protein bodies (Lending and Larkins 1989), and subsequent research where we characterized the temporal and spatial patterns of zein gene expression and zein interactions in the yeast two-hybrid system (Woo et al. 2001; Kim et al. 2002). Results showing that α-zeins reside throughout the protein body core were obtained using an antiserum raised against total α-zein proteins. However, antibodies specific to 22-kDa and 19-kDa α-zeins revealed that these proteins have distinct patterns of accumulation (Fig. 19.2C). Whereas the 19-kDa α-zein is found throughout the protein body core, the 22-kDa α-zein is found only in a discrete ring at the interface between the 19-kDa α-zein-rich core and the 27-kDa γ-zein-rich peripheral region (Holding et al. 2007). Chaperones within the RER lumen assist in zein processing (Fontes et al. 1991; Li and Larkins 1996), but direct interactions between these proteins also appear to be important for protein body formation. Experiments with the yeast two-hybrid system (Kim et al. 2002) and results from studies with heterologous systems (Geli et al. 1994; Bagga et al. 1997) suggest that the γ- and β-zeins provide the ER retention mechanism for protein body formation. There is also evidence that the cytoskeleton plays a role in directing zein mRNAs to the RER; however, the symmetry of their distribution on the RER does not appear to influence this process (Kim et al. 2002).

The mechanisms that limit the size of protein bodies are unknown, but it appears that γ-zein and perhaps β-zein organize the α- and δ-zeins and retain them within

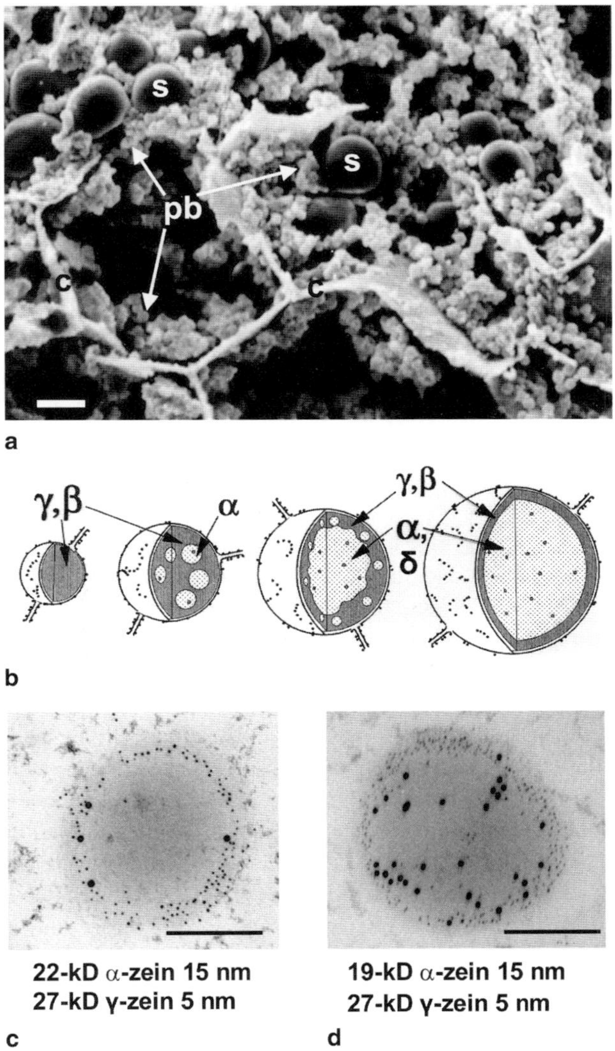

Fig. 19.2 Zein protein bodies in maize endosperm. **a** Scanning electron micrograph of endosperm cells in a developing maize kernel. The kernel was frozen and manually fractured prior to imaging; consequently, some cellular contents may have been lost. Starch grains appear as *grey spheres* and adhering protein bodies are seen as *smaller white spheres*. Representative starch grains, protein bodies and cell walls are marked with *s*, *pb* and *c*, respectively. *Scale bar* (*bottom left*) = 10 μm. **b** Diagrammatic representation of the process of protein body development showing localization of α-, β-, γ- and δ zein proteins. (Adapted from Lending and Larkins (1989) with permission from the American Society of Plant Biologists.) **c** Double immuno-gold labeling of 22-kDa α-zein (15-nm particles) and 27-kD γ-zein (5-nm gold particles) in the endosperm second sub-aleurone cell layer of an 18-DAP kernel. *Scale bar* = 200 nm. **d** Double immuno-gold labeling of 19-kDa α-zein (15-nm particles) and 27-kD γ-zein (5-nm gold particles) in the endosperm second sub-aleurone cell layer of an 18-DAP kernel. *Scale bar* = 200 nm

the ER lumen. Experiments where β- or γ-zeins were synthesized in transgenic tobacco leaves or endosperm showed they were retained within the ER (Coleman et al. 1996; Bagga et al. 1997), and it was subsequently shown that the proline-rich repeats at the N terminus of the 27-kDa γ-zein are responsible for its retention within the ER(Geli et al. 1994). When α- or δ-zeins were synthesized in petunia or tobacco tissues, both proteins appeared to be secreted and become degraded (Williamson et al. 1988; Coleman et al. 1996; Bagga et al. 1997). The α- or δ-zeins can accumulate into accretions, however, if they are co-expressed with β- and γ-zeins in transgenic tobacco. This suggests that β- and γ-zeins provide the ER retention mechanism for α- and δ-zeins. Perhaps it is the amount of β- and γ-zein in a protein body that limits its growth.

19.2.2.3 Mutations Affecting Zein Proteins and Protein Body Structure Are Associated with a Starchy, Opaque Endosperm Phenotype

Mutations that alter zein synthesis can lead to protein bodies with abnormal morphology, size or number, and they are associated with kernels that have a soft, starchy texture. At least 18 mutations have been described that cause this phenotype (Thompson and Larkins 1993), but only five have been described at the molecular level. Mutations that reduce α-zein synthesis, such as *o2*, result in small, unexpanded protein bodies (Fig. 19.3B; Geetha et al. 1991), while those that reduce γ-zein synthesis, such as *o*15 (Dannenhoffer et al. 1995), appear to cause a reduction in the number of protein bodies. Conversely, overproduction of γ-zein, as occurs in modified *o*2 mutants, appears to increase protein body number and results in the formation of more vitreous endosperm (Lopes and Larkins 1991; Moro et al. 1995). The *floury*2 (*fl*2) and *Defective endosperm B-30* (*De-B30*) mutations, which are caused by defective signal peptides on a 22-kDa and 19-kDa α-zein, respectively, disrupt protein body formation by trapping the α-zein at the surface, which leads to small, irregularly lobed protein bodies (Fig. 19.3D and E; Coleman et al. 1997; Kim et al. 2004). The small, misshapen protein bodies in the *Mucronate* (*Mc*) mutant are caused by a defective 16-kDa γ-zein that results from a deletion mutation that alters the reading frame near the 3′ end of the coding region (Kim et al. 2006). Each of the *fl2*, *De-B30* and *Mc* mutations cause the "unfolded protein response" (UPR) (Kaufman 1999), which leads to the up-regulation of genes encoding chaperones as part of the cellular mechanism for ameliorating the effects of ER stress (Hunter et al. 2002).

Although the *floury1* (*fl1*) mutant was first reported almost a century ago (Hayes and East 1915), its study received less attention than that of *o2* and *fl2*, because its nutritional quality is not substantially different from wild-type kernels. Furthermore, its opaque kernel phenotype was somewhat enigmatic, since the mutant was not apparently defective in zein accumulation. We recently determined the molecular identity of the *Fl1* gene, which encodes a novel endoplasmic reticulum protein that is specific to the membrane surrounding zein protein bodies (Holding et al. 2007). Apart from the opaque endosperm, *fl1*mutants do not share any of the phenotypes

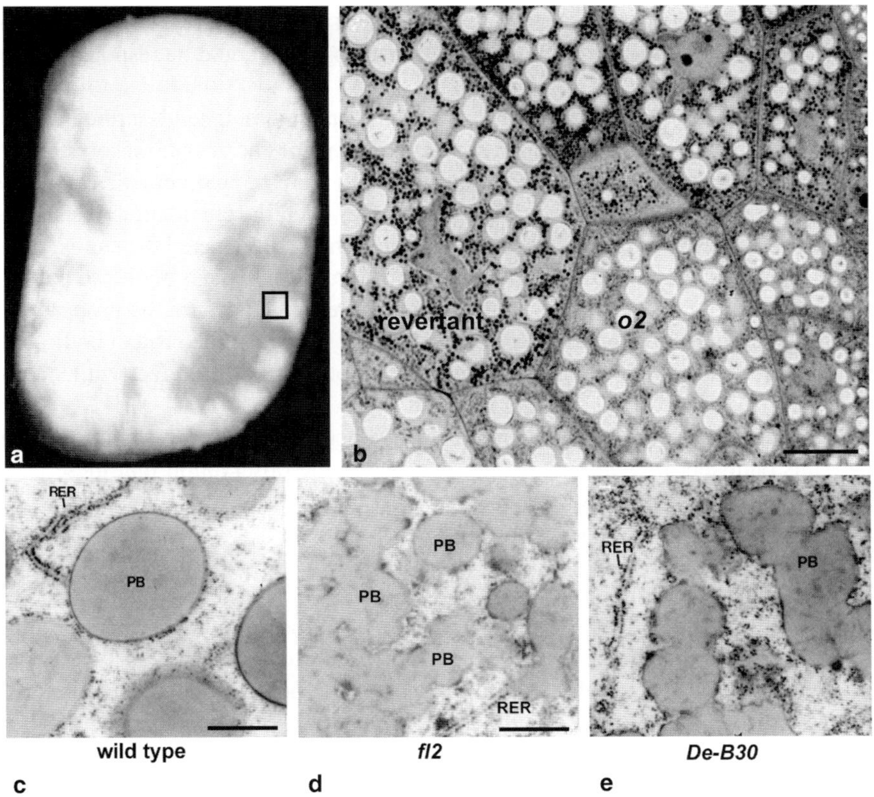

Fig. 19.3 Zein protein bodies in endosperms of opaque mutants. **a** Section of an *o2* Spm-o2o2o2-mutable kernel showing white opaque regions and sectors of yellow, revertant, vitreous endosperm. **b** TEM of a region similar to the *boxed* region in **a**. This shows the boundary between cells that constitute revertant, vitreous endosperm (with abundant protein bodies surrounding starch granules) and cells that constitute opaque endosperm (with small lightly staining protein bodies). *Scale bar* = 50 μm. **c** TEM showing wild-type protein bodies in 18-DAP endosperm cell. **d** TEM showing *fl2* protein bodies in 18-DAP endosperm cell. **e** TEM showing *De-B30* protein bodies in 18-DAP endosperm cell. In **c**, **d** and **e** (same magnification), *PB* = protein body, *RER* = rough endoplasmic reticulum and *scale bars* = 200 nm

associated with previously described opaque mutants, such as general or specific reductions in zein accumulation, a constitutive UPR, as occurs in *fl2*, DeB30 and *Mc*, or alterations in protein body size and shape. However, *fl1* mutants accumulate 22-kDa α-zeins in abnormal locations within protein bodies. In wild-type protein bodies, the 22-kDa α-zeins occupy a discrete ring at the interface between the protein body core and the γ-zein-rich periphery. In *fl1* protein bodies, 22-kDa α-zein is dispersed more randomly and is found in the center of the core as well as in the periphery, often in close association with the ER membrane. Furthermore, yeast two-hybrid experiments showed a strong interaction between the FL1 C terminus and a 22-kDa α-zein. These results support a role for FL1 in the correct placement of the hydrophobic 22-kDa α-zein and suggest that this location is essential for proper protein body formation and the generation of vitreous endosperm. They also suggest that the reduced synthesis of 22-kDa α-zeins in *o2* could be responsible for the opaque phenotype of that mutant.

The descriptions of the molecular basis of the opaque mutants described above have not resulted in a central, unifying explanation for the formation of starchy rather than vitreous endosperm, and, indeed, opacity may be symptomatic of a variety of stresses and effected in a variety of ways. However, the elucidation of a single and precise abnormality in *fl1* suggests a common theme. It appears that zeins must be packaged correctly within the protein body in order for protein bodies to participate in the formation of vitreous endosperm. These mutants share the property of ectopic accumulation of hydrophobic α-zeins at the surface of the protein body, and it is possible that this affects the way in which protein bodies interact with starch grains during desiccation and kernel maturation.

The peripheral location of γ-zeins in protein bodies and their association with a vitreous endosperm phenotype is consistent with the observation that these proteins influence the processing of maize kernels during the wet milling process. In order to separate starch grains from protein bodies, the grain must be steeped in hot sulfurous acid for 24 h to reduce disulfide bonds. Disulfide bonds also influence protein digestibility. Endosperm protein is more digestible in vitro if an agent is added that reduces disulfide bonds. Furthermore, flour from the *fl2* mutant, where protein body structure is disrupted, is more accessible to proteases than wild type (Oria et al. 2000). While several observations support the hypothesis that protein bodies, and γ-zein proteins in particular, influence endosperm texture, it is clear that other factors contribute to this trait. For example, the *o1*mutation has only a minor effect on zein synthesis and protein body size, but its endosperm is completely starchy. The *o5*, *o9* and *o11* mutants also have relatively minor effects on zein synthesis, but they also have starchy endosperms. Clearly, the relationship between zeins, protein body structure and endosperm texture is complex.

19.2.2.4 Other Endosperm Storage Proteins

Although zeins are by far the most abundant storage proteins in the endosperm, other minor storage proteins, including an 18-kDa α-globulin (Woo et al. 2001), have been identified. The α-globulin is found in smooth, vacuole-like organelles that are similar to but different from the RER-surrounded zein protein bodies. These novel protein bodies are much less abundant than zein protein bodies, have a more irregular shape and can be up to three times their size in mature cells. As their size increases, the immuno-gold labeling intensity becomes progressively weaker, suggesting that other proteins are stored within this type of protein body (Woo et al. 2001). A small amount of α-globulin is also detectable at the periphery of zein protein bodies.

A 51-kDa protein identified as legumin-1 is also found in the endosperm (Yamagata et al. 2003). Structurally, this protein resembles the 11S globulin found in the embryos of legumes and other dicots, where it is generally cleaved into 20- and

35-kDa subunits following synthesis. However in maize, it remains intact, because the canonical cleavagesequence is absent. Like the 18-kDa globulin, legumin-1 is found in protein bodies that are morphologically distinct from zein protein bodies (Yamagata et al. 2003), and it is likely to co-localize with the α-globulin.

19.3 High Lysine Corn and the Development of Quality Protein Maize

Most of the corn produced in the US ($\sim$70%) is used for food and livestock feed, and consequently the nutritional quality of the grain is an important consideration. Discovery that the *o2* mutation increases lysine content by decreasing zein synthesis and increasing the level of other types of endosperm proteins prompted a search for similar mutants in other cereal species (Munck 1992). Unfortunately, the low density and soft texture of kernels of this type of mutant were associated with a number of inferior agronomic traits, including brittleness and insect susceptibility. With only a few exceptions (Habben and Larkins 1995), these mutants were not commercially developed. However, not long after the discovery of *o2*, maize breeders began to identify genes designated "*o2* modifiers" (*mo2*) that alter the soft, starchy texture of *o2* endosperm, giving it a normal vitreous texture. The loci controlling this trait proved to be genetically complex but nevertheless effective in ameliorating the negative features of the opaque kernel phenotype (Vasal et al. 1980). By systematically introgressing *mo2* genes into *o2* germplasm, plant breeders in South Africa and at CIMMYT in Mexico were able to develop several hard endosperm *o2* mutants that they designated "Quality Protein Maize" (QPM) (Mertz 1992; Prasanna et al. 2001). QPM has the phenotype and yield of normal maize, but the high lysine content of *o2*.

The development and widespread use of QPM germplasm has been implemented slowly. In part this is because of the technical complexity of introducing multiple *mo2* loci, while maintaining a homozygous *o*2 locus and monitoring the amino acid composition (Vasal et al. 1980). This process could be greatly accelerated if we understood the mechanism(s) by which *mo2* genes create a hard, vitreous endosperm, but unfortunately relatively little is known about the number, chromosomal location or mechanism of action of *mo2* genes. Using a limited set of RFLP markers and a bulked segregant analysis (BSA) approach (Michelmore et al. 1991), Lopes et al. (1995) identified two chromosomal regions on the long arm of chromosome 7 that are associated with *o*2 endosperm modification. A locus near the centromere is linked with the gene encoding the 27-kDa γ-zein, while the other locus, identified by a single RFLP, is at the distal end of chromosome 7, near the telomere. Evaluation of the density and content of 27-kDa γ-zein in a set of recombinant inbred lines (RILs) developed from this cross revealed tight linkage of the 27-kDa γ-zein locus with the vitreous phenotype, but it was not possible to determine the role of this protein in formation of the vitreous endosperm.

Biochemical characterization of QPM showed a distinctive difference in zein composition relative to soft *o*2 mutants. Although QPMs have reduced levels of 22-kDa α-zeins, typical of *o*2 mutants, they contain approximately twice as much

27-kD γ-zein as soft *o2* genotypes (Ortega and Bates 1983; Wallace et al. 1990). Several studies demonstrated a relationship between the content of 27-kD γ-zein and *mo2* gene dosage (Geetha et al. 1991; Lopes and Larkins 1991, 1995). As described above, the 27-kDa γ-zein appears to initiate the formation of protein bodies, and endosperms with increased amounts of this protein appear to contain more protein bodies (Dannenhoffer et al. 1995). Because the 27-kDa γ-zein becomes cross-linked by disulfide bonds during kernel desiccation (Lopes and Larkins 1991), it was hypothesized that it, along with other cysteine-rich proteins, contributes to the formation of a covalently linked proteinaceous network around the starch grains (Dannenhoffer et al. 1995). Thus, the level of 27-kDa γ-zein could be an important component of endosperm modification in QPM. However, we recently obtained a 27-kDa γ-zein null mutant (in a wild-type background) and it has a vitreous phenotype. Consequently, the relationship of the 27-kDa γ-zein to *mo2* genes and the mechanism(s) of *o2* endosperm modification are unclear.

While there is good evidence that mutations affecting zein gene expression in maize endosperm influence the vitreous kernel phenotype, several observations also implicate starch synthesis and structure with this trait. One well-known example is the *waxy1* (*wx1*) mutation, which affects amylose synthesis (Nelson and Rines 1962). *wx1* is a mutation of the gene encoding granule-bound starch synthase (GBSS I), and in homozygous mutants starch grains contain amylopectin but not amylose. How this compositional change affects the molecular structure of the starch grain is poorly understood, but when placed on a light box the mutant kernels are opaque. Yet, in the region that is normally vitreous, the endosperm is congealed and hard. Mutations of some other starch genes, such as *sugary1*, which corresponds to an isoamylase-type starch debranching enzyme (James et al. 1995), and double mutants of *shrunken2* and *brittle2*, which encode subunits of ADP-glucose pyrophosphorylase (Hannah and Nelson 1976), result in kernels that are vitreous. However, the mechanisms underlying this phenotype could be quite different from those responsible for the vitreous phenotype of normal, non-starch mutant kernels. Several starch mutants, such as *sugary2*, *shrunken1*, *brittle1* and *dull1* (Hannah 1997), have essentially opaque kernel phenotypes. Many starch mutations cause a reduction in zein synthesis (Tsai et al. 1978; Giroux et al. 1994), which complicates the explanation of their phenotypes based solely on changes in starch synthesis. In any case, these observations further underscore how little we understand about the basis of the vitreous kernel phenotype.

A proteomic analysis to examine non-zein proteins that contribute to the vitreous phenotype of QPM was reported for nearly isogenic lines of CM105 wild type, *o2* and *mo2* (Gibbon et al. 2003). This study demonstrated that the starch from QPM is, in fact, different from its wild-type or *o2* counterparts. QPM starch swells in water to a larger volume than wild-type or *o2* starch, and the greater accessibility of solvent results in greater extractability of GBSS I. The starch granule swelling phenotype is associated with a change in the amylopectin branching pattern, which shows an increased proportion of short branches and decreased levels of intermediate-length branches. The most striking consequence of the altered branching pattern is that starch granules in CM105*mo2* endosperm form contacts between one another that

Fig. 19.4 Starch granule morphology in wild-type, *o2* and modified *o2* mature endosperms. **a** SEM image of wild-type (CM105+) starch granules (*SG*) surrounded by protein bodies (*PBs*). **b** SEM image of CM105*o2* starch granules. Note lack of protein bodies and lack of starch granule interconnections. **c** SEM image CM105 *mo2* starch granules. Note lack of protein bodies and the presence of multiple interconnections between starch granules. *Scale bars* = 10 μm

are not observed in wild-type or *o2* genotypes by SEM (Fig. 19.4). The arrangement of starch granules in endosperm of other QPM genotypes is similar to CM105*mo2*, which has extensive contacts between adjacent starch granules and the spaces between them are completely filled in vitreous areas of the kernel. Subtle differences in the organization of amylopectin branching, such as occurs in amorphous regions of the starch granule, could contribute to the unusual contacts observed between *mo2* starch granules.

There is evidence of starch–protein interaction influencing kernel texture in other cereals. In wheat and barley, lipid-binding puroindolin proteins associate with starch granules, causing them to be more friable, and this leads to a soft endosperm (Beecher et al. 2001; Morris 2002). However, this relationship apparently does not apply to rice and maize. Indeed, whether or not starch structure contributes to endosperm texture in rice is unclear. Studies on the physical properties of white-core (partially starchy) rice showed that physical properties are not related to starch structure (Tamaki et al. 2006), while another study of starchy endosperm cultivars showed that they have less amylose and a higher proportion of short amylopectin branches than their translucent counterparts (Patindol and Wang 2003). The importance of protein composition to kernel hardness in rice is also unclear, but increased protein content is correlated with reduced kernel breakage during milling and polishing (Leesawatwong et al. 2005).

19.4 Future Perspectives

Although zein proteins have been the subject of research for more than 100 years, many fundamentally important questions remain to be answered regarding their structure, synthesis and assembly into protein bodies, and their genetic regulation. Considering the growing importance of maize as both a feed/food and energy crop, it is likely that the structure, function and utilization of zein proteins will continue to be an active area of research in the future.

Among the developments that are likely to occur in the next few years is the creation of high protein quality maize, as was originally envisioned for the "high lysine corn" mutant *o2*. The starchy endosperm texture of *o2* was the most critical deficiency of this mutant, but this problem was ameliorated by the identification of *o2* modifier loci, which led to the development of QPM (Vasal et al. 1980). However, the creation of QPM germplasm suitable for different environments throughout the world remains a complex breeding process. First, a homozygous recessive *o2* locus must be maintained in order to reduce zein content. Second, multiple unlinked modifier loci must be introgressed to create a vitreous endosperm. Finally, the lysine content must be monitored, as this trait is also genetically variable (Moro et al. 1996). Chance contamination with wild-type pollen results in the loss of the high lysine phenotype. The need to maintain a homozygous *o2* locus in QPM is overcome by the creation of zein RNAi lines (Huang et al. 2004). These dominant mutations effectively shut down zein synthesis and increase the synthesis of other endosperm proteins, creating a kernel lysine content of between 4 and 5%. It remains to be seen whether or not *o2* modifiers can convert these mutants to a vitreous phenotype, but if this is possible, this approach could create a kernel that is valuable for both human and livestock nutrition.

Zein proteins have unique functional and biochemical properties that make them suitable for a variety of food, pharmaceutical and manufactured goods. Many of these applications are summarized in the review by Lawton (2002). As he points out, one of the major limitations to the expanded use and development of zein-based products has been the high cost of the protein, relative to other materials. However, as a consequence of the expanding utilization of corn for ethanol production, there will be large amounts of by-products, especially "distiller's grains", of which zein is a major constituent. Distiller's grains can be used directly as a source of protein for cattle, but this is likely to utilize only a portion of this by-product. As a consequence, the availability and presumably lower price of zein proteins could create additional opportunities for the utilization of these proteins.

References

Adachi M, Kanamaori J, Masuda T, Yagasaki K, Kitamura K, Mikami B, Utsumi S (2003) Crystal structure of soybean 11S globulin: glycinin A3B4 homohexamer. Proc Natl Acad Sci USA 100:7395–7400

Argos P, Pedersen K, Marks MD, Larkins BA (1982) A structural model for maize zein proteins. J Biol Chem 257:9984–9990

Bagga S, Adams HP, Rodriguez FD, Kemp JD, Sengupta-Gopalan C (1997) Coexpression of the maize delta-zein and beta-zein genes results in stable accumulation of delta-zein in endoplasmic reticulum-derived protein bodies formed by beta-zein. Plant Cell 9:1683–1696

Belanger FC, Kriz AL (1989) Molecular characterization of the major maize embryo globulin encoded by the Glb1 gene. Plant Physiol 91:636–643

Beecher B, Smidansky ED, See D, Blake TK, Giroux MJ (2001) Mapping and sequences analysis of barley hordoindolines. Theor Appl Genet 102:833–840

Carbonero P, García-Olmedo F (1999) A multigene family of trypsin/α-amylase inhibitors from cereals. In: Shewry PR, Casey R (eds) Seed proteins. Kluwer, Dordrecht, pp 617–633

Casey R (1999) Distribution and some properties of seed globulins. In: Shewry PR, Casey R (eds) Seed proteins. Kluwer, Dordrecht, pp 159–169

Chui CF, Falco SC (1995) A new methionine-rich seed storage protein from maize. Plant Physiol 107:291

Coleman CE, Larkins BA (1999) The prolamins of maize. In: Shewry PR, Casey R (eds) Seed proteins. Kluwer, Dordrecht, pp 109–139

Coleman CE, Herman EM, Takasaki K, Larkins BA (1996) The maize alpha-zein sequesters gamma-zein and stabilizes its accumulation in protein bodies of transgenic tobacco endosperm. Plant Cell 12:2335–2345

Coleman CE, Clore AM, Ranch JP, Higgins R, Lopes MA, Larkins BA (1997) Expression of a mutant alpha-zein creates the floury2 phenotype in transgenic maize. Proc Natl Acad Sci USA 94:7094–7097

Dannenhoffer JM, Bostwick DE, OR E, Larkins BA (1995) Opaque-15, a maize mutation with properties of a defective opaque-2 modifier. Proc Natl Acad Sci USA 92:931–1935

Domony C (1999) Inhibitors of legume seeds. In: Shewry PR, Casey R (eds) Seed proteins. Kluwer, Dordrecht, pp 635–655

Esen A (1987) A proposed nomenclature for the alcohol-soluble proteins (zeins) of maize (*Zea mays* L.). J Cereal Sci 5:117–128

Esen A, Stetler DA (1992) Immunocytochemical localization of the delta-zein in protein bodies of maize endosperm cells. Am J Bot 79:243–248

Garratt R, Oliva G, Caracelli I, Leite A, Arruda P (1993) Studies of the zein-like α-prolamins based on an analysis of amino acid sequences: implication for their evolution and three dimensional structures. Proteins: Structure, Function Genetics 15:88–99

Geetha KB, Lending CR, Lopes MA, Wallace JC, Larkins BA (1991) Opaque-2 modifiers increase gamma-zein synthesis and alter its spatial-distribution in maize endosperm. Plant Cell 3:1207–1219

Geli MI, Torrent M, Ludevid D (1994) Two structural domains mediate two sequential events in gamma-zein targeting – protein endoplasmic-reticulum retention and protein body formation. Plant Cell 6:1911–1922

Gibbon BC, Wang XL, Larkins BA (2003) Altered starch structure is associated with endosperm modification in quality protein maize. Proc Natl Acad Sci USA 100:15329–15334

Giroux MJ, Boyer C, Feix G, Hannah LC (1994) Coordinated transcriptional regulation of storage product genes in maize endosperm. Plant Physiol 106:713–722

Habben JE, Larkins BA (1995) Improving protein quality in seeds. In: Kijel J, Galili G (eds) Seed development and germination. Marcel Dekker, New York, pp 791–810

Hamaker BR, Larkins BA (2000) Maize food and feed: a current perspective of future possibilities. In: Khachatourians G, McHughen A, Scora R, Nip W-K, Hui YH (eds) Transgenic plants. Marcel Dekker, New York, pp 637–653

Hannah LC (1997) Starch synthesis in maize endosperm. In: Larkins BA, Vasil IK (eds) Cellular and molecular biology of plant seed development. Kluwer, Dordrecht, pp 375–405

Hannah LC, Nelson OE (1976) Characterization of ADP-glucose pyrophosphorylase from shrunken1 and brittle2 mutants of maize. Biochem Genet 14:547–560

Hayes HK, East EM (1915) Further experiments on inheritance in maize. Conn Agric Exp Stn Bull 188:1–31

Herman EM, Larkins BA (1999) Protein storage bodies and vacuoles. Plant Cell 11:601–613

Holding DR, Otegui MS, Li B, Meeley RB, Dam T, Hunter BG, Jung R, Larkins BA (2007) The maize *floury1* gene encodes a novel endoplasmic reticulum protein involved in zein protein body formation. Plant Cell 19:2569–2582

Huang SS, Adams WR, Zhou Q, Malloy KP, Voyles DA, Anthony J, Kriz AL, Luethy MH (2004) Improving nutritional quality of maize proteins by expressing sense and antisense zein genes. J Agric Food Chem 52:1958–1964

Hunter BG, Beatty MK, Singletary GW, Hamaker BR, Dilkes BP, Larkins BA, Jung R (2002) Maize opaque endosperm mutations create extensive changes in patterns of gene expression. Plant Cell 14:2591–2612

James MG, Robertson DS, Myers AM (1995) Characterization of the maize gene sugary1, a determinant of starch composition in kernels. Plant Cell 7:417–429

Kaufman RJ (1999) Stress signaling from the lumen of the endoplasmic reticulum: coordination of gene transcriptional and translational controls. Genes Dev 13:1211–1233

Kim CS, Woo YM, Clore AM, Burnett RJ, Carneiro NP, Larkins BA (2002) Zein protein interactions, rather than the asymmetric distribution of zein mRNAs on endoplasmic reticulum membranes, influence protein body formation in maize endosperm. Plant Cell 14: 655–672

Kim CS, Hunter BG, Kraft J, Boston RS, Yans S, Jung R, Larkins BA (2004) A defective signal peptide in a 19-kD alpha-zein protein causes the unfolded protein response and an opaque endosperm phenotype in the maize De*-B30 mutant. Plant Physiol 134:380–387

Kim CS, Gibbon BC, Gilikin JW, Larkins BA, Boston RS, Jung R (2006) The maize Mucronate mutation is a deletion in the 16-kDa γ-zein gene that induces the unfolded protein response. Plant J 48:440–451

Kirihara JA, Petri JB, Messing J (1988) Isolation and sequence of a gene encoding a methionine-rich 10-kDa zein protein from maize. Gene 71:359–370

Kriz AL (1989) Characterization of embryo globulins encoded by the maize Glb genes. Biochem Genetics 27:239–251

Kriz AL, Wallace NH (1991) Characterization of the maize globulin-2 gene and analysis of 2 null alleles. Biochem Genetics 29:241–254

Landry J, Moreaux T (1970) Héterogénéte des glutélines du grain du maize: extraction sélective et composition et acidés aminés des trois fractions isolíes. Bull Soc Chim Biol 52:1021–1037

Larkins BA, Vasil IK (1997) Advances in cellular and molecular biology of plant development. Vol 4. Molecular biology of plant seed development. Kluwer, Dordrecht

Larkins BA, Wallace JC, Galili G, Lending CR, Kawata E (1989) Structural analysis and modification of maize storage proteins. Dev Indust Micro 30:203–209

Lawton JW (2002) Zein: a history of processing and use. Cereal Chem 79:1–18, 31:196–200

Leesawatwong M, Jamjod S, Kuo J, Dell B, Rerkasem B (2005) Nitrogen fertilizer increases seed protein and milling quality of rice. Cereal Chem 82:588–593

Lending CR, Larkins BA (1989) Changes in the zein composition of protein bodies during maize endosperm development. Plant Cell 1:1011–1023

Li CP, Larkins BA (1996) Expression of protein disulfide isomerase is elevated in the endosperm of the maize floury-2 mutant. Plant Mol Biol 30:873–882

Lopes MA, Larkins BA (1991) Gamma-zein content is related to endosperm modification in quality protein maize. Crop Sci 31:1655–1662

Lopes MA, Larkins BA (1995) Genetic-analysis of opaque2 modifier gene activity in maize endosperm. Theor Appl Genet 91:274–281

Lopes MA, Takasaki K, Bostwick DE, Helentjaris T, Larkins BA (1995) Identification of two opaque2 modifier loci in quality protein maize. Mol Gen Genet 247:603–613

McKinney LL (1958) Zein. In: Clark GL (ed) The encyclopedia of chemistry. Reinhold, New York, pp 319–320

Mertz ET (1992) Quality protein maize. American Association of Cereal Chemists, St. Paul

Milchemore RW, Paran I, Keseli RV (1991) Identification of markers linked to disease-resistant genes by bulked segregant analysis: a rapid method to detect markers in specific genomic regions by using segregation populations. Proc Natl Acad Sci USA 88:9828–9832

Misra PS, Mertz ET, Glover DV (1975) Characteristics of proteins in single and double endosperm mutants of maize. In: High quality protein maize. CIMMYT-Purdue international symposium on protein quality in maize, El Batan, 1972. Dowden, Hutchinson & Ross, New York, pp 291–305

Moro GL, Lopes MA, Habben JE, Hamaker BR, Larkins BA (1995) Phenotypic effects of opaque2 modifier genes in normal maize endosperm. Cereal Chem 72:94–99

Moro GL, Habben JE, Hamaker BR, Larkins BA (1996) Characterization of the variability for lysine content in normal and opaque2 maize endosperm. Crop Sci 36:1651–1659

Morris CF (2002) Puroindolines: the molecular genetic basis of wheat grain hardness. Plant Mol Biol 48:633–647

Munck L (1992) The case of high-lysine barley breeding. In: Shewry P (ed) Barley, genetics, biochemistry, molecular biology and biotechnology. CAB International, Wallingford, pp 573–601

Nelson OE Jr, Rines HW (1962) The enzymatic deficiency in the waxy mutant of maize. Biochem Biophys Res Comm 9:297–300

Nielsen K, Boston RS (2001) Ribosome-inactivating proteins: a plant perspective. Annu Rev Plant Physiol Plant Mol Biol 52:785–816

Oria MP, Hamaker BR, Axtell JD, Huang CP (2000) A highly digestible sorghum mutant cultivar exhibits a unique folded structure of endosperm protein bodies. Proc Natl Acad Sci USA 97:5065–5070

Ortega EI, Bates LS (1983) Biochemical and agronomic studies of two modified hard-endosperm opaque-2 maize (*Zea mays* L.) populations. Cereal Chem 60:107–111

Osborne TB (1897) Amount and properties of the proteins of the maize kernel. J Am Chem Soc 19:525–532

Osborne TB (1908) Our present knowledge of plant proteins. Science 28:417–427

Patindol J, Wang YJ (2003) Fine structures and physicochemical properties of starches from chalky and translucent rice kernels. J Agric Food Chem 51:2777–2784

Paulis JW, Wall JS (1971) Fractionation and properties of alkylated-reduced corn glutelin proteins. Biochim Biophys Acta 251:57–69

Paulis JW, James C, Wall JS (1969) Comparison of glutelin proteins in normal and high-lysine corn. J Agric Food Chem 17:1301–1305

Pedersen K, Argos P, Naravana SVL, Larkins BA (1986) Sequence analysis and characterization of a maize gene encoding a high sulfur zein protein of Mr 15,000. J Biol Chem 261:6279–6284

Prasanna BM, Vasal SK, Kassahun B, Singh NN (2001) Quality protein maize. Curr Sci 81:1308–1319

Shewry PR (1999) Enzyme inhibitors of seeds. In: Shewry PR, Casey R (eds) Seed proteins. Kluwer, Dordrecht, pp 587–615

Shewry PR, Casey R (1999) Seed proteins. Kluwer, Dordrecht

Shewry PR, Halford NG (2002) Cereal storage proteins: structure, properties and role in grain utilization. J Exp Bot 53:947–958

Shewry PR, Tatham AS (1999) The characteristics, structures and evolutionary relationships of prolamins. In: Shewry PR, Casey R (eds) Seed proteins. Kluwer, Dordrecht, pp 11–33

Sodek L, Wilson CM (1971) Amino acid composition of proteins isolated from normal, opaque-2 and floury-2 corn endosperms by a modified Osborne procedure. J Agric Food Chem 19:1144–1150

Song R, Messing J (2002) Contiguous genomic DNA sequence comprising the 19-kD zein gene family from maize. Plant Physiol 130:1626–1635

Song R, Llaca V, Linton E, Messing J (2001) Sequence, regulation, and evolution of the maize 22-kDa alpha zein gene family. Genome Res 11:1817–1825

Tamaki M, Kurita S, Toyomaru M, Itani T, Tsuchiya T, Aramaki I, Okuda M (2006) Difference in the physical properties of white-core and non-white-core kernels of the rice varieties for sake brewing is unrelated to starch properties. Plant Prod Sci 9:78–82

Tatham AS, Field JM, Morris VJ, I'Anson KJ, Cardle L, Dufton MJ, Shewry PR (1993) Solution conformation analysis of the α-zein proteins of maize. J Biol Chem 268:26,253–26,259

Thompson GA, Larkins BA (1993) Characterization of zein genes and their regulation in maize endosperm. In: Walbot V, Freeling M (eds) The maize handbook. Springer, Heidelberg, pp 639–647

Tsai CY, Huber DM, Warren HL (1978) Relationship of kernel sink for N to maize productivity. Crop Sci 18:399–404

Vasal SK, Villegas E, Bjarnason M, Gelaw B, Goertz P (1980) Genetic modifiers and breeding strategies in developing hard endosperm opaque2 materials. In: Pollmer WG, Phillips RH (eds) Quality traits of maize for grain and silage use. Martinus Nijhoff, London, pp 37–73

Wallace JC, Lopes MA, Paiva E, Larkins BA (1990) New methods for extraction and quantitation of zeins reveal a high content of gamma-zein in modified ppaque-2 maize. Plant Physiol 92:191–196

Williamson JD, Galili G, Larkins BA, Gelvin SB (1988) The synthesis of a 19-kilodalton zein in transgenic petunia plants. Plant Physiol 88:1002–1007

Woo YM, Hu DWN, Larkins BA, Jung R (2001) Genomics analysis of genes expressed in maize endosperm identifies novel seed proteins and clarifies patterns of zein gene expression. Plant Cell 13:2297–2317

Yamagata T, Kato H, Kuroda S, Abe S, Davies E (2003) Uncleaved legumin in developing maize endosperm: identification, accumulation and putative subcellular localization. J Exp Bot 54:913–922

Young TE, Gallie DR (2000) Regulation of programmed cell death in maize endosperm by abscisic acid. Plant Mol Biol 42:397–414

Chapter 20
The Complexities of Starch Biosynthesis in Cereal Endosperms

L. Curtis Hannah and Thomas Greene

20.1 Introduction

The majority of agriculturally important crops in the world produce substantial amounts of starch. Starch, a polymer of glucose moieties, comprises approximately 70% of the dry weight of the major cereal grain grown worldwide. As such, starch provides up to 80% of the calories consumed daily by humans, and it is also the major source of energy for the animals we consume. In addition, there are many industrial uses for starch, including the growing production of ethanol as an alternative fuel. Given the growing human population, the ever-shrinking land areas available for plant production and the increasing demand for starch for ethanol and other industrial chemicals, knowing how to produce more starch and create starches of different structures are research areas of paramount importance. Because the focus of this book is on molecular and genetic approaches to maize improvement, our emphasis will be on genetic alterations – either spontaneous mutations or transgenic approaches – that alter starch quality or quantity. Genes of extant commercial use will receive particular emphasis, and discussion will be limited to genes known to be important for starch synthesis, although the reader should be aware that not all of the important genes have been identified. This chapter should be viewed as an overview, because space constraints limit detailed discussions of each topic.

L. Curtis Hannah
University of Florida, Program in Plant Molecular and Cellular Biology and Department of Horticultural Sciences, P.O. Box 110690, 2211 Fifield Hall, Gainesville, FL 32611, USA
e-mail: hannah@mail.ifas.ufl.edu

Thomas Greene
Dow AgroSciences, 9330 Zionsville Road, Indianapolis, IN 46268, USA
e-mail: twgreene@dow.com

A.L. Kriz, B.A. Larkins (eds.), *Molecular Genetic Approaches to Maize Improvement* 287
Biotechnology in Agriculture and Forestry, Vol. 63

20.2 The Starch Biosynthetic Pathway

Starch is a bioengineering marvel. While conceptually simple in structure, the non-random nature of branch points within the molecule allow the formation of insoluble granules containing tightly packed glucose moieties. In all likelihood, the cereal grain yields of today would not be possible were it not for this peculiar property of starch. In addition, because of its structure, starch synthesis and degradation can occur quickly; following appropriate cellular cues, the processes of synthesis and degradation occur so rapidly that the cycle is completed daily in leaves.

Typically two polymers, amylose and amylopectin, make up starch. Amylopectin is the major polymer, comprising approximately 75% of starch granules. Amylose is made of glucose residues linked almost exclusively through α-1, 4 linkages, whereas amylopectin has an α-1, 4 linked backbone with non-randomly placed α-1, 6 branch points. The latter linkages comprise approximately 5% of the glycosidic bonds. The more highly branched phytoglycogen, which contains 10% α-1, 6 bonds, is present in some plant mutants lacking starch debranching activity. The α-1, 6 branch points of amylopectin are clustered. It is believed that the clustered nature of the branch points allows glucan side-chains to form double-helical structures. This leads to the efficient packaging of glucose residues in an insoluble state and the synthesis of extremely large molecules.

Glucose is delivered from source leaves to maize endosperm in the form of sucrose. Two classic maize mutations, *Sh1* (*shrunken1*) and *Mn1* (*miniature seed1*) (reviewed in Koch 2004), affect the initial metabolism of sucrose. While both enzymes appear to be important for the cleavage of sucrose, multiple paralogs for both genes exist that are expressed differentially and appear to have slightly different functions. Seeds cultured in vitro with reducing sugars can synthesize sucrose (Cobb and Hannah 1988), and it appears that sucrose is made twice in the pathway, a result consistent with the early studies of Shannon (1968).

Six carbon sugars are not converted directly to starch. Rather, they undergo extensive intermediary metabolism in the cytosol. Glawischnig et al. (2002) fed developing kernels glucose or sucrose in which every labeled molecule was composed entirely of ^{13}C. The resulting starch was cleaved and the pattern of glucose labeling was examined. Label in the 1, 2 and 3 carbons or the 4, 5 and 6 carbons were the most frequent patterns observed. Extensive sugar catabolism through glycolysis, glucogenesis and the pentose phosphate pathway apparently occurred before starch synthesis. Interestingly, genetic perturbations of the starch biosynthetic pathway do not drastically affect the starch labeling pattern (Spielbauer et al. 2006). Metabolic studies clearly show that the starch biosynthetic pathway competes for carbon with other pathways in the endosperm of cereal crops. These interactions create clear challenges from a genetic engineering perspective and raise concerns for the potential to channel additional carbon into starch and away from other metabolic pathways in the seed.

Four enzymatic activities are especially important for starch biosynthesis. The first is adenosine diphosphate glucose pyrophosphorylase (AGPase) which synthesizes ADP-glucose, the primary sugar nucleotide donor of glucose residues for starch biosynthesis. A class of enzymes termed "starch synthases" elongate the glucose polymers by formation of α-1, 4 bonds. Starch debranching enzymes (DBE), which cleave α-1, 6gonds, are also required for synthesis starch. The role of these enzymes is reviewed below with particular emphasis on commercially exploited genes and enzymes. Readers are referred to a number of recent and more detailed reviews on this subject (Ball and Morell 2003; Hannah 1997, 2005, 2007; James et al. 2003; Martin and Smith 1995; Myers et al. 2000; Nelson and Pan 1995; Preiss and Levi 1980; Preiss et al. 1991; and Tetlow et al. 2004).

20.3 Adenosine Diphosphate Glucose Pyrophosphorylase (AGPase)

Because of the importance of AGPase in controlling the amount of starch synthesized, a detailed discussion of this enzyme is presented. Definitive proof that this enzyme is a rate-limiting step in starch biosynthesis was shown by five different research groups in six separate reports using four AGPase variants expressed in five different plant species. These are reviewed below.

AGPase catalyzes the first step in starch biosynthesis, the synthesis of ADP-glucose (and pyrophosphate) from glucose-1-phosphate and ATP. AGPases are tetramers composed of subunits in the 50,000-kD range. Bacterial AGPases are homotetramers, whereas the plant enzymes are $\alpha_2\beta_2$ heterotetramers composed of two small and two large subunits. In the case of maize endosperm, the small subunit is encoded by the *brittle-2* (*Bt2*) gene, whereas *shrunken-2* (*Sh2*) encodes the large subunit. Early work with maize mutants showed that alterations in *Sh2* or *Bt2* led to severe kernel phenotypes showing dramatic reductions in starch (Hannah and Nelson 1976). A similar mutational genetic approach with genes associated with the other steps in the starch biosynthetic pathway also show some level of reduction in the amount or structure of starch. However, *shrunken-2* and *brittle-2* mutants clearly show two of the most severe starch phenotypes in maize endosperm associated with single gene alterations and thus provide evidence for the prominent role of AGPase in starch biosynthesis.

20.3.1 Subunits of AGPase

The structural genes for the small and large AGPase subunits exhibit significant sequence similarity, and the evidence points to a gene duplication that occurred early in the evolution of plants. Subsequently, additional gene duplications occurred, giving rise to various coding sequences for the two subunits. Interestingly, plants typically have fewer gene copies for the small subunit compared to the large subunit, and there exists more sequence heterogeneity in the large subunit family compared to the small subunit family.

Georgelis et al. (2007) recently addressed the basis for the differential rate of sequence divergence of the two AGPase structural genes. From measurement of the rate of non-synonymous mutations per synonymous mutation in angiosperm genes, it is clear that the large subunit is under less selection pressure than the small subunit. Because investigators of potato tuber AGPase suggested that the two subunits have vastly different roles in catalysis, Georgelis and colleagues tested the hypothesis that the large subunit can tolerate a greater portion of missense mutations compared to the small subunit. Accordingly, error-prone PCR mutants created for each subunit were expressed in *E. coli* with a wildtype complementary subunit. Activity was initially measured by iodine staining of the synthesized glycogen. In contrast to the hypothesis above, these results show that each subunit is equally vulnerable to enzyme activity-altering amino acid changes.

An important difference distinguishing *E. coli* expression and plant expression of plant AGPase genes is that in plants the small subunits are typically expressed in more tissues compared to the expression pattern of the large subunit genes. This is because there are typically fewer small subunit genes than large subunit genes. To examine the significance of this difference, Georgelis et al. (2007) analyzed several small subunit mutants that are functional in combination with the wildtype *Sh2* gene. One-fourth of these enzymes were not functional with the maize embryo expressed gene, even though the wildtype BT2 protein could assemble with the embryo large subunit and produce enzyme activity in *E. coli*. In agreement with the hypothesis that expression in different tissues places extra constraints on genes, Georgelis et al. (2007) noted that in green unicellular algae the rates of divergence of the two gene sequences do not differ. The algae have only one copy of each gene. Hence, the differential rate of sequence divergence exhibited by genes encoding the two subunits is not due to different roles in catalysis; rather it is due to differences in the spectra of tissue expression of individual gene family members.

20.3.2 Subcellular localization of AGPase

A surprising feature of AGPases is that the enzyme is located in the plastid in some tissues and in the cytosol in others. Because studies of AGPase in spinach leaf and potato tuber (Kim et al. 1989; Okita et al. 1979) showed the enzyme was plastid-localized, it was assumed that this would be the case with all AGPases. The earliest indication that this might not be true came from cloning and sequencing maize and barley endosperm AGPase structural genes (Giroux and Hannah 1994; Villand and Kleczkowski 1994) which revealed the absence of transit peptides on the proteins. Additionally, proteins produced via in vitro transcription/translation experiments produced polypeptides indistinguishable in size from proteins isolated from the maize endosperm (Giroux and Hannah 1994).

Definitive evidence that the major maize endosperm AGPase is cytosolic came from studies performed by Jack Shannon and associates (Cao et al. 1995; Shannon (1996), 1998). These investigators showed that endosperms containing a null allele of *brittle-1* (*Bt1*) accumulated elevated levels of ADP-glucose. Because *Bt1* encodes a membrane-bound metabolite transporter (Sullivan et al. 1991), and loss-of-function *bt1* alleles are massively deficient in starch synthesis, the metabolite transporter encoded by *Bt1* must be important in starch synthesis. Furthermore, because ADP-glucose accumulates in this mutant, ADP-glucose synthesis must precede transport across the membrane. Importantly, a double knock-out lacking both the *Sh2*-encoded AGPase and the transporter did not accumulate ADP-glucose. This provided definitive evidence that the source of ADP-glucose must come from a cytosolic, *Sh2*-controlled AGPase. Confirmatory results for the cytosolic localization of endosperm AGPases came from an entirely different approach, i.e. cellular fractionation studies. Denyer et al. (1996) reported that greater than 95% of AGPase activity is cytosolic in maize endosperm, and Thorbjornsen et al. (1996) found that 85% of AGP activity is cytosolic in barley endosperm. Finally, Beckles et al. (2001) concluded, based on levels of ADP-glucose and UDP-glucose, that a cytosolic location for AGP is a feature of graminaceous endosperms, but not of other starch-storing organs.

The significance of the classic and seminal studies of Shannon and associates are extremely relevant to the recently proposed alternative model for starch synthesis in the cereal endosperm. It was suggested that the source of the ADP-glucose for starch synthesis is not AGPase, but rather sucrose synthase (Munoz et al. 2006; Baroja-Fernández et al. 2003). These investigators speculated that AGPase plays a role in starch synthesis by scavenging the breakdown products of starch degradation in the amyloplast. Setting aside for the moment the fact that the major source of endosperm AGPase is in the cytosol and not the amyloplast, the proposal predicts that the accumulation of ADP-glucose in the *bt1* mutant would not be reduced in the *bt1 sh2* double mutant. That ADP-glucose is reduced in a double mutant, in our view, makes the new model of starch synthesis untenable.

20.3.3 Allosteric Properties of AGPase

AGPase is an allosteric enzyme and alterations in allostery affect the amount of starch synthesized. The range of activators and inhibitors that regulate bacterial AGPase activity is diverse and reflects, for the most part, the types of sugars used to support growth of the bacterium (Preiss and Romeo 1989). Plant AGPases, in contrast, are more homogenous. Three-phosphoglyceric acid (3-PGA) and inorganic phosphate (Pi) are viewed as the “classic” activator and inhibitor, respectively, of plant enzymes.

While the activity of most plant AGPases is modulated by 3-PGA and Pi, there exist notable differences in the pattern of responses to these allosteric effectors. For example, Pi does not significantly inhibit maize endosperm AGPase in the absence of 3-PGA. Rather, Pi inhibition occurs only in the presence of 3-PGA. Hence, Pi serves as a de-activator of 3-PGA stimulation (Boehlein et al. 2005). In contrast, the closely related wheat endosperm AGPase exhibits exactly the opposite response:

3-PGA does not activate AGPase in the absence of Pi. Rather 3-PGA overcomes the inhibition caused by Pi (Gomez-Casati and Iglesias 2002). Barley endosperm AGPase exhibits yet a third pattern. It is neither activated by 3-PGA nor inhibited by Pi (Kleczkowski et al. 1993; Doan et al. 1999).

Quantitative differences in the extent of 3-PGA activation also distinguish the various plant AGPases. For example, whereas 3-PGA is effectively required for potato tuber AGPase activity (Sowokinos 1981; Iglesias et al. 1993), this is not the case for cereal endosperm AGPases. Also, while fructose-6-phosphate, glucose-6-phosphate and ribose-5-phosphate give levels of activation to the maize endosperm AGPase that are comparable to that exhibited by 3-PGA (Boehlein et al. 2005), the extent of activation of the spinach leaf chloroplast AGPase caused by ribose-5-P is only 15% of that caused by 3-PGA (Ghosh and Preiss 1966).

20.3.4 Allosteric Properties of AGPase Are Pivotal in Controlling Starch Levels

A series of investigations have shown that the allosteric properties of AGPase described above are critically important in controlling the amount of starch in agriculturally important crops. In one study (Stark et al. 1992), an allosterically enhanced, *E. coli*-derived AGPase gene was expressed in the potato tuber. This led to a greater than 35% increase in starch content. The variant, termed *glgC-16*, was isolated as a glycogen over-expresser from *E. coli*. Analysis of enzyme kinetics showed that only the allosteric properties of the enzyme were altered by the mutation.

Following the report of Stark and associates, Giroux et al. (1996) showed that a maize *Sh2* allele created by transposon mutagenesis (*Rev6*) could give rise to more than a 40% increase in seed weight (as a consequence of greater starch production), depending on the genetic background and the growth environment. Further analysis showed that this variant produced an AGPase with reduced sensitivity to inorganic phosphate, Pi. Components other than starch were also increased in this mutant, suggesting that the enhanced AGPase created a stronger carbon sink in the developing kernel.

There is a critical mass of literature (reviewed in Hannah 2007) showing that among starch biosynthetic enzymes, the endosperm AGPases are particularly susceptible to heat inactivation. To investigate this, Greene and Hannah (1998a, b) isolated stability variants of the maize endosperm enzyme by mutagenesis and expressed the maize genes in *E. coli*. Bacterial cells exhibiting increased synthesis of glycogen at high temperatures were selected for analysis. One variant, termed *HS33*, showed enhanced heat stability as a consequence of strengthened subunit interactions, supporting the hypothesis that the lability of AGPase activity is a consequence of holoenzyme stability.

A recombinant double mutant gene containing the *HS33* variant and the *Rev6* alteration was synthesized and expressed in wheat, rice and maize (Smidansky et al. 2002, 2003; Hannah and Greene, in preparation). Field-grown plants showed yield increases of 38% (wheat), 23% (rice) and 68% (maize). In maize, significant yield increases were observed under field conditions when the plants were exposed to high temperature stress during the seed filling period. Unexpectedly, the yield increase ascribed to the *Sh2HS33/Rev6* transgene is due to an increased seed number. How an alteration in the endosperm starch biosynthetic pathway affects seed number is currently under active research in a number of laboratories. A working hypothesis is that early seed abortion in the ear is blocked by increased carbon flow to the seed, thus more effectively establishing early developing kernels as a carbon sink.

Three additional reports on transgenic plants point to the importance of AGPase in controlling starch levels. Sakulsingharoja et al. (2004) placed an *E. coli*-derived AGPase variant, termed the "triple mutant", into rice. This gene variant was chosen because of the altered allosteric properties it exhibited in *E. coli*. Seed weight increases of up to 11% were obtained following its expression. Additionally, the *E. coli*-derived *glgC-16* variant used in transgene potato also increased starch content in the maize seed. Wang et al. (2007) expressed the *E. coli glgC16* gene with a zein promoter in maize seed. The resulting seed exhibited an AGPase that was less sensitive to Pi inhibition and the seed weight increased 22 to 25% in four of the eight transgenic events examined.

Finally, Obana et al. (2006) targeted expression of allosterically altered potato tuber AGPase variants (Greene et al. 1998c) to the chloroplasts of Arabidopsis leaves. Transgenic plants showed an increased leaf AGPase activity, increased levels of transient starch accumulation and increased growth properties, adding further evidence that manipulation of AGPase can impact starch levels. The particular potato tuber variants employed in these studies were more sensitive to the activator 3-PGA and less sensitive to inhibition by Pi.

20.4 Starch Synthases (SS)

Starch synthases catalyze the formation of the α D-1, 4 linked backbone of amylose and amylopectin using a glucose polymer and ADP-glucose as substrates. Five separate isoforms of this enzymatic activity are known. One form is tenaciously bound to the starch granule and is termed "granule bound starch synthase" (GBSS).

Maize GBSS and its genetic/molecular characterization are significant for at least three reasons. The very first biochemical-genetic characterization of starch biosynthesis was done with GBSS some 46 years ago by Nelson and Rines (1962), who showed that the maize *waxy* (*wx*) locus encoded this enzyme. Evidence that the *Wx* locus encoded a starch synthase was surprising since loss of *Wx* function leads to a virtual loss of amylose synthesis and little effect on amylopectin synthesis and no significant reduction in total starch accumulation. The simplest explanation for this phenotype is that amylose, surprisingly, does not serve as a precursor for amylopectin synthesis. Maize containing a homozygous mutant allele at the *wx* locus is used in commercial corn production (reviewed in Fergason 2000). For many reasons, *wx* starch is a good source of amylopectin, since it is virtually devoid of amylose. Also, amylose, relative to amylopectin, is more difficult for animals to digest, a property relevant to both the food and the feed industry.

20.4.1 Starch Synthase Isoforms

Four other synthase isoforms have been described. These enzymes, termed SSI, SSII, SSIII and SSIV/V, are sometimes referred to as "soluble" starch synthases. However, some of these isoforms can also be found in the starch granule (reviewed in James et al. 2003; Ball and Morell 2003; Tetlow et al. 2004). All five starch synthase isoforms share sequence identity, suggesting they arose from a common progenitor via gene duplication. Multiple isoforms exist for each of the enzyme classes. The results of Nelson and Rines (1962) point to the existence of a minor, starch-bound enzyme that is independent of the *Wx* locus. A recent study of rice seed reported one functional gene for SSI, three genes for SSII, and two genes each for SSIII, SSIV and GBSS (Hirose and Herao 2004).

Two classic maize mutants, *dull1* (*Du1*) and *sugary2* (*Su2*), correspond to starch synthases. Gao et al. (1998) and Zhang et al. (2004) exploited transposon mutagenesis to clone these genes and demonstrated that *Du1* encodes SSIII and *Su2* encoded SSIIa. Genetic loss of SSIII increases the frequency of branching, the relative amount of amylose and decreases the number of long amylopectin chains (Wang et al. 1993). Loss of SSIIa increases the amount of amylose and increases the proportion of short glucan side chains in amylopectin (1 to 9 glucose residues) and decreases intermediate (10 to 23 glucose residues) chain length in amylopectin. SSI mutants have not been described in maize, although one was isolated in rice (Fujita et al. 2006). Comparison of several alleles showed that none conditioned a visible starch phenotype. Molecular analysis revealed that glucan chains with 8 to 12 residues were reduced in amount, whereas chains with less (6 to 7) and more (16 to 19) moieties were increased. These results from mutant endosperm analysis are fully consistent with a model based on kinetic analysis of the various isoforms (Commuri and Keeling 2001), suggesting that SSI elongates short chain (less than 10 glucose residues) glucans, whereas the other starch synthases are involved in lengthening longer side chains.

Phosphorylation and possibly interaction with 14-3-3 regulatory proteins may control starch synthase activity. Phosphorylation of SSIIa in wheat endosperm apparently enhances its activity (Tetlow et al. 2004), and the uniquely long N terminus of the *Du1*-encoded SSIII enzyme contains a binding motif for a 14-3-3 protein. Interestingly, suppression of 14-3-3 expression via antisense technology doubles starch content in Arabidopsis leaves (Sehnke et al. 2001), suggesting that 14-3-3 binding inhibits SSIII activity. This also suggests that SSIII is quite important for starch synthesis in the leaf.

Because of their effects on starch structure, loss-of-function mutants of *Du1* and *Su2* are now used in commercial corn production. Readers are referred to an excellent review by Pamela White (2000) that describes the properties of starch from *du1* and *su2* as well as other mutants.

20.5 Starch Branching Enzymes (SBE)

The α D-1, 6 glucose linkages of amylopectin are formed by starch branching enzymes (SBEs). Acting perhaps in concert with starch synthases, SBEs cleave an α D-1, 4 bond to create a chain of glucose moieties. This chain is then attached to amylopectin through formation of an α D-1, 6 bond to a glucose residue within a linear polymer.

As judged by sequence and substrate preference, SBEs can be divided into two major classes (I and II). SBEI transfers longer glucans and makes polymers with fewer 1, 6 branches, relative to SBEII (reviewed in Yao et al. 2004). Three maize endosperm SBE isoforms have been described, SBEIa, SBEIIa and SBEIIb (reviewed in Yao et al. 2004). *Sbe1a*, *Sbe2a* and *Ae* (*amylose-extender*), respectively, encode these enzymes. *Ae* mutants give rise to an endosperm phenotype detectable by eye, and loss of *Ae* activity increases the amylose/amylopectin ratio. Sbe1a and Sbe2a were first detected via molecular/biochemical analyses (Blauth et al. 2001; Yao et al. 2004). An analysis of endosperm mutants in various cereals points to changes in functions of the various isoenzymes over evolutionary time.

SBEI mutants are known in maize (Blauth et al. 2001; Yao et al. 2004) and rice (Satoh et al. 2003). While loss of SBEI apparently has no effect on amylopectin structure in maize, Satoh and associates showed this is not the case in the rice endosperm. Chains of 12 to 21 and more than 37 glucose residues are reduced in the rice SBEI mutant, whereas chains having less than 10 and between 24 and 34 glucoses were enhanced relative to wildtype.

Transposon-induced SBEIIa mutants exist in maize (Blauth et al. 2001; Yao et al. 2004) and gene silencing constructs were made in wheat (Regina et al. 2006). Loss of SBEIIa in maize gives rise to leaf starch with little or no branching, but it has little effect on endosperm starch. Loss of SBEIIa function in wheat doubles endosperm amylose content at the expense of amylopectin, and a similar pattern is seen for SBEIIb mutants. Loss of SBEIIb via mutation at the maize *Ae* locus produces amylopectin with longer chains and fewer branches per cluster, whereas loss of SBEIIb function in wheat has no discernable effect. Relative expression of these two isoforms also differs in the two cereals, perhaps providing an explanation for the differing phenotypes. Intriguingly and in contrast to predictions, removal of SBEIIb and SBEIa in the maize endosperm decreases chain length and increases the number of branches within a cluster. Yao et al. (2004) speculated that SBEIa activity inhibits SBEIIa activity in the maize endosperm. Removal of SBEIa then uncovers SBEIIa function.

Intriguingly, ATP stimulates amylopectin synthesis but not amylose synthesis in an in vitro wheat endosperm system (Tetlow et al. 2004). This suggests that phosphorylation may be important for branching and perhaps other starch biosynthetic enzyme activities. Subsequent studies showed that all three SBE enzymes require phosphorylation for maximal activity. Phosphorylation strengthens interactions involving starch phosphorylase, SBEIIb and SBEI, and this may explain the partial reduction of SBEI activity when SBEIIb activity is lost (James et al. 2003).

20.6 Starch Debranching Enzymes (DBE)

A totally unexpected, but fundamentally important, observation made through maize mutant analysis was that cleavage of α D-1, 6 linkages by starch debranching activity is prerequisite for the synthesis of wildtype levels of starch. An analysis of mutants of the *sugary-1* (*Su1*) locus by Pan and Nelson (1984) first suggested this relationship. These investigators noted a significant reduction in DBE activity in *su1* mutants. Loss of *Su1* function leads to an increase in soluble sugars, a decrease in starch content and, uniquely, the accumulation of a highly branched glucose polymer, phytoglycogen. Phytoglycogen differs from amylopectin in being significantly more highly branched and water soluble (reviewed in Dinges et al. 2001 and James et al. 2003). The accumulation of phytoglycogen in this mutant, along with the noted altered structure of amylopectin coupled with significantly reduced DBE activity in *su1* mutants, suggested that DBE is important in the synthesis of wildtype levels of starch.

James et al. (1995) employed transposon tagging to clone *Su1*. DNA sequence analysis revealed that *Su1* encodes a debranching activity termed "isoamylase". A second major class of DBE, termed "pullulanase" was also shown to be reduced in *su1*, presumably as a downstream consequence of loss of *Su1* function. The early studies of Pan and Nelson measured pullulanase activity rather than isoamylase activity. Insight into the consequences of the reduction of pullulanase activity in a *su1* mutant came from Wu et al. (2002). These investigators showed that pullulanase activity is reduced in a *sh2* mutant. Since *su1* and *sh2* mutants contain elevated levels of soluble sugars, high sugar content may down-regulate pullulanase activity or expression of the pullulanase gene. Sugars are known to regulate gene expression. One major unanswered question concerning *su1* mutations is the cause of their associated high sucrose content. While *su1* mutants have been used in the sweet corn industry for many years, the biochemical basis for their high sugar content has escaped elucidation. The failure to efficiently package glucose into starch perhaps slows the starch biosynthetic pathway, and in turn enhances sugar levels.

Analysis of *su1* mutants has highlighted some of the complexities of this enzyme and potential interactions of the SU1 protein with other enzymes of the starch biosynthetic pathway. For example, DNA sequencing of the original loss-of-function *su1* allele, termed *sugary1-Reference* (*su1-R*), revealed three amino acid changes (Dinges et al. 2001). Another transposon-induced null allele, *su1-R4582::Mu1*, was also sequenced. A third allele, termed *su1-st* (*sugary-starchy*), conditions a leaky phenotype via alternative pre-mRNA splicing. This allele conditions an intermediate phenotype in one, two or three gene doses in the triploid endosperm. While a heterozygote involving *su1-st* and the loss-of-protein *su1-R4582::Mu1* allele has an intermediate phenotype, heterozygotes of *su1-st* with *su1-R* have extreme phenotypes. Since molecular studies have shown that the SU1 protein can exist in large aggregates (reviewed in James et al. 2003), there exists the possibility that polymers containing the SU1-R protein are enzymatically inactive. Multimers produced in heterozygotes involving the *su1-R4582::Mu1* allele do not suffer this fate, since *su1-R4582::Mu1* does not produce a protein.

Dinges et al. (2003) cloned the maize pullulanase gene (*Zpu1*) via transposon mutagenesis. Subsequent characterization showed that maize contains only one structural gene for this enzyme. Gene expression occurs in several tissues and the enzyme is involved in both starch biosynthesis and degradation. Mutational loss of this enzymic activity has pronounced effects on leaf starch; the rate of leaf starch catabolism is reduced and the percentage of amylose in leaf starch is increased. In addition, germination and early seedling growth is reduced. While endosperm starch is only moderately altered, data from double mutants (Dinges et al. 2003) suggest that the functions of isoamylase and pullulanase partially overlap. Phytoglycogen is five-fold higher in the double mutant compared with the parents, and starch levels are below either single mutant. Significantly, loss of endosperm pullulanase or isoamylase causes loss of starch branching enzyme IIa. These results/observations highlight the complexity of starch synthesis and suggest the possible existence of protein complexes containing several of these enzymatic activities.

20.6.1 Physiological Role of DBEs

Identification of the physiologically relevant substrate(s) of starch debranching enzymes is an area of active research (reviewed in Myers et al. 2000; James et al. 2003; Ball and Morell 2003; Tetlow et al. 2004). Three roles for starch debranching activity have been envisaged: (1) synthesis of glucose chains to initiate starch synthesis; (2) degradation of phytoglycogen to provide substrates for starch synthesis (the clearing model); and (3) trimming of excessive branches within amylopectin (the trimming model), thereby causing clustered branch points.

Significantly, only the trimming model predicts structurally altered amylopectin in DBE mutants, a fact consistent with chemical analysis of amylopectin from *su1* mutants. Short glucan side chains (2 to 12 glucose moieties) are elevated in these mutants and longer side chains (15 to 25 glucose moieties) are decreased. An additional important fact is that the trimming model, as originally proposed (Ball et al. 1996), provides an explanation for the clustering of α D-1, 6 linkages in amylopectin. Clustered branch points could occur if branching and debranching occur at random, but steric hindrance does not allow debranching activity to cleave closely spaced branches (Myers et al. 2000).

The clustered nature of α D-1, 6 glucose linkages gives rise to parallel glucans, which can intertwine into double helices. This is obligatory for the crystalline structure of starch and the growth of glucan side-chains that potentially are unlimited in length. The importance of this to the efficient packaging of glucose into a dense, inert starch granule cannot be overstated. Plant agriculture and perhaps life as we know it would be vastly different were it not for the ability to package glucose into starch. As noted by Myers et al. (2000), the capture by plants of bacterial genes used for glycogen synthesis (with consequent random branch points) and their modification for starch synthesis is a fantastic evolutionary success story. In our opinion, a key in the evolution of plants as we know them was the ability to synthesize α D-1, 6 glucose linkages in a non-random, clustered fashion, and debranching enzymes are essential to this process.

Acknowledgements The study was supported by NSF grants IBN-9982626 and 0444031 and USDA competitive grants 2000-01488, 2006-03034 and 2007-03575 to L.C.H. We thank Brian Larkins, Martha James, Karen Koch and Don McCarty for useful discussions and very helpful comments on an earlier version of this chapter. We thank Nick Georgelis and Brandon Futch for assistance in preparing this manuscript.

References

Ball S, Guan H, James M, Myers A, Keeling A, Mouille G, Buleon A, Colonna P, Priess J (1996) From glycogen to amylopectin: a model for the biogenesis of the plant starch granule. Cell 86:349–352

Ball SG, Morell MK (2003) From bacterial glycogen to starch: understanding the biogenesis of the plant starch granule. Annu Rev Plant Biol 54:207–233

Baroja-Fernández E, Muñoz FJ, Saikusa T, Rodríguez-López M, Akazawa T, Pozueta-Romero J (2003) Sucrose synthase catalyzes the de novo production of ADPglucose linked to starch biosynthesis in heterotrophic tissues of plants. Plant Cell Physiol 44(5):500–509

Beckles DM, Smith AM, ap Rees T (2001) A cytosolic adp-glucose pyrophosphorylase is a feature of graminaceous endosperms, but not of other starch-storing organs. Plant Physiology 125: 818–827

Blauth SL, Yao Y, Klucinec JD, Shannon JC, Thompson DB, Guiltinan MJ (2001) Identification of *Mutator* insertional mutants of the starch-branching enzyme 2a in corn. Plant Physiol 125: 1396–1405

Boehlein SK, Sewell AK, Cross J, Stewart JD, Hannah LC (2005) Purification and characterization of adenosine diphosphate glucose pyrophosphorylase from maize/potato mosaics. Plant Physiol 138(3):1552–1562

Cao H, Sullivan TD, Boyer CD, Shannon JC (1995) *Bt1*, a structural gene for the major 39–44 kD amyloplast membrane polypeptides. Physiol Plant 95:176–186

Cobb BG, Hannah LC (1988) *Shrunken-1* encoded sucrose synthase is not required for sucrose synthesis in the maize endosperm. Plant Physiol 88:1219–1221

Commuri P, Keeling PL (2001) Chain-length specificities of maize starch synthase I enzyme: studies of glucan affinity and catalytic properties. Plant J 25:475–486

Denyer K, Dunlap F, Thorbjornsen T, Keeling P, Smith AM (1996) The major form of ADP-glucose pyrophosphorylase in maize endosperm is extra-plastidal. Plant Physiol 112: 779–785

Dinges JR, Colleoni C, Myers AM, James MG (2001) Molecular structure of three mutations at the maize *sugary1* locus and their allele-specific phenotypic effects. Plant Physiol 125:1406–1418

Dinges JR, Colleoni C, James MG, Myers AM (2003) Mutational analysis of the pullulanase-type debranching enzyme in maize indicates multiple functions in starch metabolism. Plant Cell 15:666–680

Doan DN, Rudi H, Olsen OA (1999) The allosterically unregulated isoform of ADP-glucose pyrophosphorylase from barley endosperm is the most likely source of ADP-glucose incorporated into endosperm starch. Plant Physiol 121(3):965–975

Fergason V (2000) High amylose and waxy corns. In: Hallauer AR (ed) Speciality corns, 2nd edn. CRC Press, Boca Raton, pp 33–62

Fujita N, Yoshida M, Asakura N, Ohdan T, Miyao A, Hirochika H, Nakamura Y (2006) Function and characterization of starch synthase I using mutants in rice. Plant Physiol 140:1070–1084

Gao M, Wanat J, Stinard PS, James MG, Myers AM (1998) Characterization of *dull1*, a maize gene coding for a novel starch synthase. Plant Cell 10:399–412

Georgelis N, Braun EL, Shaw JR, Hannah LC (2007) The two AGPase subunits evolve at different rates in angiosperms, yet they are equally sensitive to activity altering amino acid changes when expressed in bacteria. Plant Cell 19:1458–1472

Ghosh HP, Preiss J (1966) Adenosine diphosphate glucose pyrophosphorylase. A regulatory enzyme in the biosynthesis of starch in spinach leaf chloroplasts. J Biol Chem 241:4491–4504

Giroux MJ, Hannah LC (1994) ADP-glucose pyrophosphorylase in *shrunken2* and *brittle2* mutants of maize. Mol Gen Genetics 243:400–408

Giroux MJ, Shaw J, Barry G, Cobb BG, Greene T, Okita T, Hannah LC (1996) A single gene mutation that increases maize seed weight. Proc Natl Acad Sci USA 93:5824–5829

Glawischnig E, Gierl A, Tomas A, Bacher A, Eisenreich W (2002) Starch biosynthesis and intermediary metabolism in maize kernels. Quantitative analysis of metabolite flux by nuclear magnetic resonance. Plant Physiol 130:1717–1727

Gomez-Casati DF, Iglesias AA (2002) ADP-glucose pyrophosphorylase from wheat endosperm. Purification and characterization of an enzyme with novel regulatory properties. Planta 214(3):428–434

Greene TW, Hannah LC (1998a) Assembly of maize endosperm ADP-glucose pyrophosphorylase requires motifs located throughout the large and small subunit units. Plant Cell 10:1295–1306

Greene TW, Hannah LC (1998b) Enhanced stability of maize endosperm ADP-glucose pyrophosphorylase is gained through mutants that alter subunit interactions. Proc Natl Acad Sci USA 95:13342–13347

Greene TW, Kavakli IH, Kahn M, Okita TW (1998c) Generation of up-regulated allosteric variants of potato ADP-glucose pyrophosphorylase by reversion genetics. Proc Natl Acad Sci USA 95:10322–10327

Hannah LC (1997) Starch synthesis in the maize endosperm. In: Larkins BA, Vasil IK (eds) Advances in cellular and molecular biology of plants. Cellular and molecular biology of plant seed development. Kluwer, Dordrecht, pp 375–405

Hannah LC (2005) Starch synthesis in the maize endosperm. Maydica 50:497–506

Hannah LC (2007) Starch formation in the maize endosperm. In: Olsen OA (ed) Endosperm: developmental and molecular biology. Springer, Heidelberg, pp 179–194

Hannah LC, Nelson OE (1976) Characterization of ADP-glucose pyrophosphorylase from *shrunken-2* and *brittle-2* mutants of maize. Biochem Genetics 14:547–560

Hirose T, Herao T (2004) A comprehensive expression analysis of the starch synthase gene family in rice (*Oryza sativa* L.). Planta 220:9–16

Iglesias AA, Barry GF, Meyer C, Bloksberg L, Nakata PA, Greene T, Laughlin MJ, Okita TW, Kishore GM, Preiss J (1993) Expression of the potato tuber ADP-glucose pyrophosphorylase in *Escherichia coli*. J Biol Chem 268:1081–1086

James MG, Robertson DS, Myers AM (1995) Characterization of the maize gene *sugary1*, a determinant of starch composition in kernels. Plant Cell 7:417–429

James MG, Denyer K, Myers A (2003) Starch synthesis in the cereal endosperm. Curr Opin Plant Biol 6:215–222

Kim WT, Franceschi VR, Okita TW, Robinson NL, Morell M, Preiss J (1989) Immunocytochemical localization of ADPglucose pyrophosphorylase in developing potato tuber cells. Plant Physiol 91:217–220

Kleczkowski LA, Villand P, Preiss J, Olsen OA (1993) Kinetic mechanism and regulation of ADP-glucose pyrophosphorylase from barley (*Hordeum vulgare*) leaves. J Biol Chem 268:6228–6233

Koch KE (2004) Sucrose metabolism: regulatory mechanisms and pivotal roles in sugar sensing and plant development. Curr Opin Plant Biol 7:235–246

Martin C, Smith AM (1995) Starch biosynthesis. Plant Cell 7:971–985

Munoz FJ, Zorzano MTM, Alonso-Casajus N, Baroja-Fernandez E, Etxeberria E, Pozueta-Romero J (2006) New enzymes, new pathways and alternative view on starch biosynthesis in both photosynthetic and heterotrophic tissues of plants. Biocatalysis Biotransformation 24:63–76

Myers AM, Morell MK, James MG, Ball SG (2000) Recent progress toward understanding biosynthesis of the amylopectin crystal. Plant Physiol 122:989–997

Nelson OE, Pan D (1995) Starch synthesis in maize endosperms. Annu Rev Plant Physiol Plant Mol Biol 46:475–496

Nelson OE, Rines HW (1962) The enzymatic deficiency in the *waxy* mutant of maize. Biochem Biophys Res Commun 9:297–300

Obana Y, Omoto D, Kato C, Matsumoto K, Nagai Y, Kavakli IH, Hamada S, Edwards GE, Okita TW, Matsui H, Ito H (2006) Enhanced turnover of transitory starch by expression of up-regulated ADP-glucose pyrophosphorylase in *Arabidopsis thaliana*. Plant Sci 170:1–11

Okita TW, Greenberg E, Kuhn DN, Preiss J (1979) Subcellular localization of the starch degradative and biosynthetic enzymes of spinach leaves. Plant Physiol 64:187–192

Pan D, Nelson OE (1984) A debranching enzyme deficiency in endosperms of the *sugary-1* mutants of maize. Plant Physiol 74:324–328

Preiss J, Levi C (1980) Starch biosynthesis and degradation. In: Preiss J (ed) The biochemistry of plants, vol 3. Academic, New York, pp 371–423

Preiss J, Romeo T (1989) Physiology, biochemistry and genetics of bacterial glycogen synthesis. Adv Microbiol Phys 30:183–238

Preiss J, Ball K, Hutney J, Smith-White B, Li L, Okita TW (1991) Regulatory mechanisms involved in the biosynthesis of starch. Pur Appl Chem 63:535–544

Regina A, Bird A, Topping D, Bowden S, Freeman J, Barsby T, Kosar-Hashemi B, Li Z, Rahman S, Morell M (2006) High-amylose wheat generated by RNA interference improves indices of large-bowel health in rats. PNAS 103:3546–3551

Sakulsingharoja C, Choi SB, Hwang SK, Edwards GE, Bork J, Meyer CR, Preiss J, Okita TW (2004) Engineering starch biosynthesis for increasing rice seed weight: the role of the cytoplasmic ADP-glucose pyrophosphorylase. Plant Sci 167:1323–1333

Satoh H, Nishi A, Yamashita K, Takemoto Y, Tanaka Y, Hosaka Y, Sakurai A, Fujita N, Nakamura Y (2003) Starch-branching enzyme I-deficient mutation specifically affects the structure and properties of starch in rice endosperm. Plant Physiol 133:1111–1121

Sehnke PC, Chung HJ, Wu K, Ferl RJ (2001) Regulation of starch accumulation by granule associated plant 14-3-3 proteins. Proc Natl Acad Sci USA 98:765–770

Shannon JC (1968) Carbon-14 distribution in carbohydrates of immature *Zea mays* kernels following $^{14}CO_2$ treatment of intact plants. Plant Physiol 43:1215–1220

Shannon JC, Pien FM, Liu KC (1996) Nucleotides and nucleotide sugars in developing maize endosperms: synthesis of ADP-glucose in *brittle-1*. Plant Physiol 110:835–843

Shannon JC, Pien FM, Cao H, Liu KC (1998) Brittle-1, an adenylate translocator, facilitates transfer of extraplastidial synthesized ADP-glucose into amyloplasts of maize endosperms. Plant Physiol 117:1235–1252

Smidansky ED, Clancy M, Meyer FD, Lanning SP, Blake NK, Talbert LE, Giroux MJ (2002) Enhanced ADP-glucose pyrophosphorylase activity in wheat endosperm increases seed yield. Proc Natl Acad Sci 99:1724–1729

Smidansky ED, Martin JM, Hannah LC, Fischer AM, Giroux MJ (2003) Seed yield and plant biomass increases in rice are conferred by deregulation of endosperm ADP-glucose pyrophosphorylase. Planta 216:656–664

Sowokinos JR (1981) Pyrophosphorylases in *Solanum tuberosum*. Plant Physiol 68:924–929

Spielbauer G, Margl L, Hannah LC, Romisch W, Ettenhuber C, Bacher A, Gierl A, Eisenreich W, Genschel U (2006) Robustness of central carbohydrate metabolism in developing maize kernels. Phytochemistry 67:1460–1475

Stark DM, Timmerman KP, Barry G, Preiss J, Kishore GM (1992) Regulation of the amount of starch in plant tissues by ADP-glucose pyrophosphorylase. Science 258:287–292

Sullivan TD, Strelow LI, Illingworth CA, Phillips RL, Nelson OE Jr (1991) The maize *brittle-1* locus: molecular characterization based on DNA clones isolated using the *dSpm*-tagged *brittle-1-mutable* allele. Plant Cell 3:1337–1348

Tetlow IJ, Morell MK, Emes MJ (2004) Recent developments in understanding the regulation of starch metabolism in higher plants. J Exp Bot 55:2131–2145

Thorbjornsen T, Villand P, Denyer K, Olsen O, Smith AM (1996) Distinct isoforms of ADPglucose pyrophosphorylase occur inside and outside the amyloplast in barley endosperm. Plant J 10:243–250

Villand P, Kleczkowski LA (1994) Is there an alternative pathway for starch biosynthesis in cereal seeds? Z Naturforsch 49:215–219

Wang YJ, White P, Pollak L, Jane J (1993) Characterization of starch structures of 17 maize endosperm mutant genotypes with Oh43 inbred line background. Cereal Chem 70:171–179

Wang Z, Chen X, Wang J, Liu T, Liu Y, Zhao L, Wang G (2007) Increasing maize seed weight by enhancing the cytoplasmic ADP-glucose pyrophosphorylase activity in transgenic plants. Plant Cell Tiss Organ Cult 88:83–92

White P (2000) Properties of corn starch. In: Hallauer R (ed) Speciality corns, 2nd edn. CRC Press, Boca Raton, pp 33–62

Wu C, Colleoni C, Myers AM, James MG (2002) Enzymatic properties and regulation of ZPU1, the maize pullulanase-type starch debranching enzyme. Arch Biochem Biophys 406:21–32

Yao Y, Thompson D, Guiltinan M (2004) Maize starch-branching enzyme isoforms and amylopectin structure. In the absence of starch-branching enzyme IIb, the further absence of starch-branching enzyme Ia leads to increased branching. Plant Physiol 136:3515–3523

Zhang X, Colleoni C, Ratushna V, Sirghie-Colleoni M, James MG, Myers AM (2004) Molecular characterization demonstrates that the *Zea mays* gene *sugary2* codes for starch synthase isoform SSIIa. Plant Mol Biol 54:865–879

Chapter 21
Development of a High Oil Trait for Maize

Dale L. Val, Steven H. Schwartz, Michael R. Kerns, and Jill Deikman

21.1 Introduction

Maize oil is valued as a component of animal feed and as food. Approximately half of maize consumed in the USA in 2006 was used domestically for animal feed, 31% was processed for ethanol, food or industrial use, and 19% was exported (USDA 2007). With the recent growth in use of maize for ethanol production, interest in the recovery of oil from the grain has increased (Hojilla-Evangelista and Johnson 2003).

While the composition of maize grain is suitable for feeding a variety of livestock, increasing the oil content of the grain improves its nutritional value further, due in large part to the increase in metabolizable energy (Han et al. 1987; Song et al. 2003). In addition, high oil maize typically has more protein, lysine and carotenoids than conventional maize (Han et al. 1987). Lambert (2001) reviewed a number of studies with poultry, swine, sheep and dairy cattle that demonstrated improved growth, feed efficiency and/or product quality from animals fed rations that included high oil maize.

Although oil is only a small proportion of maize grain (approximately 4%), maize oil is the second highest vegetable oil in terms of volume produced in the USA, at 2.5 billion pounds in 2005, behind soybean at 20 billion pounds (Ash and Dohlman 2006). Most of the refined maize oil is used in food (Anonymous 2006), and is composed primarily of triacylglycerol (TAG) (Weber 1987). The fatty acid profile of maize oil is desirable for food use, with a high concentration of unsaturated fatty acids, including approximately 25% oleic (18:1) and 60% linoleic (18:2) acid (Orthoefer and Sinram 1987; Weber 1987). In addition, the low level

Dale L. Val, Steven H. Schwartz, Michael R. Kerns
Monsanto Company, 1920 Fifth Street, Davis, CA 95616, USA

Jill Deikman
Monsanto Company, 1920 Fifth Street, Davis, CA 95616, USA
e-mail: jill.deikman@monsanto.com

A.L. Kriz, B.A. Larkins (eds.), *Molecular Genetic Approaches to Maize Improvement* 303
Biotechnology in Agriculture and Forestry, Vol. 63

(<1%) of linolenic acid (18:3) enhances oil stability. These values are typical of commodity maize, but can vary in maize of different genetic backgrounds (e.g. Reynolds et al. (2005) and in response to growing conditions (Weber 1987). Total oil content of a maize kernel can also vary. Grain analysis of seven dent maize hybrids typically grown in the US midwest indicated that the average oil content was 4.4%, with a range of 3.9 to 5.8% (Watson 1987). The average values for other components in this study were: 73.4% starch, 9.1% protein, 1.4% ash and 1.9% sugar.

Efforts to increase kernel oil content through breeding have had considerable success, but high oil lines usually have significantly reduced yield (Lambert 2001). One factor that may affect the yield potential of high oil maize is that oil requires 2.25 times more energy to produce than starch (Weber 1987). Another cause of reduced yield in high oil lines may be genetic linkage of yield-reducing agronomic traits, although some studies suggest that high oil grain can be produced without significant yield drag (Lambert et al. 1998).

This chapter focuses on the biology and biochemistry of oil accumulation in maize kernels, and research pertinent to increasing maize oil content. For a discussion of modified fatty acid composition, the reader is referred to recent reviews that highlight the considerable success achieved in this area of research in other crops such as canola and soy (Voelker and Kinney 2001), and the potential for similar results in maize (Kinney et al. 2002). This chapter begins with a brief summary of the development of the major high oil maize lines by recurrent selection for increased oil levels. We discuss the status of oil quantitative trait loci (QTL) mapping and use of molecular breeding methods in the development of hybrids with elevated oil levels while maintaining high grain yield. Current understanding of the physiology, biochemistry, molecular biology and transgenic engineering of oil accumulation in maize and other plants is reviewed. Finally, future prospects for enhancing maize oil content using biotechnology are discussed.

21.2 Background

21.2.1 Kernel Morphology and Lipid Content

Kernel morphology plays an important role in oil accumulation in maize kernels. The kernel contains the embryo or germ, which is typically about 10% of the kernel mass, and a starchy endosperm that comprises about 80% of the kernel mass (Watson 1987; Fig. 21.1A). The embryo includes both the embryo axis with shoot and root apices as well as the scutellum, an organ involved in nutrient storage and transport of nutrients to the growing seedling.

The mature embryo is approximately 33% lipid by weight in standard maize hybrids, and contains about 80% of the kernel lipids. The endosperm lipid concentration is only about 0.8%. The accumulation of lipids in the kernel during development parallels the growth of the embryo, beginning at about 15 days after pollination and continuing until maturity (Watson 1987).

Triacylglycerol (TAG) is the storage oil used by the seedling for energy during early seedling growth, and comprises 70–85% of embryo lipids (Tan and Morrison 1979). Other lipids in the embryo include diglycerides and steryl esters. Approximately half of the endosperm lipids are associated with starch, and these lipids are composed predominantly of free fatty acids (60%) and lysophospholipid (25%) (Weber 1987). Lipids in the aleurone and adjacent starchy endosperm layers are similar to those in the embryo, and include approximately 60% of the endosperm TAG (Weber 1987).

21.2.2 Effects of Environment and Agricultural Practices on Kernel Oil Content

Research over several decades has indicated that while environmental effects can influence oil accumulation, the primary determinant of maize kernel oil content is genetic makeup (Weber 1987; Lambert 2001). For example, an early field study showed that location and year of planting had a statistically significant effect on oil content, but the relative oil content among hybrids was similar in each test (Jellum and Marion 1966).

A number of studies have investigated the impact of nitrogen on maize oil content, and found either no effect (Jellum et al. 1973) or small effects from treatments that caused much larger changes in grain yield and/or protein content (Genter et al. 1956; Welch 1969; Duarte et al. 2005; Miao et al. 2006).

Several studies have looked at the effects of severe abiotic stresses on grain quality. Mixed results were reported for the effects of high temperature (Thompson et al. 1973; Wilhelm et al. 1999), and severe drought reduced oil content (Jurgens et al. 1978). However, growth at high plant density had only small effects on kernel oil concentration (Maddonni and Otegui 2006), and complete defoliation of plants that reduced grain yield by 29% did not significantly alter oil concentration (Mangen et al. 2005).

The relative stability of the oil trait in response to environmental conditions is an advantage to researchers interested in modification of oil content, since small differences between different varieties can be reliably measured.

21.2.3 Inheritance of Oil Phenotype

Inheritance of the high kernel oil trait is a function of both maternal and paternal genetics (Curtis et al. 1956; Letchworth and Lambert 1998). Reciprocal crosses demonstrated that the oil percentage of a kernel is approximately midway between a high oil and low oil parent. This effect is not observed for kernel protein or starch contents, which are determined primarily by the genetics of the maternal parent (Letchworth and Lambert 1998). This pattern of inheritance suggests that oil accumulation in the kernel is controlled in large part by expression of genes in the embryo and endosperm.

21.3 Breeding for High Oil

21.3.1 High Oil Sources

Maize lines with increased kernel oil have been selected through breeding. Three well-characterized high oil sources are Illinois High Oil (IHO), Alexho Synthetic and Beijing High Oil (BHO).

IHO was developed in the Illinois Long-Term Selection Experiment, which also developed low oil and high and low protein lines (Dudley and Lambert 2004; Moose et al. 2004). In 1896, 24 high oil ears were selected from a population of an open-pollinated variety called Burr's White, and selection for high oil was conducted every year for more than 100 years, except for 3 years during World War II. The oil content was initially 4% oil, and attained 20% by the 100th cycle of selection (Dudley and Lambert 2004). It has not yet reached a plateau, suggesting further increases are still possible.

The Alexho Synthetic population, also from the University of Illinois, was created from 53 open-pollinated populations and synthetics, including an IHO selection (Lambert et al. 2004). A single kernel selection strategy was used after cycle 5. After 27 cycles of selection, the population reached 21% oil.

Song et al. (1999) developed the BHO high oil population from the Zhongzong no. 2 synthetic, and also used a single kernel selection method. The initial oil concentration was 4.7% and reached 15.6% after 18 cycles of selection (Song and Chen 2004).

The usefulness of these high oil populations has been in the accumulation of alleles favorable for oil in single sources. However, a number of negative traits are present in these populations, including reduced grain yield, plant and ear height, and kernel size (Dudley et al. 1977; Misevic and Alexander 1989). Kernel oil also has negative correlation with starch content (Song and Chen 2004; Clark et al. 2006).

Understanding the mechanisms of enhanced oil accumulation in high oil maize may provide insights into approaches for biotech modification of oil content. Several studies have demonstrated that the high oil trait is typically associated with an increase in embryo size and elevated oil concentration in the embryo (Curtis et al. 1968; Lambert et al. 1997; Lambert 2001; Dudley and Lambert 2004). High oil kernels may also have smaller endosperms, which is undesirable since this phenotype may contribute to decreased grain yield. Figure 21.1 illustrates the morphology of high oil kernels from three sources compared to a control with standard oil content.

21.3.2 Oil QTL Analysis

A number of studies have been conducted to identify QTL that affect the oil content of maize (Table 21.1). Dudley et al.'s estimate that 54 factors, or QTL, affect oil (Dudley et al. 1977) was supported using molecular markers on a random-mated

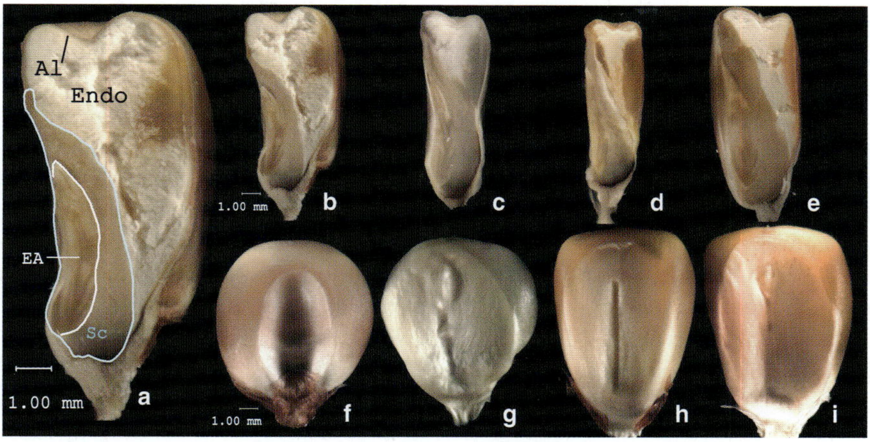

Fig. 21.1 Wild type and high oil kernels in cross section (**a–e**) or whole kernels (**f–i**). **a** Mo17 kernel labeled to show endosperm (*Endo*); *Al* aleurone; *EA* embryo axis; *Sc* scutellum. **b–i** Mature kernels of wild type and high oil mutant lines. Genotypes are indicated with oil concentration in parentheses: **b** and **f**, Mo17 (3.8%); **c** and **g**, IHO (17.8%); **d** and **h**, Alexho Cycle 20 (12.1%); and **e** and **i**, BHO (14%)

population (Clark et al. 2006). Other studies identified from 7 to 63 oil QTL (Table 21.1). The variation in the number of QTL in different experiments may be explained in part by the number of markers used in the studies. It is interesting that similar numbers of oil QTL are found in IHO × ILO populations and populations selected for variation in protein content (Dudley et al. 2007). This fact is consistent with the ability to reverse low oil populations and select for high oil lines (Dudley and Lambert 2004).

A uniform observation from these QTL studies is that many loci are involved in oil accumulation, and that these QTL each contribute a small effect. These findings

Table 21.1 Summary of marker-QTL studies for maize oil

Study	Population	Oil source	Number of QTL
Dudley 1977	IHO × ILO	IHO	54
Goldman et al. 1994	IHP × ILO	None	25
Alrefai et al. 1995	IHO × ILO	IHO	12
Berke and Rocheford 1995	IHO × ILO	IHO	31
Séne et al. 2001	F-2 × Io	None	7
Lazic-Jancic et al. 2003	YuSSSu, DS7u	None	8
Dudley et al. 2004	IHP × ILP	None	8
Laurie et al. 2004	IHO × ILO RM	IHO	> 50
Mangolin et al. 2004	L-20-01 × L-02-03	None	13
Song et al. 2004	By804 × B73	BHO	20
Clark et al. 2006	IHO × ILO	IHO	51/54[a]
Dudley et al. 2007	IHP × ILP	None	70/ 63[a]

[a]Number of QTL found in per se and test-cross, respectively

agree with the gradual and continuous response of increasing oil content in the Illinois Long-Term Selection Experiment over 100 years of selection (Dudley and Lambert 2004; Laurie et al. 2004).

It was recently demonstrated that the causal gene for a major maize high oil/high oleic QTL on chromosome 6 is a *DGAT1* gene (Zheng et al. 2008). The *DGAT1-2* allele from the high oil line encodes an enzyme with a phenylalanine insertion at position 469 (F469), missing in *DGAT1-2* from conventional maize. This high oil variant was shown to have greater DGAT activity, and while over-expression of either *DGAT1-2* allele in transgenic maize resulted in increased kernel oil and oleic acid, the high oil allele resulted in greater increases in both these traits. It was further shown that the *DGAT1-2* allele from high oil corn is likely to be the ancestral *DGAT1* allele, and is not common in commercially relevant germplasm.

21.3.3 TopCross™ Strategy for High Oil Hybrids

One strategy that has been developed to produce high oil grain without large negative effects on yield is to plant a mixture consisting of a high oil pollinator and a high yielding male sterile hybrid (Berquist et al. 1998). Research supporting this approach includes a 3-year study which found that hybrids pollinated with high oil or normal oil pollinators had comparable yield, while oil content was 19 g kg^{-1} greater when a high oil pollinator was used (Lambert et al. 1998). Evaluation of TopCross^{TM}blend hybrids demonstrated that oil content of the harvested grain was increased 31 g kg^{-1} compared to the controls (Thomison et al. 2003), but yield was reduced by 8% on average (Thomison et al. 2002). The yield decrease was likely due in part to the reduced number of high yielding plants in the field compared to controls, although the higher energy cost of oil synthesis could also be a factor.

21.3.4 Contemporary Strategies for High Oil Breeding

Current breeding programs focus on introgression of high oil traits from high oil selections into elite germplasm. The genetic correlation of oil content with grain yield is often neutral, so it may be possible to increase oil while maintaining yield at commercially viable levels (Rocheford et al. 2003). The use of molecular markers will accelerate development of high oil lines with high yield potential.

21.4 Synthesis of Oil in the Kernel

Knowledge of the oil biosynthetic pathway in maize is necessary for genetic engineering to elevate oil content of the kernel. Current understanding of this pathway is derived primarily from studies conducted with model species such as Arabidopsis and with oil seed crops. The pathway is outlined in Fig. 21.2 and described below,

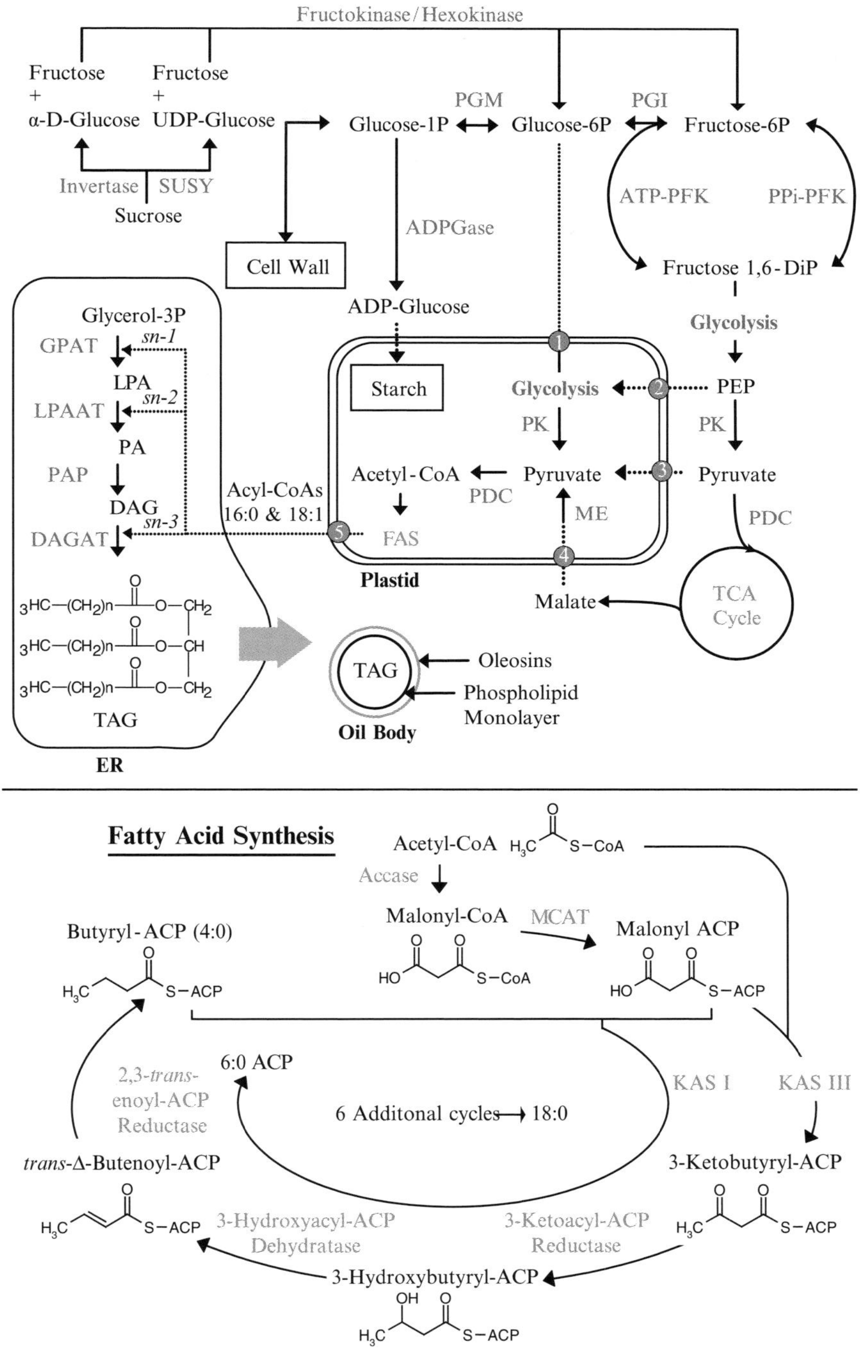

Fig. 21.2 The key enzymes and intermediates in the synthesis and storage of TAG beginning with sucrose hydrolysis. Abbreviations for the enzymes are defined in the text. Plastidial transporters/translocators are indicated by *numbered circles*: ❶ glucose-6-phosphate translocator; ❷ PEP translocator; ❸ pyruvate translocator; ❹ malate translocator; ❺ long chain acyl CoA synthetases. *Inset* represents the core reactions of de novo fatty acid synthesis

with particular attention paid to maize and to transgenic experiments to alter oil content.

21.4.1 Synthesis and Plastid Import of the Carbon Precursor(s) of Acetyl-CoA

Sucrose enters the kernel through the pedicel and basal endosperm transfer layer, where it is hydrolyzed by a cell wall invertase to produce glucose and fructose (Taliercio et al. 1999). The transport, uptake and utilization of sugars for starch synthesis in the endosperm have been extensively studied. For oil synthesis, the partitioning of assimilates into the embryo is of particular interest, but this process is not as well understood. Since there are no symplastic connections between the embryo, the endosperm and the maternal tissues, uptake of sugars into the embryo must occur from the apoplast. Griffith et al. (1987) showed that isolated maize embryos can assimilate both sucrose and hexose sugars. When ^{14}C-labeled glucose was fed to isolated embryos, a significant amount of the label was incorporated into sucrose. Sucrose resynthesis from hexose sugars could create a gradient that would facilitate passive uptake. The uptake of sucrose in isolated embryos is pH dependent, indicating the involvement of a sucrose transporter as well (Griffith et al. 1987). A number of sucrose and monosaccharide transporters have been identified in plants (Lalonde et al. 2004), but their functions in assimilate partitioning to the developing embryo have not yet been described.

The embryo surrounding region (ESR) is a specialized area of the endosperm that is defined by unique gene transcripts (Opsahl-Ferstad et al. (1997), including an invertase inhibitor that could affect carbohydrate partitioning to the embryo (Bate et al. 2004). Further characterization of ESR localized genes should be helpful in determining its role in the development and assimilation of carbon to the embryo.

The intracellular metabolism of sucrose and its utilization for the synthesis of oil are illustrated in Fig. 21.2. The different hexose phosphates (Glu-1-P, Glu-6-P and Fru-6-P) are readily interconverted and their utilization for starch, oil or other metabolites is determined largely by the relative flux into each pathway. Fru-6-P is the immediate precursor to glycolysis, which is necessary to generate the carbon skeleton for fatty acid synthesis.

Glycolysis is the sequential oxidation of Fru-6-P to produce two pyruvates. Minimally, there are seven enzymatic steps in the pathway, which yield three ATPs and two reducing equivalents for each Fru-6-P (Plaxton and Podesta 2006). The phosphorylation of Fru-6-P to fructose-1, 6-bisphosphate (Fru-1, 6-P_2) is the entry point into glycolysis and is considered an important regulatory step in many organisms. In higher plants, this reaction may be catalyzed by an ATP-dependent phosphofructokinase (ATP-PFK) or pyrophosphate-dependent phosphofructokinase (PPi-PFK), an enzyme found only in plants, a few bacteria and Archaea. Higher activities for fructokinase, glucokinase and ATP-PFK have been associated with the period of oil deposition in maize (Doehlert 1990).

Attempts to alter flux through glycolysis by altered expression of PFKs have met with limited success in plants (Burell et al. 1994; Paul et al. 1995). Expression levels alone are not sufficient to explain the activity of the enzymes or flux through glycolysis (Gibon et al. 2004). Concentrations of allosteric effectors, pH and covalent modifications have been shown to regulate the activity of various glycolytic enzymes. The activities of the plant PFKs are controlled by several allosteric effectors. Phosphoenolpyruvate (PEP), a downstream intermediate in the pathway, is an inhibitor of both PFKs. This type of allosteric regulation allows for the feedback regulation of glycolysis to accommodate changes in demand for metabolic precursors and energy.

In higher plants, glycolysis may occur in the cytosol or the plastid. In a number of plant species, the period of oil deposition is marked by an increase in expression and activity for enzymes in the cytosolic pathway of glycolysis (Eastmond and Rawsthorne 2000; Ruuska (2002)). Lee et al. (2002) analyzed a subset of metabolic genes during maize embryo development and found that genes encoding enzymes for the cytosolic pathway of glycolysis and the TCA cycle were coordinately regulated with maximal levels of expression between 10 and 20 days after pollination. If the precursors for fatty acid synthesis are being generated in the cytosol, they must be translocated across the inner envelope of the plastid. A phosphate translocator superfamily has been identified that is able to import specific metabolites into the plastid stroma using inorganic phosphate as a counter ion (Flugge 1998). In Brassica and Arabidopsis, the primary precursors of fatty acid synthesis are likely to be PEP and pyruvate (Eastmond and Rawsthorne 2000; Ruuska (2002); Schwender (2004)). The PEP translocator, which was first identified in maize (Fischer et al. 1997), is a member of the plastidic phosphate translocator family. The expression of a homologous PEP translocator in Arabidopsis is coordinate with the expression of genes encoding enzymes for fatty acid synthesis in developing seeds (Ruuska (2002)).

In castor bean, malate is imported directly into the plastid where it is decarboxylated to pyruvate by the malic enzyme (ME). This reaction also provides reducing equivalents for later steps in FAS (Eastmond et al. 1997). A similar malate translocating system has been described in plastids isolated from maize embryos (Lara-Nunez and Rodriguez-Sotres 2004). Gene expression and steady state flux analyses will be helpful in further elucidating the relative contribution of plastidial glycolysis and the import of various intermediates into the plastid for fatty acid synthesis.

Fatty acid synthesis also requires several cofactors. ATP is produced through plant respiration, and some ATP may be produced in the plastid by glycolysis. Additional ATP may be imported into the plastid by a specific ADP/ATP translocator (Mohlmann et al. 1998). The reducing equivalents that are generated from glycolysis are probably not sufficient for all of the metabolic needs of the plastid. In green seeds, such as Brassica, photosynthesis provides a significant source of reducing equivalents (Schwender (2004)). However, there is no indication that photosynthesis is occurring in the maize embryo. Presumably, a significant amount of the reducing equivalents for fatty acid synthesis are supplied by the oxidative pentose phosphate pathway within the plastid.

21.4.2 Synthesis of Acetyl-CoA in the Plastid

Acetyl-CoA is the membrane impermeable building block that supplies all of the carbon for de novo fatty acid synthesis, a process that occurs within the plastids. In vivo labeling studies have indicated that acetyl-coA is derived from plastidial pyruvate in Arabidopsis and Brassica (Bao et al. 2000; Schwender (2004)), most likely by the action of the plastidial pyruvate dehydrogenase complex (PDC) (Ke et al. 2000). Although similar studies have not yet been conducted with maize, it is likely that the immediate precursor of plastidial acetyl-CoA in maize kernels is also pyruvate. Like its mitochondrial counterpart, plastidial PDC catalyzes the oxidative decarboxylation of pyruvate to yield NADH and acetyl-CoA, but comparatively little is known about the regulation of the plastidial isoform (reviewed in Tovar-Mendez et al. (2003) and Patel and Korotchkina (2006)).

21.4.3 Plastidial de novo Fatty Acid Synthesis

Fatty acid synthesis involves the action of two enzyme complexes, acetyl-CoA carboxylase (ACCase) and fatty acid synthase (FAS). ACCase catalyzes the first committed step in fatty acid synthesis, an ATP-dependent carboxylation of acetyl-CoA to form the 3-carbon precursor, malonyl-CoA. Most plants have two kinds of ACCases: the plastid-localized heteromeric type II and the cytosolic homomeric type I (reviewed in Sasaki and Nagano (2004)). However, maize and other grasses lack a type II ACCase and instead have homomeric ACCase isoforms in both the cytosol and the plastid (Egli et al. 1993; Ashton et al. 1994). The plastidial isoform is responsible for most of the ACCase in developing germ and endosperm (Somers et al. 1993), and has different kinetic properties from the cytosolic isoform (Herbert et al. 1996). By using specific inhibitors to generate minor alterations in ACCase activity in isolated chloroplasts, Page et al. (1994) found that plastidial ACCase exerts a significant ($\sim$50%) control over flux through the fatty acid synthesis pathway in leaves of maize and barley.

At present there is little published research that addresses the importance of maize plastidial ACCase in controlling fatty acid synthesis flux to TAG in developing kernels. However, transgenic over-expression of a homomeric ACCase in plastids has been reported to lead to small increases in oil accumulation in rapeseed (Roesler et al. 1997) and potato tubers (Klaus et al. 2004).

All of the reactions required for the de novo synthesis of 16:0 and 18:0 fatty acids from malonyl-CoA are carried out by FAS. Plants, like most bacteria, have dissociable multi-subunit (type II) FAS enzymes. Kinetic and protein interaction studies suggest that the plant type II component enzymes assemble together in a complex (Roughan and Ohlrogge 1996; Roughan 1997; Brown et al. 2006), functionally analogous to the FAS complexes of non-plant eukaryotic type I enzymes (Jenni et al. 2006; Maier et al. 2006). The basic sequence of reactions involved in the type II dissociable FAS process in plants is shown schematically in Fig. 21.2, is described below and has been reviewed extensively (e.g. Harwood (1996); Voelker and Kinney (2001)).

Malonyl-CoA: acyl carrier protein transacylase (MCAT) catalyzes the thioester transfer of the 3-carbon malonyl unit from malonyl-CoA to the terminal SH of the flexible prosthetic group of acyl carrier protein (ACP), forming malonyl-ACP. During the whole process of fatty acid synthesis, ACP functions as the mobile carrier of the growing fatty acid chain between the different active sites of the component enzymes in the complex.

All the carbon–carbon bond formation steps (condensation steps) during elongation of the fatty acid chains are catalyzed by enzymes known as condensing enzymes or β-ketoacyl-ACP synthases (KAS). KAS III catalyzes the condensation between acetyl-CoA and malonyl-ACP, generating the first "primer" (4-carbon) for fatty acid synthesis. KAS I catalyzes all subsequent 2-carbon additions during the elongation cycles until the fatty acid is 16 carbons in length, and KAS II is responsible for the 2-carbon addition from 16 to 18 carbons in the final elongation cycle.

By catalyzing the irreversible chain lengthening steps, KAS enzymes are viewed as key rate and chain length specificity "gate-keepers" of the whole process. Consistent with its proposed "chain length gatekeeper" role, the crystal structure of the first KAS enzyme (*E. coli* KAS II) and subsequent mutagenesis studies confirmed that the size of a hydrophobic fatty-acid binding pocket at the enzyme's active site determines the chain length specificity toward the acyl-ACP substrates (Moche et al. 1999; Val et al. 2000).

After the initiating KAS step, each cycle of elongation is completed by three additional enzymatic steps resulting in the regeneration of a saturated fatty acid attached to ACP that is lengthened by 2-carbons, and is potentially ready for the next elongation cycle (Fig. 21.2). Considering that the component enzymes of fatty acid synthesis form a complex, transgenically engineering higher oil levels in maize will likely be more challenging than simply over-expressing one key enzyme component (Thelen and Ohlrogge 2002; Kinney 2006).

21.4.4 Fatty Acid Termination, Export and Transfer to the ER

Any 18:0-ACPs produced by FAS in the plastid are desaturated to 18:1-ACP (oleic) by a soluble plastidial Δ^9 desaturase. The 16:0 and 18:1 fatty acids are then terminated from the FAS process by fatB- and fatA-type thioesterases (TEs) respectively (Voelker and Kinney 2001) and subsequently exported from the plastid to the ER by the action of long chain acyl-CoA synthetase(s) (LACSs) localized in the outer plastid envelope. Although the export of fatty acid from the plastids to the ER represents the largest trans-membrane flux of lipids in plants, to date many of the molecular details of this process have yet to be determined (Benning et al. 2006; Jouhet et al. 2007).

21.4.5 ER Membrane Glycerolipid Synthesis and Fatty Acid Desaturation

In plants TAGs and ER membrane glycerolipids are synthesized in the ER by the sequential attachment of fatty acids onto a glycerol backbone by a series of membrane localized enzymes in the Kennedy pathway. The glycerol backbone is generated by the action of glycerol-3-phosphate (gly-3P) dehydrogenase (Gly3PDH), reducing the glycolytic intermediate dihydroxyacetone phosphate. Transgenic elevation of Gly3PDH activity has been shown to increase oil levels in Brassica, suggesting that the gly-3P supply was limiting TAG accumulation (Vigeolas et al. 2007).

Glycerol-3-phosphate acyltransferase (GPAT) transfers the first fatty acid chain from an acyl-CoA to the *sn*-1 position of glycerol-3-phosphate, producing lysophoshatidic acid (LPA). Lysophosphatidic acid acyltransferase (LPAAT) catalyzes the acyl transfer of the second fatty acid onto the glycerol backbone, again from an acyl-CoA, onto the *sn*-2 position of LPA, generating phosphatidic acid (PA). LPAAT from maize, and many other plants, has a high preference for 18:1 (Oo and Huang 1989) and is the most substrate-specific of the three acyltransferases. Kinetic labeling studies in maize (Sun et al. 1988; Oo and Huang 1989) and other plants suggest that LPAAT activity is relatively abundant and does not appear to limit the overall rate of glycerolipid synthesis.

Phosphatidic acid phosphatase (PAP) cleaves the phosphate group off the glycerol backbone of PA producing s*n*-1,2-diacylglycerol (DAG), the branch-point intermediate for either ER membrane lipid or TAG synthesis. In order to generate the linoleic (18:2) and linolenic (18:3) fatty acids present in the seed storage TAGs of maize and other plants, the DAG intermediate is first converted into the membrane glycerolipid phosphatidyl-choline (PC) and the fatty acids are subsequently desaturated by 18:1 (FAD2) and 18:2 (FAD3) desaturases, respectively. The unsaturated 18:2 and 18:3 fatty acids can then re-enter the pathway, leading to TAG formation by a number of mechanisms.

21.4.6 Synthesis of TAG from ER Membrane Lipids

The final step of TAG synthesis involves the transfer of a third fatty acid chain onto the glycerol backbone. This last step of the Kennedy pathway is the only reaction unique to TAG synthesis and it is carried out by the enzyme diacylglycerol acyltransferase (DGAT), which transfers a fatty acid from acyl-CoA onto the *sn*-3 position of DAG. Like GPAT, DGAT has much broader substrate specificity than LPAAT. Nevertheless, the DGAT activity in maize embryo utilizes 18:2 approximately twice as efficiently as 18:1, in keeping with the fact that maize oil typically has more 18:2 than 18:1 (Cao and Huang 1987).

There are two separate gene families of ER membrane-localized DGATs. DGAT1, which shares similarity to mammalian acyl-CoA:cholesterol acyltransferase (ACAT) genes, and the unrelated DGAT2 family originally cloned from the oleaginous fungus *Mortierella ramaniana* (Mr) (Lardizabal et al. 2001). Genes encoding DGAT1 enzymes have been cloned in several plants species (Lung and Weselake 2006), and DGAT2 has been cloned and characterized from castor beans (Kroon et al. 2006).

Being located at the branch point of TAG synthesis, DGAT has long been considered to be a potentially important enzyme for regulating TAG synthesis in plants (reviewed in Lung and Weselake 2006). *DGAT1* is highly expressed during seed development in canola and Arabidopsis (Hobbs et al. 1999), and reduced expression of *DGAT1* in Arabidopsis or tobacco resulted in a significant decrease in seed oil accumulation (Routaboul et al. 1999; Zou et al. 1999; Zhang et al. 2005). In castor, the expression of *DGAT2* rather than *DGAT1* was correlated with seed oil deposition (Kroon et al. 2006). The importance of DGATs in enhancing oil accumulation has been confirmed transgenically. Over-expression of Arabidopsis *DGAT1* was found to elevate oil levels in Arabidopsis (Jako et al. 2001), and DGAT2 from *M. ramaniana* elevated oil levels in soybean (Lardizabal et al. 2003). Recently in maize, transgenic over-expression of the maize DGAT1 enzyme encoded by the ancestral/high oil *DGAT1-2* allele led to kernel oil increases of up to 41%, while over-expression of the less active allele from conventional lines led to more moderate increases in kernel oil (Zheng et al. 2008).

In addition to the DGAT reaction, phospholipid:diacylglycerol acyltransferase (PDAT) also synthesizes TAG. PDAT transfers a fatty acid from PC to DAG, generating TAG and LPA (Dahlqvist et al. 2000) in an acyl-CoA-independent reaction. However, neither over-expression (Stahl et al. 2004) nor gene knockout (Mhaske et al. 2005) of the Arabidopsis *PDAT* had any significant effect on seed oil accumulation. The physiological role of PDAT in plants is still an area of active research, and its function may involve editing unwanted fatty acids out of the membrane glycerolipids.

Following its synthesis, TAG is deposited between the phospholipid bilayer of the ER membrane (Hsieh and Huang 2004). Oil bodies, which are approximately 1–2 μm in size, are formed when the oil-rich deposits bud off of the ER membrane. These oil bodies contain a core of TAG, a phospholipid monolayer and an outer layer of abundant proteins known as oleosins. The proposed function of the oleosins is to control the oil body size and prevent oil bodies from coalescing during seed desiccation. In the maize embryo and aleurone there are three isoforms of oleosin that are coordinately expressed, with peak levels occurring at 25 days after pollination (Ting et al. 1996). These genes encode the low molecular weight 16-kD isoform OLE16, and two high molecular weight isoforms, OLE17 and OLE18. The two high molecular weight isoforms are very similar to each other and they appear to be functionally equivalent in maize (Lee et al. 1995). Although electron microscopy revealed that the oil bodies from IHO kernels were significantly larger than those from ILO kernels, accumulation of oleosin proteins was similar in these lines (Ting et al. 1996). These authors suggested that over-expression of oleosin genes would not likely impact oil accumulation in maize.

21.5 Regulation of Oil Biosynthesis

The plant hormone abscisic acid (ABA) plays an important role in promoting synthesis and accumulation of storage proteins and lipids during seed maturation (reviewed in Kermode 2005). The role of ABA in storage deposition may be mediated in large part by the transcription factor VIVIPAROUS1 (VP1), which has been shown to regulate the expression of oleosin genes (Paiva and Kriz 1994; Suzuki et al. 2003). *VP1* is the maize ortholog of *ABI3*, and can complement most aspects of the Arabidopsis *abi3* mutant phenotype (Suzuki et al. 2001), indicating that there is some conservation in the regulation of embryo development and oil deposition between maize and Arabidopsis. A number of additional transcription factors important in the control of embryo development and oil accumulation have been identified in Arabidopsis. These include *LEC1*, *LEC2* and *FUS3* genes, which were initially identified as mutants impaired in embryo development (reviewed in To et al. 2006), as well as repressors of these transcription factors, *PICKLE* (Rider et al. 2003) and *VP1/ABI3-like* (Suzuki et al. 2007; Tsukagoshi et al. 2007). Ectopic expression of *LEC2* resulted in increased oil accumulation in Arabidopsis (Mendoza et al. 2005), and ectopic expression of *FUS3* was shown to cause increased expression of fatty acid biosynthetic genes (Wang et al. 2007).

The Arabidopsis *WRINKLED1* gene encodes a putative transcription factor of the *AP2/EREB* family that has an important role in regulating oil biosynthesis in seeds (Cernac and Benning 2004). Comparison of gene expression in *wri1* and wild-type seeds indicated that many genes encoding enzymes involved in central lipid and carbohydrate metabolism are controlled by *WRI1* (Ruuska (2002); Baud et al. 2007). The *wri1* mutant has a wrinkled seed phenotype and an 80% reduction in TAG accumulation in seeds, but early embryo development appears normal (Focks and Benning 1998). Over-expression of *WRI1* in Arabidopsis resulted in 10–20% more oil in seeds (Cernac and Benning 2004). *WRI1* is regulated by *LEC2*, and these two genes work together to influence oil accumulation (Baud et al. 2007).

Shen et al. (2006) identified ten low-oil mutants and one high-oil mutant from Arabidopsis with a screen for altered density. The high oil mutant was disrupted in the *GLABRA2* gene, which encodes a homeobox protein required for leaf trichome and root hair development. While *GL2* is expressed during early seed development in wild type, it is not understood how *GL2* regulates oil accumulation in seeds.

It is likely that additional transcription factors will be identified that are involved in control of lipid accumulation in seeds (Ruuska (2002)). In the meantime, it is important to begin to understand whether genes orthologous to these Arabidopsis regulators control oil accumulation in maize.

21.6 Conclusions

The selection of high oil lines through breeding has demonstrated that it is possible to increase oil content of maize kernels. This achievement was facilitated by the high heritability of the oil phenotype and the relative insensitivity of the trait to environmental variation. Unfortunately, these high oil selections tend to have poor yield. While the biochemical cost of oil biosynthesis relative to that of starch may make some negative yield impact inevitable, there are experiments that suggest it should be possible to develop high oil lines with competitive yield potential (Lambert et al. 1998). Source production has been shown to be responsive to sink demand (Paul and Foyer 2001), so a plant could theoretically accommodate an increased energy requirement in the kernel.

The oil increase of high oil maize is primarily a consequence of an increase in embryo size and an increase in the concentration of oil in the embryo, suggesting that these traits could be productive targets of transgenic approaches to increasing oil content. An increase in oil accumulation in the endosperm could also be attempted, as a novel mechanism. Transgenic modification of embryo size will likely require greater understanding of the molecular mechanisms controlling embryo growth in maize. The rice giant embryo mutant may provide some insight into the control of this process (Hong et al. 1996).

The pathways involved in oil biosynthesis have been fairly well characterized, particularly in model plants such as Arabidopsis. Nevertheless, even in Arabidopsis the details of how metabolic flux through the pathways to TAG is controlled are not fully known. Furthermore, studies with other complex multi-step pathways suggest that the flux control is likely to be distributed among the enzymes in the pathway rather than focused in one "rate-determining step" (Fell 1997), adding an additional layer of complexity. Indeed, QTL analysis indicates that there are at least 50 loci important for oil accumulation in the Illinois high oil population, and that each QTL has only a small effect (Laurie et al. 2004). The recent successful demonstration of an increase in maize kernel oil (up to 41%) by overexpression of *DGAT1* is encouraging (Zheng et al. 2008). However, it remains to be seen whether greater oil increases can be achieved by further manipulation of this step alone, or whether additional transgenes will be required to achieve larger oil increases.

Despite the inherent complexity of successfully generating higher oil levels in maize by breeding or biotechnology, the value and heritability of the trait continue to make it an attractive target. Application of powerful new technologies, such as transcription profiling, metabolite profiling and flux analysis, should prove valuable to accomplishing this goal. In addition, completion of the maize genomic sequence will provide improved tools for both breeding and biotechnology. This work will facilitate marker assisted breeding of higher oil lines that have competitive yield, as well as the positional cloning of additional causal genes for oil QTLs. Identification of transcription factors (Cernac and Benning 2004) or other regulatory proteins that exert higher level control of oil biosynthesis or embryo development will be particularly attractive candidates for biotechnology approaches in the future.

Acknowledgements We gratefully acknowledge the contribution of Jan Oakes in producing Fig. 21.1 and Tom Carlson for providing the kernels for that figure. We thank Beth Savidge and Benjamin Moll for constructive review of the manuscript.

References

Alrefai R, Berke TG, Rocheford TR (1995) Quantitative trait locus analysis of fat acid concentrations in maize. Genome 38:894–901

Anonymous (2006) Corn oil, 5th edn. Corn Refiners Association, Washington, DC

Ash M, Dohlman E (2006) USDA oil crops situation and outlook yearbook. Market and Trade Economics Division, Economic Research Service, US Department of Agriculture, Washington, DC

Ashton AR, Jenkins CLD, Whitfeld PR (1994) Molecular cloning of two different cDNAs for maize acetyl CoA carboxylase. Plant Mol Biol 24:35–49

Bao XM, Focke M, Pollard M, Ohlrogge J (2000) Understanding in vivo carbon precursor supply for fatty acid synthesis in leaf tissue. Plant J 22:39–50

Bate NJ, Niu XP, Wang YW, Reimann KS, Helentjaris TG (2004) An invertase inhibitor from maize localizes to the embryo surrounding region during early kernel development. Plant Physiol 134:246–254

Baud S, Mendoza MS, To A, Harschet E, Lepiniec L, Dubreucq B (2007) *WRINKLED1* specifies the regulatory action of *LEAFY COTYLEDON2* towards fatty acid metabolism during seed maturation in Arabidopsis. Plant J 50:825–838

Benning C, Xu C, Awai K (2006) Non-vesicular and vesicular lipid trafficking involving plastids. Curr Opin Plant Biol 9:241–247

Berke TG, Rocheford TR (1995) Quantitative trait loci for flowering, plant and ear height, and kernel traits in maize. Crop Sci 35:1542–1549

Berquist RR, Nubel DS, Thompson DL (1998) Production method for corn with enhanced quality grain traits. Patent 5706603

Brown AP, Affleck V, Fawcett T, Slabas AR (2006) Tandem affinity purification tagging of fatty acid biosynthetic enzymes in *Synechocystis* sp PCC6803 and *Arabidopsis thaliana*. J Exp Bot 57:1563–1571

Burell MM, Mooney PJ, Blundy M, Carter D, Wilson F, Green J, Blundy KS, Rees TA (1994) Genetic manipulation of 6-phosphofructokinase in potato tubers. Planta 194:95–101

Cao YZ, Huang AHC (1987) Acyl coenzyme A preference of diacylglycerol acyltransferase from the maturing seeds of *Cuphea*, maize,rapeseed, and canola. Plant Physiol 84:762–765

Cernac A, Benning C (2004) WRINKLED1 encodes an AP2/EREB domain protein involved in the control of storage compound biosynthesis in Arabidopsis. Plant J 40:575–585

Clark D, Dudley JW, Rocheford TR, LeDeaux JR (2006) Genetic analysis of corn kernel chemical composition in the random mated 10 generation of the cross of generations 70 of IHO × ILO. Crop Sci 46:807–819

Curtis J, Brunson A, Hubbard J, Earle F (1956) Effect of the pollen parent on oil content of the corn kernel. Agron J 48:551–555

Curtis PE, Leng ER, Hageman RH (1968) Developmental changes in oil and fatty acid content of maize strains varying in oil content. Crop Sci 8:689–693

Dahlqvist A, Stahl U, Lenman M, Banas A, Lee M, Sandager L, Ronne H, Stymne S (2000) Phospholipid:diacylglycerol acyltransferase: an enzyme that catalyzes the acyl-CoA-independent formation of triacylglycerol in yeast and plants. Proc Natl Acad Sci USA 97:6487–6492

Doehlert D (1990) Distribution of enzyme activities within the developing maize (*Zea mays*) kernel in relation to starch, oil and protein accumulation. Physiol Plant 78:560–567

Duarte AP, Mason SC, Jackson DS, Kiehl J de C (2005) Grain quality of Brazilian maize genotypes as influenced by nitrogen level. Crop Sci 45:1958–1964

Dudley JW (1977) Seventy-six generations of selection for oil and protein percentage in maize. In: Pollak E, et al. (eds) Proc Int Conf on Quantitative Genetics, Ames, 16—21 August 1976, pp. 459–473. Iowa State University Press, Ames

Dudley JW, Lambert RJ (2004) 100 generations of selection for oil and protein in corn. Plant Breed Rev 24:79–110

Dudley JW, Lambert RJ, de la Roche IA (1977) Genetic analysis of crosses among corn strains divergently selected for percent oil and protein. Crop Sci 17:111–117

Dudley JW, Dijkhuizen A, Paul C, Coates ST, Rocheford TR (2004) Effects of random-mating on marker-QTL associations in the cross of the Illinois High Protein × Illinois Low Protein maize strains. Crop Sci 44:1419–1428

Dudley JW, Clark D, Rocheford TR, LeDeaux JR (2007) Genetic analysis of corn kernel chemical composition in the random mated 7 generation of the cross of generations 70 of IHP × ILP. Crop Sci 47:45–57

Eastmond PJ, Rawsthorne S (2000) Coordinate changes in carbon partitioning and plastidial metabolism during the development of oilseed rape embryos. Plant Physiol 122:767–774

Eastmond PJ, Dennis DT, Rawsthorne S (1997) Evidence that a malate/inorganic phosphate exchange translocator imports carbon across the leucoplast envelope for fatty acid synthesis in developing castor seed endosperm. Plant Physiol 114:851–856

Egli MA, Gengenbach BG, Gronwald JW, Somers DA, Wyse DL (1993) Characterization of maize acetyl-coenzyme A carboxylase. Plant Physiol 101:499–506

Fell DA (1997) Understanding the control of metabolism. Plenum Press, London

Fischer K, Kammerer B, Gutensohn M, Arbinger B, Weber A, Hausler RE, Flugge UI (1997) A new class of plastidic phosphate translocators: a putative link between primary and secondary metabolism by the phosphoenolpyruvate/phosphate antiporter. Plant Cell 9:453–462

Flugge UI (1998) Metabolite transporters in plastids. Curr Opin Plant Biol 1:201–206

Focks N, Benning C (1998) *wrinkled1*: A novel, low-seed-oil mutant of Arabidopsis with a deficiency in the seed-specific regulation of carbohydrate metabolism. Plant Physiol 118:91–101

Genter C, Eheart J, Linkous W (1956) Effects of location, hybrid, fertilizer, and rate of planting on the oil and protein contents of corn grain. Agron J 48:63–67

Gibon Y, Blaesing OE, Hannemann J, Carillo P, Hohne M, Hendriks JHM, Palacios N, Cross J, Selbig J, Stitt M (2004) A robot-based platform to measure multiple enzyme activities in Arabidopsis using a set of cycling assays: comparison of changes of enzyme activities and transcript levels during diurnal cycles and in prolonged darkness. Plant Cell 16:3304–3325

Goldman IL, Rocheford TR, Dudley JW (1994) Molecular markers associated with maize kernel oil concentration in an Illinois high protein × Illinois low protein cross. Crop Sci 34:908–915

Griffith SM, Jones RJ, Brenner ML (1987) In vitro sugar transport in *Zea mays* L. Kernels: II. Characteristics of sugar absorption and metabolism by isolated developing embryos. Plant Physiol 84:472–475

Han Y, Parsons CM, Alexander DE (1987) Nutritive value of high oil corn for poultry. Poultry Sci 66:103–111

Harwood JL (1996) Recent advances in the biosynthesis of plant fatty acids. Biochim Biophys Acta Lipids Lipid Metab 1301:7–56

Herbert D, Price LJ, Alban C, Dehaye L, Job D, Cole DJ, Pallett KE, Harwood JL (1996) Kinetic studies on two isoforms of acetyl-CoA carboxylase from maize leaves. Biochem J 318: 997–1006

Hobbs DH, Lu CF, Hills MJ (1999) Cloning of a cDNA encoding diacylglycerol acyltransferase from *Arabidopsis thaliana* and its functional expression. FEBS Lett 452:145–149

Hojilla-Evangelista MP, Johnson LA (2003) Sequential extraction processing of high-oil corn. Cereal Chem 80:679–683

Hong SK, Kitano H, Satoh H, Nagato Y (1996) How is embryo size genetically regulated in rice? Development 122:2051–2058

Hsieh K, Huang AHC (2004) Endoplasmic reticulum, oleosins, and oils in seeds and tapetum cells. Plant Physiol 136:3427–3434

Jako C, Kumar A, Wei YD, Zou JT, Barton DL, Giblin EM, Covello PS, Taylor DC (2001) Seed-specific over-expression of an Arabidopsis cDNA encoding a diacylglycerol acyltransferase enhances seed oil content and seed weight. Plant Physiol 126:861–874

Jellum MD, Marion JE (1966) Factors affecting oil content and oil composition of corn (*Zea mays* L.) grain. Crop Sci 6:41–42

Jellum MD, Boswell FC, Young CT (1973) Nitrogen and boron effects on protein and oil of corn grain. Agron J 65:330–331

Jenni S, Leibundgut M, Maier T, Ban N (2006) Architecture of a fungal fatty acid synthase at 5 Å resolution. Science 311:1263–1267

Jouhet J, Marechal E, Block MA (2007) Glycerolipid transfer for the building of membranes in plant cells. Progr Lipid Res 46:37–55

Jurgens SK, Johnson RR, Boyer JS (1978) Dry matter production and translocation in maize subjected to drought during grain fill. Agron J 70:678–682

Ke J, Behal RH, Back SL, Nikolau BJ, Wurtele ES, Oliver DJ (2000) The role of pyruvate dehydrogenase and acetyl-coenzyme A synthetase in fatty acid synthesis in developing Arabidopsis seeds. Plant Physiol 123:497–508

Kermode AR (2005) Role of abscisic acid in seed dormancy. J Plant Growth Regul 24:319–344

Kinney AJ (2006) Metabolic engineering in plants for human health and nutrition. Curr Opin Biotechnol 17:130–138

Kinney AJ, Cahoon EB, Hitz WD (2002) Manipulating desaturase activities in transgenic crop plants. Biochem Soc Trans 30:1099–1103

Klaus D, Ohlrogge JB, Neuhaus HE, Doermann P (2004) Increased fatty acid production in potato by engineering of acetyl-CoA carboxylase. Planta 219:389–396

Kroon JTM, Wei WX, Simon WJ, Slabas AR (2006) Identification and functional expression of a type 2 acyl-CoA: diacylglycerol acyltransferase (DGAT2) in developing castor bean seeds which has high homology to the major triglyceride biosynthetic enzyme of fungi and animals. Phytochemistry 67:2541–2549

Lalonde S, Wipf D, Frommer WB (2004) Transport mechanisms for organic forms of carbon and nitrogen between source and sink. Annu Rev Plant Biol 55:341–372

Lambert RJ (2001) High-oil corn hybrids. In: Hallauer AR (ed) Specialty corns, 2nd edn. CRC Press, Boca Raton, pp 131–154

Lambert RJ, Alexander DE, Mollring EL, Wiggens B (1997) Selection for increased oil concentration in maize kernels and associated changes in several kernel traits. Maydica 42: 39–43

Lambert RJ, Alexander DE, Han ZJ (1998) A high oil pollinator enhancement of kernel oil and effects on grain yields of maize hybrids. Agron J 90:211–215

Lambert RJ, Alexander DE, Mejaya IJ (2004) Single kernel selection for increased grain oil in maize synthetics and high-oil hybrid development. Plant Breed Rev 24:153–175

Lara-Nunez A, Rodriguez-Sotres R (2004) Characterization of a dicarboxylate exchange system able to exchange pyrophosphate for L-malate in non-photo synthetic plastids from developing maize embryos. Plant Sci 166:1335–1343

Lardizabal KD, Mai JT, Wagner NW, Wyrick A, Voelker T, Hawkins DJ (2001) DGAT2 is a new diacylglycerol acyltransferase gene family – purification, cloning, and expression in insect cells of two polypeptides from *Mortierella ramanniana* with diacylglycerol acyltransferase activity. J Biol Chem 276:38862–38869

Lardizabal KD, Thompson GA, Hawkins D (2003) Diacylglyerol transferase proteins. Patent US23115632A1

Lazic-Jancic V, Ignajatovic-Micic D, Markovic K, Coric T, Kovacevic D, Saratlic G, Quarrie S (2003) Changes in RFLP allele frequencies during recurrent selection for high oil content in maize. Maize Genetics Coop Newsl 77:4–5

Laurie CC, Chasalow SD, LeDeaux JR, McCarroll R, Bush D, Hauge B, Lai CQ, Clark D, Rocheford TR, Dudley JW (2004) The genetic architecture of response to long-term artificial selection for oil concentration in the maize kernel. Genetics 168:2141–2155

Lee JM, Williams ME, Tingey SV, Rafalski JA (2002) DNA array profiling of gene expression changes during maize embryo development. Funct Integr Genomics 2:13–27

Lee K, Ratnayake C, Huang AHC (1995) Genetic dissection of the coexpression of genes encoding the 2 isoforms of oleosins in the oil bodies of maize kernel. Plant J 7:603–611

Letchworth MB, Lambert RJ (1998) Pollen parent effects on oil, protein, and starch concentration in maize kernels. Crop Sci 38:363–367

Lung SC, Weselake RJ (2006) Diacylglycerol acyltransferase: a key mediator of plant triacylglycerol synthesis. Lipids 41:1073–1088

Maddonni GA, Otegui ME (2006) Intra-specific competition in maize: contribution of extreme plant hierarchies to grain yield, grain yield components and kernel composition. Field Crops Res 97:155–166

Maier T, Jenni S, Ban N (2006) Architecture of mammalian fatty acid synthase at 4.5 Å resolution. Science 311:1258–1262

Mangen TF, Thomison PR, Strachan SD (2005) Early-season defoliation effects on TopCross high-oil corn production. Agron J 97:823–831

Mangolin CA, de Souza CL Jr, Garcia AAF, Garcia AF, Sibov ST, de Souza AP (2004) Mapping QTLs for kernel oil content in a tropical maize population. Euphytica 137:251–259

Mendoza MS, Dubreucq B, Miguel M, Caboche M, Lepiniec L (2005) LEAFY COTYLEDON 2 activation is sufficient to trigger the accumulation of oil and seed specific mRNAs in *Arabidopsis* leaves. FEBS Lett 579:4666–4670

Mhaske V, Beldjilali K, Ohlrogge J, Pollard M (2005) Isolation and characterization of an *Arabidopsis thaliana* knockout line for phospholipid: diacylglycerol transacylase gene (At5g13640). Plant Physiol Biochem 43:413–417

Miao Y, Mulla DJ, Robert PC, Hernandez JA (2006) Within-field variation in corn yield and grain quality reponses to nitrogen fertilization and hybrid selection. Agron J 98:129–140

Misevic D, Alexander DE (1989) Twenty-four cycles of phenotypic recurrent selection for percent oil in maize. I. Per se and test-cross performance. Crop Sci 29:320–324

Moche M, Schneider G, Edwards P, Dehesh K, Lindqvist Y (1999) Structure of the complex between the antibiotic cerulenin and its target, beta-ketoacyl-acyl carrier protein synthase. J Biol Chem 274:6031–6034

Mohlmann T, Tjaden J, Schwoppe C, Winkler HH, Kampfenkel K, Neuhaus HE (1998) Occurrence of two plastidic ATP/ADP transporters in *Arabidopsis thaliana* L. – molecular characterisation and comparative structural analysis of similar ATP/ADP translocators from plastids and *Rickettsia prowazekii*. Eur J Biochem 252:353–359

Moose SP, Dudley JW, Rocheford TR (2004) Maize selection passes the century mark: a unique resource for 21st century genomics. Trends Plant Sci 9:358–364

Oo KC, Huang AHC (1989) Lysophosphatidate acyl transferase activities in the microsomes from palm endosperm, maize scutellum, and rapeseed cotyledon of maturing seeds. Plant Physiol 91:1288–1295

Opsahl-Ferstad HG, LeDeunff E, Dumas C, Rogowsky PM (1997) ZmEsr, a novel endosperm-specific gene expressed in a restricted region around the maize embryo. Plant J 12: 235–246

Orthoefer FT, Sinram RD (1987) Corn oil: composition, processing, and utilization. In: Watson SA, Ramstad PE (eds) Corn chemistry and technology. American Association of Cereal Chemists, St. Paul, pp 535–551

Page RA, Okada S, Harwood JL (1994) Acetyl-CoA carboxylase exerts strong flux control over lipid synthesis in plants. Biochim Biophys Acta 1210:369–372

Paiva R, Kriz AL (1994) Effect of abscisic acid on embryo-specific gene exprsession during normal and precocious germination in normal and viviparous maize (*Zea mays*) embryos. Planta 192:332–339

Patel MS, Korotchkina LG (2006) Regulation of the pyruvate dehydrogenase complex. Biochem Soc Trans 34:217–222

Paul M, Foyer C (2001) Sink regulation of photosynthesis. J Exp Bot 52:1383–1400

Paul M, Sonnewald U, Hajirezaei M, Dennis D, Stitt M (1995) Transgenic tobacco plants with strongly decreased expression of pyrophosphate-fructose-6-phosphate 1-phosphotransferase do not differ significantly from wild-type in photosynthate partitioning, plant-growth or their ability to cope with limiting phosphate, limiting nitrogen and suboptimal temperatures. Planta 196:277–283

Plaxton WC, Podesta FE (2006) The functional organization and control of plant respiration. Crit Rev Plant Sci 25:159–198

Reynolds TL, Nemeth MA, Glenn KC, Ridley WP, Astwood TD (2005) Natural variability of metabolites in maize grain: differences due to genetic background. J Agric Food Chem 53:10061–10067

Rider SD, Henderson JT, Jerome RE, Edenberg HJ, Romero-Severson J, Ogas J (2003) Coordinate repression of regulators of embryonic identity by PICKLE during germination in Arabidopsis. Plant J 35:33–43

Rocheford T, Dudley JW, Lambert R, Widholm J, Endgeseth N (2003) Development of high oil, high oleic value-added maize hybrids. Final Report for Illinois Missouri Biotechnology Alliance Project 98-4, http://www.imba.missouri.edu/report/1998_4.php

Roesler K, Shintani D, Savage L, Boddupalli S, Ohlrogge J (1997) Targeting of the Arabidopsis homomeric acetyl-coenzyme A carboxylase to plastids of rapeseeds. Plant Physiol 113: 75–81

Roughan PG (1997) Stromal concentrations of coenzyme A and its esters are insufficient to account for rates of chloroplast fatty acid synthesis: evidence for substrate channelling within the chloroplast fatty acid synthase. Biochem J 327:267–273

Roughan PG, Ohlrogge JB (1996) Evidence that isolated chloroplasts contain an integrated lipid-synthesizing assembly that channels acetate into long-chain fatty acids. Plant Physiol 110:1239–1247

Routaboul JM, Benning C, Bechtold N, Caboche M, Lepiniec L (1999) The TAG1 locus of Arabidopsis encodes for a diacylglycerol acyltransferase. Plant Physiol Biochem 37:831–840

Ruuska SA, Girke T, Benning C, Ohlrogge JB (2002) Contrapuntal networks of gene expression during Arabidopsis seed filling. Plant Cell 14:1191–1206

Sasaki Y, Nagano Y (2004) Plant acetyl-CoA carboxylase: structure, biosynthesis, regulation, and gene manipulation for plant breeding. BioSci Biotechnol Biochem 68:1175–1184

Schwender J, Ohlrogge J, Shachar-Hill Y (2004) Understanding flux in plant metabolic networks. Curr Opin Plant Biol 7:309–317

Séne M, Thévenot C, Hoffmann D, Bénétrix F, Causse M, Prioul J-L (2001) QTLs for grain dry milling properties, composition and vitreousness in maize recombinant inbred lines. Theor Appl Genet 102:591–599

Shen B, Sinkevicius KW, Selinger DA, Tarczynski MC (2006) The homeobox gene GLABRA2 affects seed oil content in Arabidopsis. Plant Mol Biol 60:377–387

Somers DA, Keith RA, Egli MA, Marshall LC, Gengenbach BG, Gronwald JW, Wyse DL (1993) Expression of the Acc1 gene-encoded acetyl-coenzyme A carboxylase in developing maize (*Zea mays* L.) kernels. Plant Physiol 101:1097–1101

Song GL, Li DF, Piao XS, Chi F, Wang JT (2003) Comparisons of amino acid availability by different methods and metabolizable energy determination of a Chinese variety of high oil corn. Poultry Sci 82:1017–1023

Song TM, Chen SJ (2004) Long term selection for oil concentration in five maize populations. Maydica 49:9–14

Song TM, Kong F, Li C, Song G (1999) Eleven cycles of single kernel phenotypic recurrent selection for percent oil in Zhongzong no. 2 maize synthetic. J Genet Breed 53:31–35

Song XF, Song TM, Dai JR, Rocheford TR, Li JS (2004) QTL mapping of kernel oil concentration with high-oil maize by SSR markers. Maydica 49:41–48

Stahl U, Carlsson AS, Lenman M, Dahlqvist A, Huang BQ, Banas W, Banas A, Stymne S (2004) Cloning and functional characterization of a phospholipid: diacylglycerol acyltransferase from Arabidopsis. Plant Physiol 135:1324–1335

Sun C, Cao YZ, Huang AHC (1988) Acyl coenzyme A preference of the glycerol phosphate pathway in the microsomes from the maturing seeds of palm, maize and rapeseed. Plant Physiol 88:56–60

Suzuki M, Kao CY, Cocciolone S, McCarty DR (2001) Maize VP1 complements Arabidopsis abi3 and confers a novel ABA/auxin interaction in roots. Plant J 28:409–418

Suzuki M, Ketterling MG, Li Q-B, McCarty DR (2003) Viviparous1 alters global gene expression patterns through regulation of abscisic acid signaling. Plant Physiol 132:1664–1677

Suzuki M, Wang HHY, McCarty DR (2007) Repression of the LEAFY COTYLEDON 1/B3 regulatory network in plant embryo development by VP1/ABSCISIC ACID INSENSITIVE 3-LIKE B3 genes. Plant Physiol 143:902–911

Taliercio EW, Kim JY, Mahe A, Shanker S, Choi J, Cheng WH, Prioul JL, Chourey PS (1999) Isolation, characterization and expression analyses of two cell wall invertase genes in maize. J Plant Physiol 155:197–204

Tan SL, Morrison WR (1979) The distribution of lipids in the germ endosperm pericarp and tip cap of amylomaize Lg-11 hybrid maize and waxy maize. J Am Oil Chem Soc 56:531–535

Thelen JJ, Ohlrogge JB (2002) Metabolic engineering of fatty acid biosynthesis in plants. Metab Eng 4:12–21

Thomison PR, Geyer AB, Lotz LD, Siegrist HJ, Dobbels TL (2002) TopCross high-oil corn production: agronomic performance. Agron J 94:290–299

Thomison PR, Geyer AB, Lotz LD, Siegrist HJ, Dobbels TL (2003) TopCross high oil corn production: select grain quality attributes. Agron J 95:147–154

Thompson DL, Jellum M, Young C (1973) Effect of controlled temperature environments on oil content and on fatty acid composition of corn oil. J Am Chem Soc 50:540–542

Ting JT, Lee K, Ratnayake C, Platt KA, Balsamo RA, Huang AH (1996) Oleosin genes in maize kernels having diverse oil contents are constitutively expressed independent of oil contents. Size and shape of intracellular oil bodies are determined by the oleosins/oils ratio. Planta 199:158–165

To A, Valon C, Savino G, Guilleminot J, Devic M, Giraudat J, Parcy F (2006) A network of local and redundant gene regulation governs Arabidopsis seed maturation. Plant Cell 18:1642–1651

Tovar-Mendez A, Miernyk JA, Randall DD (2003) Regulation of pyruvate dehydrogenase complex activity in plant cells. European J of Biochemistry 270:1043–1049

Tsukagoshi H, Morikami A, Nakamura K (2007) Two B3 domain transcriptional repressors prevent sugar-inducible expression of seed maturation genes in Arabidopsis seedlings. Proc Natl Acad Sci USA 104:2543–2547

USDA (2007) Feed Grains Database: Yearbook Tables. In. United States Department of Agriculture, Economic Research Service

Val D, Banu G, Seshadri K, Lindqvist V, Dehesh K (2000) Re-engineering ketoacyl synthase specificity. Structure 8:565–566

Vigeolas H, Waldeck P, Zank T, Geigenberger P (2007) Increasing seed oil content in oil-seed rape (*Brassica napus* L.) by over-expression of a yeast glycerol-3-phosphate dehyrogenase under the control of a seed-specific promoter. Plant Biotechnology J 5:431–441

Voelker T, Kinney AJ (2001) Variations in the biosynthesis of seed-storage lipids. In Annual Review of Plant Physiology and Plant Molecular Biology, pp 335–361

Wang H, Gui J, Lambert KN, Lin Y (2007) Developmental control of Arabidopsis seed oil biosynthesis. Planta 226:773–783

Watson SA (1987) Structure and Composition. In SA Watson, PE Ramstad, eds, Corn Chemistry and Technology. American Association of Cereal Chemists, Inc., St. Paul, MN, pp 53–82

Weber EJ (1987) Lipids of the kernel. In SA Watson, PE Ramstad, eds, Corn Chemistry and Technology. American Association of Cereal Chemists, Inc., St. Paul, MN, pp 311–349

Welch L (1969) Effect of N, P, and K on the percent and yield of oil in corn. Agron J 61:890–891

Wilhelm EP, Mullen R, Keeling PL, Singletary G (1999) Heat stress during grain filling in maize: effects on kernel growth and metabolism. Crop Sci 39:1733–1741

Zhang F-Y, Yang M-F, Xu Y-N (2005) Silencing of *DGAT1* in tobacco causes a reduction in seed oil content. Plant Sci 169:689–694

Zheng P, Allen WB, Roesler K, Williams ME, Zhang S, Li J, Glassman K, Ranch J, Nubel D, Solawtz W, Bhattramakki D, Llaca V, Deschamps S, Zhong G-Y, Tarczynski MC, Shen B (2008) A phenylalanine in DGAT is a key determinant of oil content and composition in maize. Nature Genetics 40: 367-372

Zou JT, Wei YD, Jako C, Kumar A, Selvaraj G, Taylor DC (1999) The Arabidopsis thaliana TAG1 mutant has a mutation in a diacylglycerol acyltransferase gene. Plant J 19:645–653

Chapter 22
Chloroplasts

Delene J. Oldenburg and Arnold J. Bendich

22.1 Introduction

The introduction of a transgene to alter the properties of the chloroplast raises the question of whether the transgene should be integrated into the nuclear or plastid genome (Daniell et al. 2004; Grevich and Daniell 2005; Maliga 2002, 2004). For the nucleus, we need to consider a plastid targeting sequence, gene silencing, and cell- and development-specific expression. For both locations, regulation of gene expression is a concern, but different mechanisms predominate in each location. Transcriptional regulation is the most important issue to address in the nucleus, whereas post-transcriptional regulation is primary in the plastid. Success in this endeavor may be further affected by the presence of multiple copies of the genome per plastid and multiple plastids per cell.

Another consideration is the structure of the plastid chromosome (Bendich 2004). Since we require that the transgene be present in all cells derived from the cell containing the initial transformed plastid, it is important to target a plastid DNA molecule capable of acting as a chromosome, a segregating genetic unit. Thus, we need to know what a plastid chromosome looks like and where in the plant to find such a chromosome. The concept of the circular chloroplast chromosome has impeded progress toward an understanding of the process by which chloroplast (cp) DNA is replicated and inherited. The "ploidy paradox" illustrates the problem: there is a small number of segregating genetic units, but a high level of ploidy (computed as the mass of DNA per plastid divided by its genome size) (Birky 1994; Gillham 1994). If the chromosome were comprised of a multigenomic structure of

Delene J. Oldenburg
Biology Department, University of Washington, Box 355325, Seattle, WA 98195-5325, USA
e-mail: delene@u.washington.edu

Arnold J. Bendich
Biology Department, University of Washington, Box 355325, Seattle, WA 98195-5325, USA
e-mail: bendich@u.washington.edu

A.L. Kriz, B.A. Larkins (eds.), *Molecular Genetic Approaches to Maize Improvement*
Biotechnology in Agriculture and Forestry, Vol. 63

replicating cpDNA, this paradox would be resolved. Furthermore, we could then aim our transgene at cells containing bona fide plastid chromosomes and avoid cells no longer containing cpDNA able to serve as a plastid chromosome.

The focus of this chapter is the chromosomal DNA in maize plastids. We will review recent advances in DNA sequence analysis, structure and replication and discuss the factors affecting cpDNA copy number. Some information has been included on the regulation and expression of proteins involved in plastid biogenesis and photosynthesis, but most of this information is found elsewhere (Barkan and Goldschmidt-Clermont 2000; Daniell and Chase 2004; Lopez-Juez and Pyke 2005; Stern et al. 2004). Finally, we will discuss strategies for plastid genetic engineering.

22.2 Size, Form, and Genomic Map of cpDNA

22.2.1 Highlights of the Genome Sequence

In 1995, maize joined what was at that time an exclusive group of plant species for which the complete chloroplast genome sequence was known (Maier et al. 1995; accession X86563). As of July 2008, the list reached 114 species (http://megasun.bch.umontreal.ca/ogmp/projects/other/cp_list.html). Prior to sequencing, restriction analysis of maize cpDNA predicted a genome size of 129 kb and two single copy regions separated by a pair of inverted repeats (Bedbrook and Bogorad 1976). The sequenced genome is 140,387 base pairs (bp) in length (Maier et al. 1995). The large (LSC) and small (SSC) single copy regions are 82,355 and 12,536 bp, respectively, and the inverted repeats (IR_A and IR_B) are each 22,748 bp. Following the convention used for previously sequenced plastid genomes, nucleotide numbering for the maize genome begins with nucleotide (nt) position 1 in the LSC and ends in IR_A at nt 140,387 (Fig. 22.1b, map 1).

There are 107 genes of known function encoded in the maize plastome (Maier et al. 1995; Matsuoka et al. 2002). Most of these can be grouped into two categories: a set of housekeeping genes necessary for gene expression and comprised of 60 genes for RNAs, ribosomal proteins, and RNA polymerase (61 if translational initiation factor 1, *infA*, is included); and 43 genes encoding proteins for photosynthesis and energy production. In addition, *matK* codes for a maturase, *clpP* for a protease and *cemA* for an envelope membrane protein. Eighteen unidentified open reading frames (ORFs) were found (Maier et al. 1995). A comparison of hypothetical ORFs in maize, rice, tobacco, liverwort and black pine reveals that eight sequences (*ycf3–10*) are common to all five land plants, two sequences (*ycf1* and *ycf2*) are present in the non-vascular, gymnosperm and dicot plants but absent in the two monocots, and nine ORFs are exclusive to the monocots.

In maize as in other land plants, with the exception of the liverwort *Marchantia polymorpha*, RNA editing of plastid transcripts is evident and shows a strong bias towards modification of the second codon position from C to U. The total number of edited sites, however, is low (0.13%, 25 verified editing sites) (Maier et al. 1995).

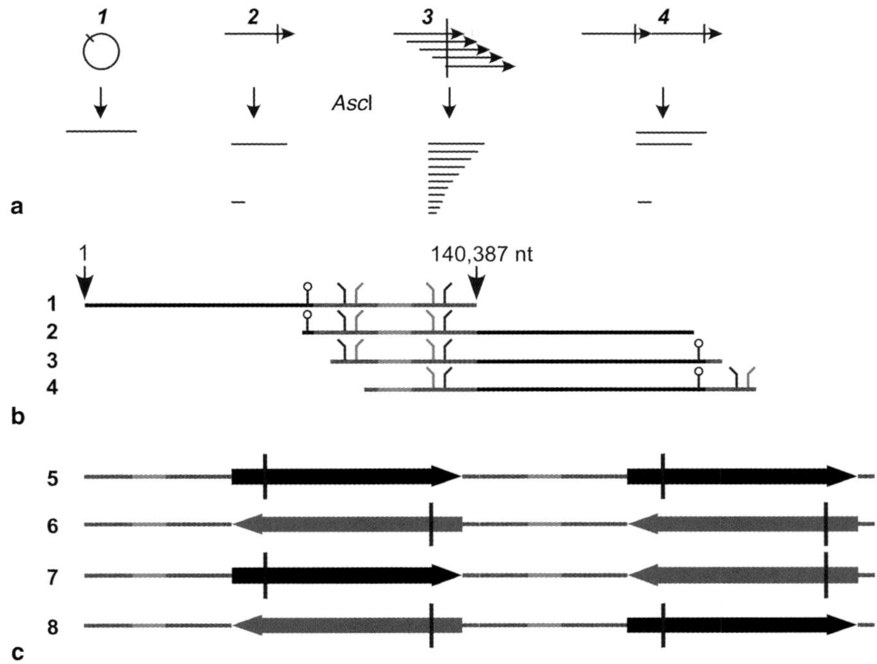

Fig. 22.1 Model of maize cpDNA structure: linear molecules with defined ends. **a** Digestion with an enzyme that cuts once per genome generates: a genome-sized linear from a circular monomer (*1*); two subgenomic fragments from a linear monomer with defined ends (*2*); a smear of subgenomic fragments from a collection of circularly permuted linear monomers (*3*); and three fragments (one genome-sized and two subgenomic) from a head-to-tail (h-t) linear concatemer (*4*). *Asc*I digestion of maize cpDNA yields patterns *2* and *4*. **b** For *map 1*, the end of maize cpDNA is arbitrarily designated as nt 1 (reference sequence X86563). Ends predicted from restriction digestion and blot hybridization are: nt 78,000 for *map 2*, 88,000 for *3* and 100,000 for *4*. The standard deviation is ±5,000 for the location of the ends. The putative origins are indicated by *line-with-circle* for the ori near *rpl16* and by the *black and gray angled lines* for the orthologs of *Oenothera ori*A and *ori*B, respectively. **c** The four h-t dimeric (280-kb) isomers with the LSC in the standard and inverted orientations. A diagnostic larger-than-genome-sized fragment is produced by *Sfi*I digestion (*tick marks*) of isomer 7. The LSC is indicated by *black lines* in **b** and *black and gray arrows* in **c**, SSC by *gray lines* and IRs by *stippled lines* in both **b** and **c**. (Adapted from Oldenburg and Bendich (2004a))

22.2.2 The Traditional Model: The Circular Form of Maize cpDNA

The chromosomal DNA in chloroplasts of maize, as in other plants, is typically depicted as a genome-sized circle. The circular depiction is based, in part, on the assumptions that the ancestral bacterial chromosome was circular and that linear forms result from breakage of circular molecules (the Broken Circles theory) (Bendich 2004; Deng et al. 1989). Restriction analysis, using DNA prepared by in-solution methods and enzymes with several cutting sites per genomic unit, predict a circular map of cpDNA from maize (Bedbrook and Bogorad 1976) and other plants (Palmer 1985). Linear head-to-tail (h-t) concatemers or a collection of circular permuted linear forms, however, yield the same fragments and thus can also generate a circular map (Bendich 1993; Oldenburg and Bendich 2001). Estimates of the amount of circular DNA in chloroplasts vary greatly and depend on the methods of DNA preparation, which often include fractionation of the cpDNA prior to analysis (Bendich 1991, 2004; Bendich and Smith 1990; Kolodner and Tewari 1972, 1975a; Lilly et al. 2001). They also depend on whether the quantification is based on the number of cpDNA molecules or the fraction of cpDNA mass found in circular form. An even more important variable is the tissue used for extracting the cpDNA (Sects. 22.4.2 and 22.4.3). The circular concept has persisted for over 30 years, in part because no information concerning the ends of the linear molecules has been available. It is time to revise this outdated model in light of the recent advances described below.

22.2.3 The Revised Model: Linear and Complex Forms of Maize cpDNA

In-gel preparation of cpDNA avoids two problems associated with in-solution methods: shearing of large molecules and discarding a substantial fraction of the cpDNA before analysis. Restriction enzymes with a single site per genomic unit can be used to distinguish between linear and circular forms and determine whether a population of linear molecules has defined or random ends (Fig. 22.1a; Oldenburg and Bendich 2001, 2004a). Pulsed-field gel electrophoresis (PFGE) of unfractionated, uncut, in-gel-prepared maize cpDNA revealed linear molecules of subgenomic and unit genome sizes, but much of the cpDNA remained immobile (well-bound) (Oldenburg and Bendich 2004a). Fluorescence microscopy of the ethidium-stained well-bound cpDNA revealed that this fraction was primarily comprised of complex, branched, multigenomic forms (Figs. 22.2d and e), with only a minor circular DNA component. Based on these two methods (PFGE and fluorescence microscopy), coupled with *Asc*I digestion (single site at nt 42,131), it was estimated that 3.5% of the cpDNA mass was in circular, 50% in linear and 46.5% in complex forms. These values are for chloroplasts from young maize seedlings, 10–14 days old and grown in the light. However, the amount of cpDNA in each molecular form varies, depending on the developmental stage of the tissue and growth conditions (Sect. 22.4). Linear and complex forms have also been reported for cpDNAs of tobacco, Arabidopsis, pea, *Medicago truncatula* and watermelon (Bendich 1991; Bendich and Smith 1990; Lilly et al. 2001; Rowan et al. 2004; Shaver et al. 2006). The failure of early investigators to report cpDNA in complex forms of multigenomic size has been attributed to cpDNA fractionation by centrifugation prior to analysis, dismissal of such forms as nuclear contamination or uninterpretable structures, and a strong expectation that such complex forms would not be found in chloroplasts (Bendich 2004).

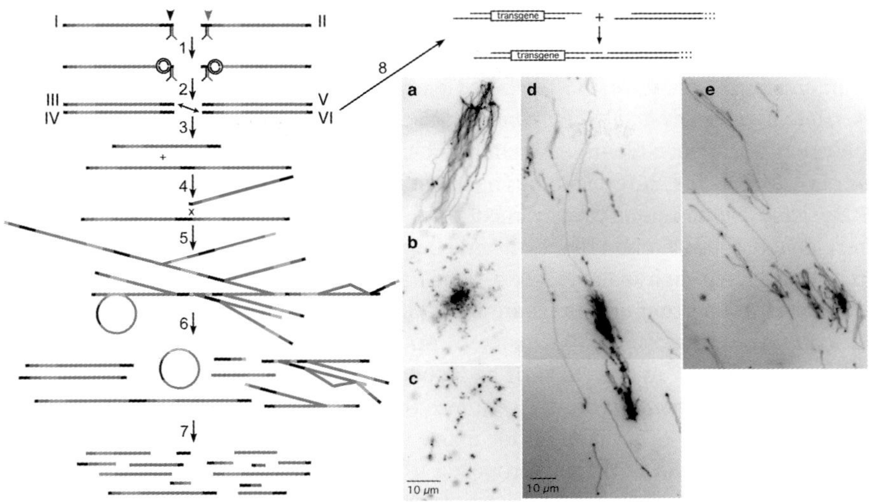

Fig. 22.2 Model for cpDNA replication: OPaLI-RDR and images of cpDNA molecules. *Step 1*: Initiation. Note that *ori*A (*black*) and *ori*B (*gray*) are spaced by about 4 kb on each 140-kb isomer. *Step 2*: Elongation. *Step 3*: Annealing of 3′ overhangs to form head-to-tail concatemers. *Step 4*: Recombination. *Step 5*: Amplification. *Step 6*: Resolution. *Step 7*: Degradation. *Step 8*: A linear transgene vector could be designed to anneal to a 3′ overhang on one of the nascent monomers. *Steps 1–7* are described in Sect. 22.3.2 and *step 8* in Sect. 22.5.2. Fluorescence microscopic images of ethidium-stained cpDNA molecules from individual chloroplasts after in-gel lysis are depicted in **a–c** and from the well-bound fraction following PFGE in **d** and **e**. Three classes of cpDNA molecules were described (Oldenburg and Bendich 2004b). Class I contains multigenomic complex structures representing replicating DNA produced in *step 5* (**a**). Class II contains smaller complex structures with many simple molecules produced by resolution in *step 6* (**b**). Class III contains simple molecules without complex forms, corresponding to degraded molecules produced in *step 7* (**c**). The multigenomic branched forms in **d** and **e** are likely due to recombination-dependent replication, *step 5*. *Scale bar* in **c** applies to **a–c** and in **d** to **d** and **e**. (Adapted from Oldenburg and Bendich (2004a, b))

PFGE of in-gel, *Asc*I-digested maize cpDNA shows a genome-sized fragment and discrete subgenomic fragments, as expected for linear molecules with defined ends (Oldenburg and Bendich 2004a). Four major and six minor fragments are present and grouped in pairs, with each pair approximately equal to the genome size of 140 kb. For example, one of the major pairs was 94 and 44 kb (fragments a2 and a5 in Oldenburg and Bendich 2004a). Thus, there are two major and three minor populations of linear isomers with different left and right ends. The location of the ends was determined using hybridization probes to regions within the LSC, SSC and IRs. The ends of the major isomers are predicted to be at nt 88,000 and 100,000 (Fig. 22.1b, maps 3 and 4, respectively) within IR_B. It is noteworthy that the restriction and hybridization data are equally compatible with ends in the corresponding IR_A sequences (nt 134,000 and 122,000, respectively). The end of one minor isomer (Fig. 22.1b, map 2) is predicted to be at nt 78,000 within the LSC (fragment pair a8/a9 in Oldenburg and Bendich 2004a). The terminal locations of the other two minor isomers were not determined. Linear forms with defined ends are also predicted for cpDNAs of tobacco (Scharff and Koop 2006, 2007) and *M. truncatula* (Shaver et al. 2008).

Interestingly, some of the predicted ends for maize (Fig. 22.1b), tobacco and *M. truncatula* cpDNAs also lie near putative origins of replication. Although there is no single sequence identified unambiguously as a consensus origin of replication for plastid genomes, several studies have implicated sequences near the 16S and 23S rRNA genes, and in some cases pairs of origins 4–7 kb apart are proposed (Heinhorst and Cannon 1993; Kunnimalaiyaan and Nielsen 1997). These characteristics are exemplified by two such sequences, *ori*A and *ori*B in *Oenothera* (Chiu and Sears 1993; Sears et al. 1996), and orthologs are found near the ends of maize (Fig. 22.1b), *M. truncatula* and tobacco cpDNA isomers. Another origin in tobacco, *ori*B (T*ori*B), is also located near an end (Scharff and Koop 2006), but this sequence is not found in maize or *M. truncatula* plastid genomes (our BLAST search). Gold et al. (1987) proposed an origin of replication for maize cpDNA near the *rpl16* gene within the LSC and it is similar to origins proposed for *Chlamydomonas* and soybean (Kunnimalaiyaan and Nielsen 1997). The right end of the maize isomer shown as map 2 in Fig. 22.1b also lies near the *rpl16* gene. A corresponding end also maps near the *rpl16* gene in tobacco. These results suggest that the ends are near sites of replication initiation and common mechanism(s) of replication for plant cpDNAs, both those with (maize and tobacco) and those without (*M. truncatula*) IR sequences.

In light of the evidence for linear forms with defined ends, two other features of the circular model should be re-evaluated: (1) two equimolar populations of circular molecules differing only in the relative orientation (inversion) of the single copy regions; and (2) intramolecular recombination ("flipping") between IR_A and IR_B generating the two isoforms (Palmer 1983). The restriction fragments indicative of an inversion, however, may result from either linear or circular isomers. In addition, a larger-than-genome-sized restriction fragment was identified that could only be produced from h-t concatemers containing LSCs in both the "standard" (reference sequence) and inverted orientation (Fig. 22.1c, isomer 5; Oldenburg and Bendich 2004a). Considering the scarcity of circular molecules, a more likely model is a set of four equimolar, h-t linear isomers with the LSC either in the same orientation within each genomic unit or in opposite orientations (Fig. 22.1c). This model is analogous to that of the 152-kb genome of herpes simplex virus (HSV) (Lehman and Boehmer 1999; Oldenburg and Bendich 2004a; Sandri-Goldin 2003). Flipping between the IRs within an h-t linear concatemer could produce isomers with an inverted LSC, although inversion may arise via a recombination-dependent mechanism of replication (Sect. 22.3.2). The latter process is more likely for two reasons: (1) the ends of the molecules are near putative replication origins; and (2) homologous recombination in *E. coli* is stimulated both by double-stranded DNA ends and by DNA replication (Kuzminov and Stahl 1999).

It is time to replace the circular model with one that more accurately represents most of the DNA in chloroplasts – linear and complex multigenomic molecules. We will now consider the mechanism of replication of the plastid DNA in light of this revised structural model.

22.3 Replication of cpDNA

Three stages during replication of any DNA molecule can be recognized: initiation, elongation and resolution (separation) of the nascent molecules. For plastids, mitochondria and viruses, however, an additional stage of genome amplification produces multigenomic molecules. The replication of DNA in the nucleus does not include multigenomic amplification, at least for the production of chromosomes to be partitioned at cell division. We will also address the degradation of cpDNA, a phenomenon not generally appreciated.

22.3.1 The Traditional Model: D-Loop-to-Theta-to-Rolling Circle Replication

The traditional model for cpDNA replication involves a genome-sized circular molecule, initiation of replication at two sites forming D-loops, merging of the D-loops to form a theta or Cairns replication intermediate, and conversion to a rolling circle mechanism (Heinhorst and Cannon 1993; Kolodner and Tewari 1975b; Kunnimalaiyaan and Nielsen 1997). In this model, the start site for the rolling circle mechanism is 180° from the D-loop origin of replication, at the terminus for theta replication where the replication forks converge. One problem with this model is the assumption, despite the lack of evidence, of a circular molecule as the primary or only form of DNA in the chloroplast. Another is that the end sequence of the rolling circle "tail" would not be near the origin(s) of replication. A rolling circle mechanism could be used, however, if initiation began at a single origin and replication was unidirectional with strand displacement around the circle to generate an h-t concatemeric tail with an end that corresponds to the site of initiation. Such a mechanism has been suggested for maize, tobacco and other plastid DNAs (Oldenburg and Bendich 2004a; Scharff and Koop 2006), but requires the circular form as the template for replication. Given the abundance of linear, not circular, cpDNA forms, what other mechanism(s) might be used for cpDNA replication?

22.3.2 The Revised Model: OPaLI-RDR

In developing a model for cpDNA replication, it is useful to consider mechanisms that have been well characterized for other replicons. In the case of maize cpDNA, similarities with DNAs from T7 bacteriophage and HSV were used to develop the OPaLI-RDR (origin paired linear isomers – recombination-dependent replication) model (Fig. 22.2; Oldenburg and Bendich 2004a, b). Although this model may not be the only mechanism used to replicate cpDNAs, it does account for the two closely spaced origins, the location of ends near replication origins, and the presence of LSC inversion isomers for cpDNAs with IRs. It also applies to cpDNAs with and without IRs. Finally, it accounts for the myriad of structural forms observed including circular, linear and branched multigenomic complexes.

22.3.2.1 Steps 1 and 2 – Initiation and Elongation

Replication begins near the ends of two linear isomers. Each isomer has unique terminal sequences and two origins in close proximity forming a "pair of origins". A key feature is the initiation at a different origin on each of the two isomers: replication begins on *ori*A of isomer I and on *ori*B of isomer II (arrowheads in Fig. 22.2), and replication is bidirectional. Replication is incomplete at one end of each nascent molecule, as occurs with any linear DNA molecule due to the "end problem" (Kornberg and Baker 1992).

22.3.2.2 Step 3 – Annealing to Create an h-t Concatemer

A partial solution to the end problem is formation of an h-t concatemer by annealing compatible single-strand regions between *ori*A and *ori*B on two of the four nascent molecules ($3'$ overhangs of isomers III and VI in Fig. 22.2). This step is analogous to that proposed for T7 DNA, where h-t linear concatemers are formed by annealing of terminally redundant ends of identical molecules (Kornberg and Baker 1992).

22.3.2.3 Steps 4 and 5 – Recombination and Genome Amplification

Two of the four newly replicated molecules do not have compatible $3'$ overhangs but still need their terminal sequences filled in (isomers IV and V in Fig. 22.2). To solve this problem, the single-strand $3'$ overhang can invade a homologous region of another molecule. Additional rounds of strand invasion and RDR, as proposed for T7, would amplify the plastid genome and create the branched multigenomic molecules that are the major form of cpDNA found in developing plastids (Sects. 22.4.1 and 22.4.2). This process also generates LSC inversion isomers, because strand invasion can occur at either IR_A or IR_B.

22.3.2.4 Step 6 – Resolution

In both T7 and HSV, multigenomic replicative forms of DNA are processed to genome-sized molecules for packaging into as many viral particles as possible before release from the host cells (Kornberg and Baker 1992; Sandri-Goldin 2003). Since plastids remain in the cell after division, there is no apparent requirement for processing cpDNA to genome-sized units (Bendich 2004; Oldenburg and

Bendich 2004a). Branched multigenomic chromosomes would be partitioned to daughter plastids following resolution of the replicative form as forks reach the ends of their template strands and through breakage-rejoining of DNA strands at Holliday junctions. Resolution also produces additional forms of maize cpDNA including unit-genome and subgenomic linears, small branched forms and, in some instances, genome-sized circles that may result from incidental recombination between direct repeats on an h-t concatemer (Oldenburg and Bendich 2004a, b).

22.3.2.5 Step 7 – Degradation

For light-grown maize seedlings, most of the DNA is found in the well-bound fraction and the less-than-genome-sized smear plus a faint, sometimes barely detectable, band of linear monomeric cpDNA. Higher oligomers, common in other plants (Backert et al. 1995; Bendich and Smith 1990; Deng et al. 1989), are rarely found in maize. The paucity of unit-genome-sized linear forms may be due to rapid degradation, without or concomitant with resolution of branched forms, in the absence of reinitiation of replication and/or repair. This progressive degradation leads to the less-than-genome-sized molecules that may then be lost entirely in mature chloroplasts (Oldenburg and Bendich 2004b; Oldenburg et al. 2006). Linear oligomers and circular forms would be expected under conditions where replication ceases, resolution proceeds and degradation is minimal (Sect. 22.4.3).

22.4 Plastid Development in Maize

22.4.1 Progressive Leaf and Plastid Development

In grasses such as maize, development progresses from undifferentiated tissue at the base of the sheath or stalk to differentiated tissue at the tip of the blade (Stern et al. 2004; Sylvester et al. 1990). This progression makes maize well suited for studying changes in the developing plastid, including cpDNA structure and genome copy number. Plastid development may be divided into three stages that correlate with stages of cell development (Fig. 22.3). Stage I represents undifferentiated proplastids in dividing cells at or near the meristematic region. Stage II includes dividing and developing plastids in dividing and expanding cells. Stage III consists of the mature, green chloroplasts in differentiated leaf cells. Three stages of leaf growth (preligule, blade growth and sheath growth) have also been identified (Sylvester et al. 1990). In the studies described below, the age of the plants (day, D) and leaf number (L) are given, where L1 is the first leaf to emerge after the coleoptile. An appreciation of the developmental program will be important for engineering plastids with transgenes.

22.4.2 Changes in Genome Copy Number and cpDNA Molecular Form

As cells develop from the basal meristem to the expanded green leaf, there is a concomitant change in both size and appearance of plastid nucleoids (Fig. 22.3), as well as individual ethidium-stained cpDNA molecules for maize (Figs 22.2a–c; Oldenburg and Bendich 2004b; Oldenburg et al. 2006). Such changes indicate that plastid development is accompanied by changes in the structure of cpDNA molecules of predominantly multigenomic size, not just in copy number of genome-sized circles.

Stage I proplastids have a relatively high average genome copy number per plastid (165 for D8 and 36 for D13) and per cell (~1600 for D14), as determined by DAPI-DNA fluorescence and real-time quantitative PCR, respectively (Oldenburg

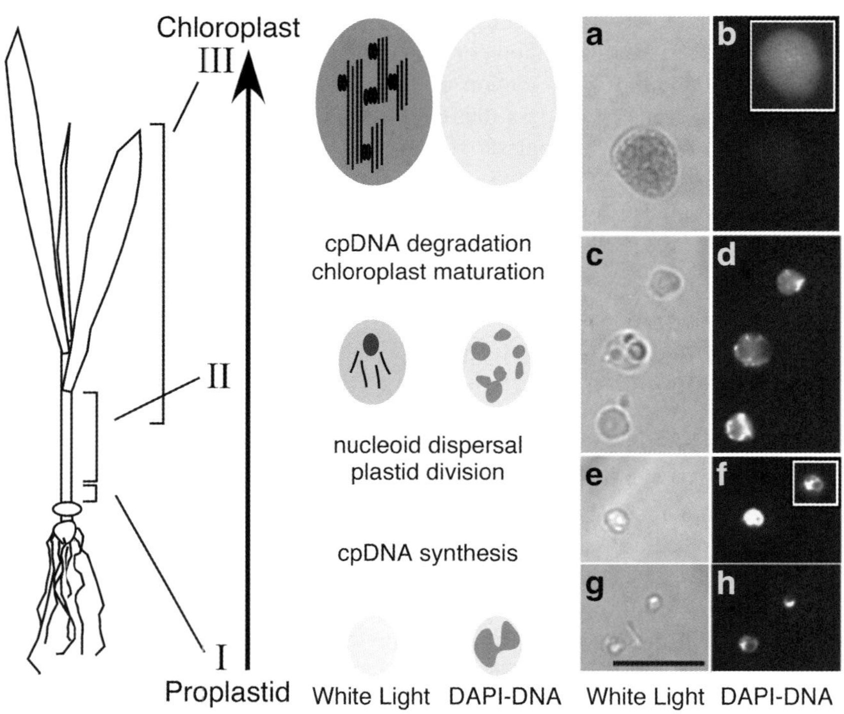

Fig. 22.3 Stages of chloroplast development and representative images of plastids. Stage *I* (**e–h**): small, colorless proplastids with one or two bright DAPI-DNA nucleoids in dividing cells at the base of the stalk. Stage *II* (**c** and **d**): developing plastids with several bright DAPI-DNA nucleoids in elongating cells above the basal meristem. Prolamellar bodies are evident in some plastids at this stage. Stage *III* (**a** and **b**): mature, green chloroplasts in differentiated cells of the upper stalk and leaf blade. DAPI-DNA fluorescence is weak or undetectable in most chloroplasts at this stage. Bright-field (**a**, **c**, **e** and **g**) and DAPI-DNA fluorescence images (**b**, **d**, **f** and **h**) are shown. Plastid developmental stages are described in Sect. 22.4. *Scale bar* is 10 μm and is the same for all images. (Adapted from Oldenburg and Bendich (2004a, b))

and Bendich 2004b; Oldenburg et al. 2006). Typically, there are one or two nucleoids per plastid, usually at the periphery, with fainter fluorescence throughout the plastid (Fig. 22.3f and h). The highest copy number per plastid (225 for D8 and 96 for D13) and per cell ($\sim$2100 for D14) is found above the basal meristem. These developing stage II plastids have intense DAPI-DNA fluorescence throughout the plastid, with 5–10 discrete, brighter nucleoids (Fig. 22.3d). In the mature green chloroplasts (stage III), cpDNA per cell is greatly reduced (100–200 copies for D14). The average cpDNA per plastid is also reduced (106 for D8 and 28 for D13) and DAPI-DNA fluorescence is undetectable in 53–88% of the chloroplasts (Fig. 22.3b). It should be noted that the DNA content per plastid varies greatly among individual plastids at all stages of development for maize.

In both stages I and II, the cpDNA is found mostly in complex structures that contain many genome copies (Fig. 22.2a). These complexes are likely the products of the recombination-dependent replication process described above. At stage II, however, smaller branched structures and more simple individual molecules are found (Fig. 22.2b), indicating that the complexes are beginning to disassemble as replication ceases and forks reach the ends of template strands. With increasing leaf age and long before the start of leaf senescence, cpDNA is degraded completely or to subgenome-sized molecules (Fig. 22.2c) and dispersed, so that DAPI-DNA fluorescence is undetectable in most of the mature chloroplasts. The changes in molecular form are apparent by both fluorescence microscopic imaging of individual molecules and PFGE-blot hybridization.

It may seem counterintuitive that retention of cpDNA is not necessary for photosynthesis. Nonetheless, degradation can benefit the plant in two ways. First, once an adequate stock of mRNAs is produced to last the growing season, DNA nucleotides may be recycled for their nutritive value. Second, the elimination of cpDNA relieves the cell of the burden of maintaining and repairing the many copies of the chloroplast genome that are no longer needed for their coding function.

22.4.3 Effects of Light on cpDNA

Light exerts the major environmental influence on whole plant development (Wada et al. 2005). At the subcellular level, proplastids develop into etioplasts in the dark, with exposure to light triggering chloroplast maturation (Daniell and Chase 2004). As might be expected, light stimulates DNA synthesis as the chloroplast develops. Surprisingly, light also triggers the destruction of cpDNA in expanded leaf cells of maize.

As discussed above, in the light the cpDNA content declines with development from the base to the leaf tip. In dark-grown plants, however, cpDNA levels remain high in the leaf blade (Oldenburg et al. 2006). PFGE-blot hybridization also shows differences in cpDNA structure. A band of circular cpDNA is present only in dark-grown tissue, and linear oligomers that are rarely, if ever, seen in the light are abundant in the dark. These circular and linear forms are detectable probably because replication slows and resolution of the multigenomic complexes proceeds, but without the concomitant degradation of the cpDNA that occurs in the light. The cpDNA levels decline rapidly after transfer of dark-grown plants to the light. Within 3 days, cpDNA levels in dark-to-light transferred plants are the same as in light-grown plants. Our recent unpublished data show that cpDNA declines to 50% of the dark level by 6 h after transfer, and by 24 h the small amount remaining is equivalent to that of light-grown plants. These results suggest that a light signal-transduction pathway is involved in the regulation of cpDNA maintenance in maize.

22.4.4 Genes that Influence Plastid Development and cpDNA Levels

Chloroplast development requires coordination between the activities of genes found in both the nucleus and the plastid (Koussevitzky et al. 2007; Lopez-Juez and Pyke 2005; Nott et al. 2006). We would like to identify genes involved in this coordination, since they are probably involved in regulating plastid gene expression at the posttranscriptional level, as well as the replication, repair and degradation of cpDNA. Knowledge of such genes will also be important for genetically engineering the plastid (Sect. 22.5).

Plastid-encoded gene expression is largely regulated by post-transcriptional mechanisms including RNA editing, stabilization and intron splicing, as well as by controlling translation (Barkan and Goldschmidt-Clermont 2000; Daniell and Chase 2004). Studies of maize mutants, such as Mu transposon mutants from the Photosynthetic Mutant Library (PML, http://pml.uoregon.edu), have begun to elucidate such regulation. Examples include mRNA translation by ATP1 (McCormac and Barkan 1999) and mRNA abundance and splicing by HCF38 (Taylor et al. 1987). The pentatricopeptide repeat (PPR) family includes many proteins targeted to chloroplasts and mitochondria that affect diverse processes controlling organellar gene expression (Lurin et al. 2004), such as CRP1and PPR2 for plastid mRNA translation and splicing in maize (Schmitz-Linneweber 2005; Williams and Barkan 2003).

A limited screen of maize mutants (including seven of the >2000 PML lines available) defective in RNA processing, protein import and photomorphogenesis indicates that nuclear genes can affect cpDNA retention (or loss) (Oldenburg et al. 2006). The cpDNA levels are similar for wild-type proplastids and undeveloped plastids of ivory mutants and higher than for mature chloroplasts. These results and those from dark-grown seedlings suggest that the signal to degrade cpDNA is effective only in developed chloroplasts. CpDNA retention and loss, however, appears to be regulated independently of chloroplast maturation, since two mutants, *hcf38* and *csr1*, have higher cpDNA levels than wild type and yet display some degree of normal chloroplast development, as indicated by size and appearance using light microscopy. Nevertheless, some components of the regulatory pathways are probably involved in both cpDNA maintenance and chloroplast biogenesis. For optimal expression of plastid transgenes, it is necessary to identify the genes controlling cpDNA maintenance and RNA translation and stability (Sect. 22.5).

It is clear that nuclear genes can affect the level and persistence of maize cpDNA, a conclusion that should surprise nobody. It is equally clear, however, that we are a long way from elucidating the genetically defined steps by which light leads to the demise of cpDNA in mature maize chloroplasts.

22.4.5 Mesophyll and Bundle Sheath Cell-Specific Processes

The final events in leaf development involve differentiation to specific cell and plastid types. In C4 plants such as maize, there is a "division of labor" with respect to photosynthetic functions between mesophyll (MS) and bundle sheath (BS) cells, and physical differences in chloroplast morphology reflect this distinction. MS chloroplasts contain grana and stroma thylakoids and have higher photosystem II (PSII) mRNA and protein levels, whereas BS chloroplasts lack grana and have higher photosystem I (PSI) activity (Darie et al. 2006; Hahnen et al. 2003; Sheen 1999). A recent proteomic study by Majeran et al. (2005) gives a detailed profile of the variation in protein abundance and function in MS and BS cells of maize. This information will be useful for plastid genetic engineering to improve photosynthetic performance and to identify regulatory sequences to use for transcriptional regulation and transgene expression in a cell-specific manner. For example, proteins G2 and ZmGLK1 are implicated as transcriptional regulators for cell-type differentiation of BS and MS cells, respectively (Rossini et al. 2001). Thus, a nuclear-located transgene with a ZmGLK1 promoter might be used to alter PSII activity in MS plastids or, after determining the factors controlling cpDNA degradation, to inhibit degradation in MS, but not BS, plastids.

At present, we have only limited information on the differences between cpDNA levels in MS and BS cells. Lindbeck et al. (1989) reported that the average DNA content per plastid was similar for MS and BS cells from young seedlings. Recent reports of a cpDNA decline during development did not distinguish between isolated plastids from MS and BS cells (Oldenburg and Bendich 2004b; Oldenburg et al. 2006), although plastids in tissue sections from young and old leaves did not show any difference in the presence or absence of DAPI-DNA signal depending on cell type (Shaver et al. 2006). Additional studies comparing cpDNA levels in MS and BS cells, however, may provide insight into the processes controlling retention and loss. For instance, cpDNA degradation may be due, in part, to DNA damage from reactive oxygen species without subsequent repair and may occur more frequently in MS plastids where PSII is higher. Thus, genes regulating cpDNA maintenance could be identified by differential expression and targeting of nuclear-encoded transgenic proteins to MS- and BS-specific plastids.

22.5 Strategies for Engineering the Chloroplast

The modification of chloroplast function through genetic engineering can be achieved by introducing a transgene into either the nuclear or the plastid genome, and the relative merits for choosing either as a target have been discussed previously (Grevich and Daniell 2005; Maliga 2004). Here we will address a few key points in light of the recent advances in cpDNA structure, replication and stability described above.

22.5.1 Nuclear-Encoded Plastid-Targeted Transgenes

Reasons for targeting transgenes to the nucleus include eliminating, replacing or modifying resident genes involved in plastid function and adding a novel function to the plastid. Some examples are described above for altering cell-specific processes in maize. Nuclear integration could also elucidate signal transduction pathways that affect plastid development.

22.5.2 Plastid-Encoded Transgenes

Transgene integration into plastid DNA could avoid problems associated with nuclear transgene expression, such as gene silencing, the spread of the transgene to non-target plants and sequestration of potentially cytotoxic metabolites (Daniell 2007; Grevich and Daniell 2005; Maliga 2004). For some crops (tobacco, carrot, cotton and soybean) successful plastid transformation has led to the introduction of useful agronomic traits, such as pathogen resistance and the production of pharmaceutical proteins (vaccines) (Grevich and Daniell 2005). While success has been achieved with many dicots, monocots such as maize and wheat have proved recalcitrant to plastid transformation, and stable plastid transformation with transmission of the transgenes to the progeny has only recently been reported for rice (Lee et al. 2006). Plant regeneration via tissue culturing (required for agronomically useful plastid transformation) has also been more difficult for maize than for dicots such as tobacco. Nonetheless, regeneration of maize can be accomplished using somatic embryogenesis and seedling meristematic tissue (Al-Abed et al. 2006; Huang and Wei 2004). We now consider the retention of cpDNA, cpDNA structure and mRNA stability as factors that may influence successful transgene expression in maize.

22.5.2.1 Retention and Loss of cpDNA in Mature Chloroplasts and Transgene Expression

An unanticipated problem with plastid transformation and transgene expression is the decline of DNA in mature chloroplasts of maize (Sects. 22.4.2–22.4.4). This loss of cpDNA may also contribute to the recalcitrant nature of maize plant regeneration from leaf tissue, since cpDNA is almost certainly required for totipotency. Thus, we need to elucidate the genetic and environmental factors that affect the retention of cpDNA, such as light-mediated signal transduction in maize (Oldenburg et al. 2006) and nutrient stress in *Chlamydomonas* (Yehudai-Resheff et al. 2007).

Another consideration is the type of tissue or cell where the transgene is to be expressed, because cpDNA may be retained in some tissues but not others. For example, if the goal is to increase photosynthetic efficiency or produce a useful biomaterial, such as *p*-hydroxybenzoic acid (Grevich and Daniell 2005), and we know that cpDNA is lost in the mature maize leaf blade, then improving transgene RNA stability would be one way to maintain production of the transgene product. On the other hand, if the goal is to increase the content of seed oil or insect protection in the root – assuming that the cpDNA is retained in these tissues – then integration and transcription of the transgene is all that is required.

22.5.2.2 Structure and Replication of cpDNA and Integration of a Transgene

The molecular structure and integrity of the cpDNA may affect the success of plastid transformation. What is the form of cpDNA that can function as a chromosome, as a template for replication and as a template for transcription? Are there particular tissues in which cpDNA replication is high and degradation low? For maize, the preferred tissue would be located at or slightly above the basal meristem. Alternatively, it may be possible to stimulate cpDNA replication or suppress its degradation by light or chemical signaling. Since maize cpDNA is retained in the dark, growing the plants in the dark, perhaps followed by a brief light treatment, may stimulate cpDNA replication and facilitate transgene integration. Further study of the environmental (light) and genetic factors that affect the retention or loss of cpDNA will be invaluable in determining optimal conditions for successful plastid transformation.

For tobacco and other plants, Grevich and Daniell (2005) reported that 22 out of 35 successfully integrated transgenes were located in the *trn*I/*trn*A spacer region and suggested that this site "allows highly efficient transgene integration and expression". Is there a reason for this "hotspot" for transgene integration? This region lies within the IR and close to *ori*B in several plastid genomes, including maize. Furthermore, the *trn*I/*trn*A region is near the end of one of the linear isomers (Fig. 22.1b, map 4). We suggest that since the ends of DNA molecules are recombinagenic (Kuzminov and Stahl 1999), this terminal location may facilitate transgene integration.

The OPaLI-RDR mechanism for maize cpDNA includes the joining of 3′ complementary overhangs to promote replication (Fig. 22.2). Perhaps a transgene could be created with such flanking sequences that could be integrated either by annealing to the ends of a linear molecule (Fig. 22.2, *Step 8*) or by strand invasion of complementary sequences (Fig. 22.2, *Step 5*). The RDR process would then amplify the transgenic cpDNA, increasing the likelihood of achieving homoplasmy in a regenerated transgenic plant. In order to employ this strategy, we first need to determine the terminal sequences of the linear cpDNA isomers.

22.5.2.3 Regulation of Plastid Transgene Expression, mRNA Stability and Translation

If most of the cpDNA carrying the transgene is no longer present in the mature maize chloroplast, is plastid transformation nevertheless a viable option for expression of the transgene product? We will consider three factors that relate to this question. First, what forms of cpDNA are transcriptionally competent? One would expect the multigenomic and monomeric molecules to be used for transcription, but in the leaf blade of maize these forms are rarely, if ever, found in mature chloroplasts. Functional transgene expression could still be maintained in mature chloroplasts if the less-than-genome-sized molecules, although reduced in abundance, are not degraded completely and are still transcriptionally competent. Second, it is likely that sufficient plastid mRNAs are retained to maintain photosynthesis during the growing season for maize. It is known that transcript stability varies among plastid genes of maize (Cahoon et al. 2004), and in barley one of the most stable is the *psb*A mRNA for D1 protein (Kim et al. 1993). It is interesting to note that the most common 5′ and 3′ UTRs used for plastid transgene expression are for *psb*A (Grevich and Daniell 2005). Thus, in order to design a vector for transgene expression in mature maize chloroplasts, the use of flanking sequences that stabilize mRNA would be prudent. Finally, forthcoming knowledge of nuclear-encoded plastid proteins that influence mRNA stability and translation, such as the PPR proteins (Sect. 22.4.4), will aid in the design of transgenes.

One of the advantages of plastid transformation over nuclear transgene integration is the potential for high copy number and expression levels of the transgene in the plastid (Grevich and Daniell 2005; Maliga 2002). Plastid transformation has been highly successful in some crops, but not in maize and other cereals. The loss of the transgenic cpDNA in mature chloroplasts may contribute to this lack of success. Thus, it is imperative that a greater understanding of the factors influencing cpDNA maintenance is obtained in order to make plastid transformation a useful agronomic tool for maize and other cereals.

Acknowledgements We thank Jeff Shaver and Beth Rowen for reading the manuscript. We thank the Botany Greenhouse and Comparative Genomics Center at the University of Washington, Department of Biology for technical assistance. The project was supported in part by the National Research Initiative of the USDA Cooperative State Research, Education and Extension Service, grant number 2002-35301-12021, and Public Health Service, National Research Service Award, T32 GM07270, from the National Institute of General Medical Sciences.

References

Al-Abed D, Rudrabhatla S, Talla R, Goldman S (2006) Split-seed: a new tool for maize researchers. Planta 223:1355–1360

Backert S, Dörfel P, Börner T (1995) Investigation of plant organellar DNAs by pulsed-field gel electrophoresis. Curr Genet 28:390–399

Barkan A, Goldschmidt-Clermont M (2000) Participation of nuclear genes in chloroplast gene expression. Biochimie 82:559–572

Bedbrook JR, Bogorad L (1976) Endonuclease recognition sites mapped on *Zea mays* chloroplast DNA. Proc Natl Acad Sci USA 73:4309–4313

Bendich AJ (1991) Moving pictures of DNA released upon lysis from bacteria, chloroplasts, and mitochondria. Protoplasma 160:121–130

Bendich AJ (1993) Reaching for the ring: the study of mitochondrial genome structure. Curr Genet 24:279–290

Bendich AJ (2004) Circular chloroplast chromosomes: the grand illusion. Plant Cell 16:1661–1666

Bendich AJ, Smith SB (1990) Moving pictures and pulsed-field gel electrophoresis show linear DNA molecules from chloroplasts and mitochondria. Curr Genet 17:421–425

Birky CW Jr (1994) Relaxed and stringent genomes: why cytoplasmic genes don't obey Mendel's laws. J Hered 85:355–365

Cahoon AB, Harris FM, Stern DB (2004) Analysis of developing maize plastids reveals two mRNA stability classes correlating with RNA polymerase type. EMBO 5:801–806

Chiu WL, Sears BB (1993) Plastome–genome interactions affect plastid transmission in *Oenothera*. Genetics 133:989–997

Daniell H (2007) Transgene containment by maternal inheritance: effective or elusive? Proc Natl Acad Sci USA 104:6879–6880

Daniell H, Chase C (eds) (2004) Molecular biology and biotechnology of plant organelles: chloroplasts and mitochondria. Springer, Dordrecht

Daniell H, Cohill PR, Kumar S, Dufourmantel N (2004) Chloroplast genetic engineering. In: Daniell H, Chase C (eds) Molecular biology and biotechnology of plant organelles: chloroplasts and mitochondria. Springer, Dordrecht, pp 443–490

Darie CC, De Pascalis L, Mutschler B, Haehnel W (2006) Studies of the Ndh complex and photosystem II from mesophyll and bundle sheath chloroplasts of the C4-type plant *Zea mays*. J Plant Physiol 163:800–808

Deng X-W, Wing RA, Gruissem W (1989) The chloroplast genome exists in multimeric forms. Proc Natl Acad Sci USA 86:4156–4160

Gillham NW (ed) (1994) Organelle genes and genomes. Oxford University Press, New York

Gold B, Carrillo N, Tewari K, Bogorad L (1987) Nucleotide sequence of a preferred maize chloroplast genome template for in vitro DNA synthesis. Proc Natl Acad Sci USA 84:194–198

Grevich JJ, Daniell H (2005) Chloroplast genetic engineering: recent advances and future perspectives. Crit Rev Plant Sci 24:83–107

Hahnen S, Joeris T, Kreuzaler F, Peterhansel C (2003) Quantification of photosynthetic gene expression in maize C(3) and C(4) tissues by real-time PCR. Photosynth Res 75:183–192

Heinhorst S, Cannon GC (1993) DNA replication in chloroplasts. J Cell Sci 104:1–9

Huang XQ, Wei ZM (2004) High-frequency plant regeneration through callus initiation from mature embryos of maize (*Zea mays* L.). Plant Cell Rep 22:793–800

Kim M, Christopher DA, Mullet JE (1993) Direct evidence for selective modulation of *psbA*, *rpoA*, *rbcL*, and 16S RNA stability during barley chloroplast development. Plant Mol Biol 22:447–463

Kolodner R, Tewari KK (1972) Molecular size and conformation of chloroplast deoxyribonucleic acid from pea leaves. J Biol Chem 247:6355–6364

Kolodner R, Tewari KK (1975a) The molecular size and conformation of the chloroplast DNA from higher plants. Biochim Biophys Acta 402:372–390

Kolodner RD, Tewari KK (1975b) Chloroplast DNA from higher plants replicates by both the Cairns and the rolling circle mechanism. Nature 256:708–711

Kornberg A, Baker TA (1992) DNA replication. W.H. Freeman, New York

Koussevitzky S, Nott A, Mockler TC, Hong F, Sachetto-Martins G, Surpin M, Lim J, Mittler R, Chory J (2007) Signals from chloroplasts converge to regulate nuclear gene expression. Science 316:715–719

Kunnimalaiyaan M, Nielsen BL (1997) Chloroplast DNA replication: mechanism, enzymes and replication origins. J Plant Biochem Biotechnol 6:1–7

Kuzminov A, Stahl FW (1999) Double-strand end repair via the RecBC pathway in *Escherichia coli* primes DNA replication. Genes Dev 13:345–356

Lee SM, Kang K, Chung H, Yoo SH, Xu XM, Lee SB, Cheong JJ, Daniell H, Kim M (2006) Plastid transformation in the monocotyledonous cereal crop, rice (*Oryza sativa*) and transmission of transgenes to their progeny. Mol Cells 21:401–410

Lehman IR, Boehmer PE (1999) Replication of herpes simplex virus DNA. J Biol Chem 274:28059–28062

Lilly JW, Havey MJ, Jackson SA, Jiang J (2001) Cytogenomic analyses reveal the structural plasticity of the chloroplast genome in higher plants. Plant Cell 13:245–254

Lindbeck AGC, Rose RJ, Lawrence ME, Possingham JV (1989) The chloroplast nucleoids of the bundle sheath and mesophyll cells of *Zea mays*. Physiol Plant 75:7–12

Lopez-Juez E, Pyke KA (2005) Plastids unleashed: their development and their integration in plant development. Int J Dev Biol 49:557–577

Lurin C, Andres C, Aubourg S, Bellaoui M, Bitton F, Bruyere C, Caboche M, Debast C, Gualberto J, Hoffmann B, Lecharny A, Le Ret M, Martin-Magniette ML, Mireau H, Peeters N, Renou JP, Szurek B, Taconnat L, Small I (2004) Genome-wide analysis of Arabidopsis pentatricopeptide repeat proteins reveals their essential role in organelle biogenesis. Plant Cell 16:2089–2103

Maier RM, Neckermann K, Igloi GL, Kössel H (1995) Complete sequence of the maize chloroplast genome: gene content, hotspots of divergence and fine tuning of genetic information by transcript editing. J Mol Biol 251:614–628

Majeran W, Cai Y, Sun Q, van Wijk KJ (2005) Functional differentiation of bundle sheath and mesophyll maize chloroplasts determined by comparative proteomics. Plant Cell 17: 3111–3140

Maliga P (2002) Engineering the plastid genome of higher plants. Curr Opin Plant Biol 5: 164–172

Maliga P (2004) Plastid transformation in higher plants. Annu Rev Plant Biol 55:289–313

Matsuoka Y, Yamazaki Y, Ogihara Y, Tsunewaki K (2002) Whole chloroplast genome comparison of rice, maize, and wheat: implications for chloroplast gene diversification and phylogeny of cereals. Mol Biol Evol 19:2084–2091

McCormac DJ, Barkan A (1999) A nuclear gene in maize required for the translation of the chloroplasts *atpB/E* mRNA. Plant Cell 11:1709–1716

Nott A, Jung HS, Koussevitzky S, Chory J (2006) Plastid-to-nucleus retrograde signaling. Annu Rev Plant Biol 57:739–759

Oldenburg DJ, Bendich AJ (2001) Mitochondrial DNA from the liverwort *Marchantia polymorpha*: circularly permuted linear molecules, head-to-tail concatemers, and a 5′ protein. J Mol Biol 310:549–562

Oldenburg DJ, Bendich AJ (2004a) Most chloroplast DNA of maize seedlings in linear molecules with defined ends and branched forms. J Mol Biol 335:953–970

Oldenburg DJ, Bendich AJ (2004b) Changes in the structure of DNA molecules and the amount of DNA per plastid during chloroplast development in maize. J Mol Biol 344:1311–1330

Oldenburg DJ, Rowan BA, Zhao L, Walcher CL, Schleh M, Bendich AJ (2006) Loss or retention of chloroplast DNA in maize seedlings is affected by both light and genotype. Planta 225: 41–55

Palmer JD (1983) Chloroplast DNA exists in two orientations. Nature 301:92–93

Palmer JD (1985) Comparative organization of chloroplast genomes. Annu Rev Genet 19: 325–354

Rossini L, Cribb L, Martin DJ, Langdale JA (2001) The maize *Golden2* gene defines a novel class of transcriptional regulators in plants. Plant Cell 13:1231–1244

Rowan BA, Oldenburg DJ, Bendich AJ (2004) The demise of chloroplast DNA in *Arabidopsis*. Curr Genet 46:176–181

Sandri-Goldin RM (2003) Replication of the herpes simplex virus genome: does it really go around in circles? Proc Natl Acad Sci USA 100:7428–7429

Scharff LB, Koop HU (2006) Linear molecules of tobacco ptDNA end at known replication origins and additional loci. Plant Mol Biol 62:611–621

Scharff LB, Koop HU (2007) Targeted inactivation of the tobacco plastome origins of replication *A* and *B*. Plant J 50(5):782–794

Schmitz-Linneweber C, Williams-Carrier R, Barkan A (2005) RNA immunoprecipitation and microarray analysis show a chloroplast pentatricopeptide repeat protein to be associated with the 5′ region of mRNAs whose translation it activates. Plant Cell 17:2791–2804

Sears BB, Stoike LL, Chiu W-L (1996) Proliferation of direct repeats near the *Oenothera* chloroplast DNA origin of replication. Mol Biol Evol 13:850–863

Shaver JM, Oldenburg DJ, Bendich AJ (2006) Changes in chloroplast DNA during development in tobacco, *Medicago truncatula*, pea, and maize. Planta 224:72–82

Shaver JM, Oldenburg DJ, Bendich AJ (2008) The structure of chloroplast DNA molecules and the effects of light on the amount of chloroplast DNA during development in *Medicago truncatula*. Plant Physiol 146:1064–1074

Sheen J (1999) C4 gene expression. Annu Rev Plant Physiol Plant Mol Biol 50:187–217

Stern DB, Hanson MR, Barkan A (2004) Genetics and genomics of chloroplast biogenesis: maize as a model system. Trends Plant Sci 9:293–301

Sylvester AW, Cande WZ, Freeling M (1990) Division and differentiation during normal and *liguleless-1* maize leaf development. Development 110:985–1000

Taylor WC, Barkan A, Martienssen RA (1987) Use of nuclear mutants in the analysis of chloroplast development. Dev Genet 8:305–320

Wada M, Shimazaki K, Iino M (eds) (2005) Light sensing in plants. Botanical Society of Japan, Yamada Science Foundation and Springer, Tokyo

Williams PM, Barkan A (2003) A chloroplast-localized PPR protein required for plastid ribosome accumulation. Plant J 36:675–686

Yehudai-Resheff S, Zimmer SL, Komine Y, Stern DB (2007) Integration of chloroplast nucleic acid metabolism into the phosphate deprivation response in *Chlamydomonas reinhardtii*. Plant Cell 19:1023–1038

Part VI
Biomass and Energy

Chapter 23
Ethanol Production from Maize

Stefan Schwietzke, Youngmi Kim, Eduardo Ximenes, Nathan Mosier, and Michael Ladisch

Abstract The production of fuel ethanol from corn grain is widely carried out in the US, with total current production at 7 billion gallons. This may soon reach 10 billion gallons or more. This chapter addresses the potential of fuel ethanol as an additional source of product based on utilization of the cellulosic (non-food) portions of maize, and in particular the pericarp, cobs, stalks and leaves of the corn plant. An analysis of the composition of corn, and possible processing schemes that transform the cellulosic portions to ethanol are addressed. Technologies for the bioprocessing of cellulose to ethanol, as well as the impact of cellulose utilization on supplementing corn ethanol, are presented.

Stefan Schwietzke
Laboratory of Renewable Resources Engineering, Purdue University, Potter Engineering Center, 500 Central Drive, West Lafayette, IN 47907, USA

Youngmi Kim, Eduardo Ximenes
Department of Agricultural and Biological Engineering, Purdue University, 225 South University Street, West Lafayette, IN 47907, USA

Nathan Mosier
Laboratory of Renewable Resources Engineering, Purdue University, Potter Engineering Center, 500 Central Drive, West Lafayette, IN 47907, USA
Department of Agricultural and Biological Engineering, Purdue University, 225 South University Street, West Lafayette, IN 47907, USA

Michael Ladisch
Laboratory of Renewable Resources Engineering, Purdue University, Potter Engineering Center, 500 Central Drive, West Lafayette, IN 47907, USA
Department of Agricultural and Biological Engineering, Purdue University, 225 South University Street, West Lafayette, IN 47907, USA
Weldon School of Biomedical Engineering, Purdue University, 206 S. Intramural Drive, West Lafayette, IN 47907, USA
e-mail: ladisch@purdue.edu

A.L. Kriz, B.A. Larkins (eds.), *Molecular Genetic Approaches to Maize Improvement* 347
Biotechnology in Agriculture and Forestry, Vol. 63

23.1 Introduction

Ethanol is commonly derived from biological feedstocks utilizing fermentation processes. During these processes, monosaccharides are fermented to ethanol by yeast or bacteria. There are a variety of carbohydrate-containing feedstocks that yield monosaccharides for fermentation, such as corn grain, sugarcane, wheat, sugar beet and other biomass. Brazil produces similar volumes of ethanol as the USA, using sugarcane as the main feedstock, which requires less processing than corn since the sugar is present in a soluble form, rather than as a structural polysaccharide.

The USA and Brazil are the world's largest producers of fuel ethanol, with outputs of 4.9 and 4.5 billion gallons, respectively, in 2006. US production reached 4.65 billion gallons by September, 2007, while demand was 4.85 billion gallons. The demand for fuel ethanol in the USA more than doubled between 2000 and 2004 and has increased to 7 billion gallons in response to the Energy Act of 2005. In comparison, Brazil used about 4 billion gallons of ethanol in 2006 and produced sufficient quantities of ethanol from sugarcane to satisfy its own demand and export 0.43 billion gallons to the USA. In Brazil, the price of ethanol is tied to the price of gasoline (Energy Information Administration 2007; Renewable Fuels Association 2007). Ethanol production in the USA in 2006 and 2007 increased significantly as new plants started up and increased production by an estimated 2 billion gallons to levels now approaching 7 billion gallons (Renewable Fuels Association 2007).

23.2 Maize as a Feedstock for Ethanol Production

In the USA, maize is a key candidate for ethanol production. It yields corn grain which is converted to ethanol. The potential for ethanol from maize lies not only in converting the grain to ethanol, but also in applying cellulose conversion technology to the pericarp that covers the grain. Cellulose conversion technology, consisting of pretreatment and hydrolysis, offers the prospect of extending conversion to other parts of the corn plant, such as corn stover (cobs, stalks, and leaves). Both corn grain and corn stover are discussed in this chapter. Significant increases in the ethanol yield per acre of maize harvested is possible if biomass from the maize residue is utilized for ethanol production. A quantitative analysis of mass balance has been carried out to address this issue.

The corn cob, stalks, and leaves can be converted to fermentable sugars with cellulose processing technology that consists of pretreatment, hydrolysis, and fermentation using yeast or other microorganisms. In contrast to grain-based feedstocks, cellulose-based ethanol production requires microorganisms that are capable of producing ethanol from both glucose and xylose.

Corn grain contains high amounts of starch, which is readily convertible to monosaccharides upon pretreatment (i.e., cooking in water) and hydrolysis. Glucan is also present in the cob, the stalk, and the leaves, but in a different form, i.e., cellulose, and at lower amounts compared to corn grain. The other major structural

Table 23.1 Compositions of corn grain, corn cob, and corn stover in percentage of total. The lower portion of the table shows the sugar and ethanol yields from these components. n/m indicates not measured

Type of material	Grain[a]	Cob[b]	Stover[c]
Starch	71.7	n/m	n/m
Cellulose	2.4	42.0	36.0
Hemicellulose	5.5	33.0	26.0
Protein	10.3	n/m	5.0
Oil	4.3	n/m	n/m
Lignin	0.2	18.0	19.0
Ash	1.4	1.5	12.0
Other	4.2	5.5	2.0
Total	100.0	100.0	100.0
Maximum yield of monosaccharides (lb/ton, 100% efficiency)	1778	1684	1392
Calculated best case ethanol yield (gal/ton, 100% efficiency)	135	128	105
Dry weight (%)[d]	52.4	(9.5)	47.6
Dry weight (kg/acre)[e]	4000	(725)	3630

[a]Gulati et al. (1996)
[b]Corn cob composition was measured at LORRE
[c]US DOE (2007)
[d]Pordesimo et al. (2005). Reference data are based on maize harvest at 151 days after planting
[e]Absolute weight for corn grain is based on corn grain data provided by the USDA National Agricultural Statistics Service (2005) which were used for calculation of ethanol yields. Absolute weights for corn cob and corn stover are derived from the given weight percentages based on the absolute weight of corn grain

polysaccharide is hemicellulose, which is predominantly xylan. The conversion of the xylan fraction and the ability to ferment it to ethanol is therefore important to improve the efficiency and the economics of ethanol production from maize.

The main components of maize are corn grain and corn stover (corn cob, stalk, and leaves). The ratio of grain:stover is about 1:1, with the grain accounting for slightly more weight than stover. The corn cob represents approximately 20% of the weight of corn stover. Corn grain includes the starchy endosperm, germ, and corn hull. Corn stover contains principally cellulose and hemicellulose, which are the main sources of fermentable sugars for ethanol production. The corn cob contains more cellulose and hemicellulose than stalks and leaves. This enables a higher potential yield of ethanol. The compositions of the three major components of maize are summarized in Table 23.1. Compositional analysis of cob is based on an analysis carried out in LORRE (Laboratory of Renewable Resources Engineering – Purdue University) for cobs obtained in the Delphi, Indiana, area. The data in the last three rows give the mass composition for grain, cob, and stover.

Starch is a type of glucan that is readily convertible to glucose. Cellulose and hemicellulose are also sugar polymers, and the cellulose is more difficult to convert into fermentable sugars than starch due to its crystalline structure and close association with lignin and hemicellulose. The efficiency of starch hydrolysis is a reflection of the long experience in industrial processing of starch, while the well-known recalcitrance of the cellulose itself forms an impediment to hydrolysis. Furthermore,

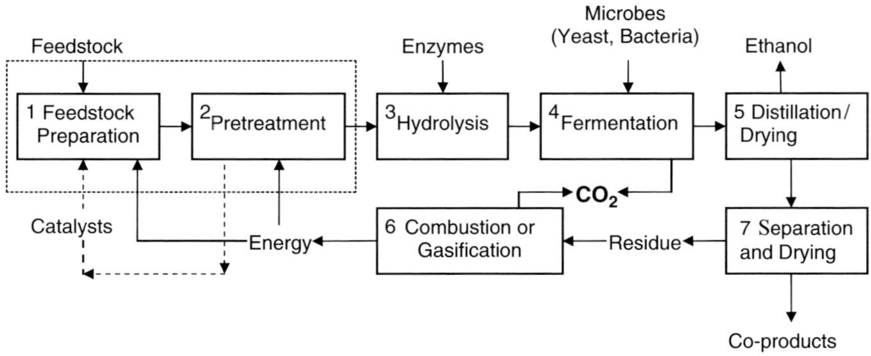

Fig. 23.1 The seven major unit operations of a biorefinery. Biorefinery as represented here is viewed as being energy self-sufficient through combustion or gasification of residual lignocellulose. If only corn grain is processed, the remaining solids that are high in protein are recovered as co-products and sold as animal feed, which has a higher value than use as a boiler fuel. CO_2 is recycled into plant matter through production agriculture. (Adapted from Eggeman and Elander (2005))

a large portion of the hemicellulose (xylan and arabinan) consists of 5-carbon sugars which cannot be fermented by natural yeast. As a result, only the starch part of the maize is used for ethanol conversion in current industrial processes. Since starch makes up less than half of the weight of the entire plant, only about 40–50% of the theoretical yield of ethanol of a maize plant is obtained from starch. A higher yield of ethanol for every acre of harvested corn can be achieved if the cellulose and hemicellulose fractions in corn stalks, cobs, and leaves are used to produce ethanol in addition to starch.

Processing technology consisting of pretreatment and hydrolysis is necessary to make use of the glucan, xylan, and arabinan portions of the plant. Enzymes are required for the conversion of cellulose and hemicellulose into monomeric sugars, and recombinant yeast or bacteria are needed for subsequent fermentation of both 5-carbon and 6-carbon sugars into ethanol. Figure 23.1 is a schematic representation of the key steps in the sequence of pretreatment, hydrolysis, and fermentation that make up a biorefinery for converting both starch and lignocelluloses to ethanol.

23.3 Ethanol Production from Corn Grain

Ethanol is produced when yeast ferments 6-carbon sugars (mainly glucose) via the glycolytic pathway. In the USA, starch from maize is used as a feedstock for hydrolysis, where the glucan is converted to glucose by enzyme hydrolysis and fermentation of the glucose to ethanol by yeast. The mash is fermented using natural yeast and bacteria in a process that takes up to 40 hours. The fermented mash is separated into ethanol and residues (for feed production) via distillation. A simplified process scheme for ethanol production from starch is shown in Fig. 23.2. Hydrolysis is carried out using amylase and glucoamylase, i.e., starch and maltose

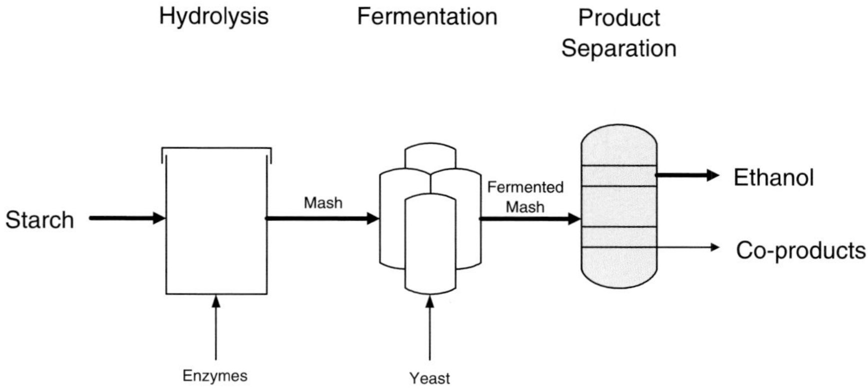

Fig. 23.2 Process flow diagram for ethanol production from starch

hydrolyzing enzymes. The substrate, in this case, is starch that has been gelatinized (i.e., pretreated) by cooking it in water.

Either wet or dry milling processes may be used for ethanol production from corn starch. A dry-grind process is simpler and typically will require less capital than wet milling. The capital costs for large-scale dry milling plants have ranged from about \$1.50 to \$2.20 per annual gallon of capacity (2006). A dry-grind process entails grinding the corn into a fine powder, which is then cooked, hydrolyzed, and fermented. In a wet-milling plant, the number of co-products is higher and more flexible, with processing consisting of steeping and separation of the corn kernel into germ, starch, and other components. Currently, most new US fuel ethanol production facilities utilize dry-grind processes. The expansion in the industry in 2005 to 2007 was attained through construction of new or an increase in existing dry-grind facilities. The key steps involved in both technologies are shown in Fig. 23.3 (wet milling) and Fig. 23.4 (dry milling).

23.3.1 Wet Milling

The wet milling process fully fractionates the corn grain into carbohydrates, lipids, and protein. These can be efficiently recovered and purified for the production of value-added products. When the starch is converted to fuel ethanol, the processing steps of saccharification, fermentation, and recovery are similar to those in a dry-grind operation.

The first step in the wet milling process is steeping, where the corn kernel is placed in an aqueous solution of 0.1–0.2% SO_2 and allowed to cook at 48–52°C for 30–50 hours. This facilitates downstream fractionation by hydrolyzing disulfide bonds in proteins so that they are more soluble. The corn is then ground in its wet state and oil, fiber, and gluten are separated from the starch for further processing into value-added co-products. During saccharification, enzymes break down the starch into glucose. In the fermentation step, yeast grown in seed tanks is added to

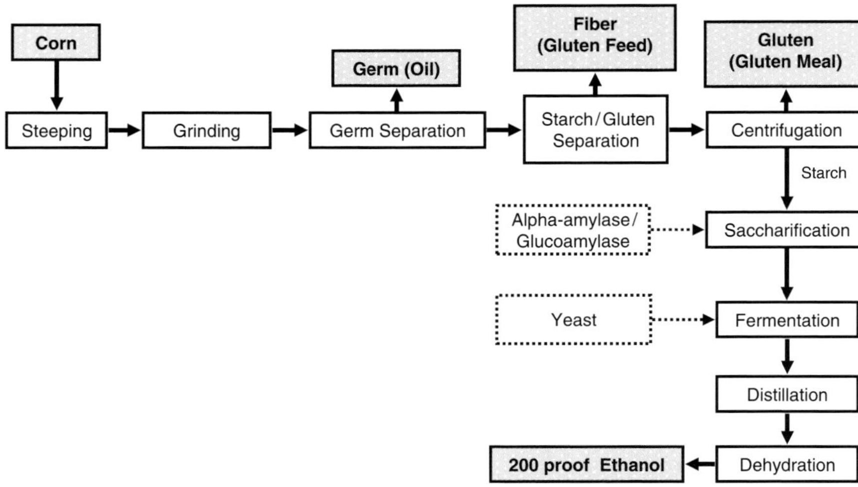

Fig. 23.3 Schematic flow diagram of wet milling process (steps for concurrent production of high fructose corn syrup, dextrose, and dry starch, which are found in many wet mill facilities, are omitted from this diagram that shows ethanol production only.)

the corn mash to ferment the simple sugars (glucose) to ethanol. Finally, ethanol is separated from the water by means of distillation and dehydration.

Corn fiber contains cellulose and hemicellulose (Table 23.2) which cannot be used for producing ethanol in wet mill facilities, since these facilities do not

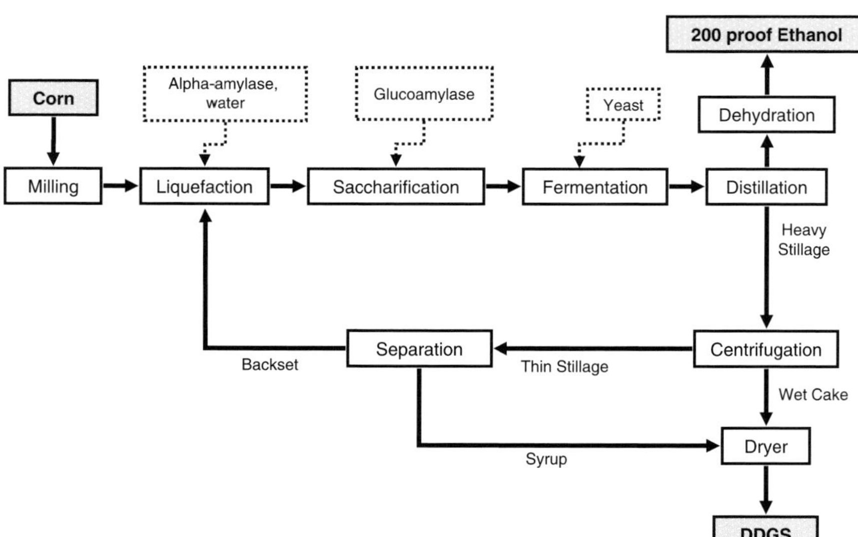

Fig. 23.4 Schematic flow diagram of current dry milling process. (Adapted from Kim et al. (2008b))

Table 23.2 Compositions of corn fiber, DG, and DDGS in percentage of dry matter. n/m indicates not measured

Type of material	Corn fiber[a]	DG[b]	DDGS[b]
Glucan (cellulose and starch)	38.0	18.5	21.2
Xylan (hemicellulose)	28.0	20.4	13.5
Protein	12.0	37.0	25.0
Oil	n/m	10.0	12.0
Lignin	8.0	n/m	n/m
Ash	0.4	2.0	5.0
Other	13.6	12.1	23.3
Total	100.0	100.0	100.0
Weight (kg/acre)[c]	1135	n/m	1510

[a]Mosier et al. (2005b)
[b]Kim et al. (2008b)
[c]Graboski (2002). Corn grain yield per acre is based on corn grain data provided by the USDA National Agricultural Statistics Service (2005) which were used for calculation of ethanol yields in Table 23.3

currently incorporate cellulose conversion technology. Fiber is a potential feedstock for additional ethanol production if lignocellulosic conversion technologies are applied. Each acre of harvested corn that is processed by wet milling yields approximately 1135 kg of fiber (Graboski 2002).

23.3.2 Dry Milling

Dry milling technology produces high ethanol yields at lower capital investment than wet milling. However, the only major co-products, other than CO_2, are the fermentation residuals which are sold as animal feed. These products are commonly known as distillers' grains (DG) and dried distillers' grains with solubles (DDGS). Compositions of DG and DDGS are given in Table 23.2.

In a dry mill, cleaned corn is first ground in hammer mills, which breaks the tough outer coating of the seed and grinds the corn into a fine powder. During the liquefaction process, water and enzymes are added to the ground corn in order to create a slurry. The gelatinized starch feedstock is easier to hydrolyze into monomeric sugars than uncooked corn, although processes that avoid the cooking step are being considered for ethanol plants. Saccharification and fermentation are similar to the processes performed in a wet mill. Ethanol is obtained from the water slurry via a number of complex steps including distillation and dehydration. A co-product of the dry milling process, heavy stillage, leaves the bottom of the first distillation column. The heavy stillage is centrifuged to remove the majority of the solids. The thin stillage is partly recycled to the liquefaction step. The centrifuged solids are referred to as wet cake or wet distiller's grains (35–40% solids). These are further dried to give DDGS.

23.3.3 Enzymes

Starch processing is a mature technology that makes use of enzymes for liquefaction and saccharification. These produce a relatively clean glucose stream which can be further fermented to ethanol by yeast (*Saccharomyces*). In the liquefaction step, a thermostable α-amylase is added before the heat treatment (105–110°C) for 5–7 min. Taking into consideration that the starch-slurry is then flash-cooled to 95°C and kept at that temperature for 60–120 min to complete the enzymatic liquefaction, a highly thermostable enzyme is required that will be active during the entire procedure. In addition to the originally used enzymes from *Bacillus stearothermophilus* or *B. licheniformis*, there are several other enzyme preparations available and marketed for this purpose (for example, Valley "Ultra-thin™" from Valley Research/Diversa, Multifect AA 21L® from Genencor, and Termamyl® and Liquozyme® from Novozymes). Ideally, the enzyme should be active and stable at a low pH (~4.5) and not demand calcium for stability (Van der Veen et al. 2006; Turner et al. 2007). Some engineered enzymes have been reported to fulfill these desired properties. Genetic engineering has been extensively used for cloning α- and gluco-amylase genes from different microbial sources, in order to express enzymes with such desirable properties in appropriate hosts. The development of low-pH α-amylases that simplify starch processing and reduce chemical costs as well as improve ethanol production (Gray et al. 2006) is an example. Another example is that of a glucoamylase from a thermoacidophilic microorganism (*Thermoplasma acidophilum*) which has been cloned and successfully expressed in *Escherichia coli*; the amylase activity of the recombinant enzyme is reported to be maximal at 75°C and pH 5.0 (Dock et al. 2008) .

The saccharification step involves hydrolysis of remaining oligosaccharides into either maltose syrup (using β-amylase) or glucose/glucose syrups (using glucoamylase) (Pandey 1995). The process conditions (pH 4.2–4.5 and 60°C) are consistent with the range over which *Aspergillus niger* glucoamylase is stable. However, the liquefied corn slurry must be cooled down after liquefaction and the pH must be adjusted in order to achieve the best conditions for the glucoamylase action. Perhaps less costly would be to utilize an enzyme that is active in the same pH and temperature range as the liquefaction enzymes. It is also possible to increase the efficiency in saccharification by use of a debranching enzyme. Pullulanases have been added to the process for this purpose (Turner et al. 2007).

23.3.4 DDGS

With increasing dry mill capacities, the use of DDGS as a feed may be limited (Belyea et al. 2004). DDGS contain large amounts of fiber that contain cellulose and hemicellulose as well as protein, some starch, and oil (Table 23.2). Integrating lignocellulosic conversion technologies in dry-grind facilities might further increase the value of DDGS by lowering its fiber content and increasing its relative protein content, while increasing the overall ethanol yield per bushel (Mosier et al. 2005b; Kim et al. 2008c). In a dry grind mill, each acre of harvested corn yields approximately 1510 kg of DDGS, which can be used for ethanol conversion (Graboski 2002).

Previous studies (Mosier et al. 2005b; Kim et al. 2008a) have shown that the fiber-containing co-products of dry milling and wet milling processes, such as DDGS and corn fiber, are easily hydrolyzed upon cooking in liquid hot water and addition of cellulase enzymes. Processing of the co-products, such as DDGS and corn fiber, of dry or wet milling processes for additional fermentable sugars and ethanol is achieved by releasing fermentable sugars from glucans and other polymeric sugars left in these co-products. Due to the recalcitrant nature of the cellulose, these materials need pretreatment prior to saccharification.

Kim et al. (2008c) have described several dry mill process alternatives in which the DDGS is recycled and hydrolyzed to produce more ethanol. The modified dry grind process involves recycling the sugar-rich liquid stream in the process, which, in turn, results in accumulation of various fermentation inhibitors. A simulated material balance model by Kim et al. (2008c) shows that two- to five-times higher concentrations of by-products and inhibitory components during the fermentation step of the modified processes may accumulate due to water recycling as compared to the conventional dry grind process. Nonetheless, ready fermentation of the sugars to ethanol is still achieved. The water balance for the modified dry grind processes also shows that the proposed processes require less fresh water input for the liquefaction than a conventional dry grind process. The water input due to additional unit operations required to extract and ferment sugars from the by-products can be compensated for by the water saved for the liquefaction in the modified process.

Processing the co-products for enhanced ethanol yield in the dry milling or wet milling process also affects the compositions of the final co-product exiting the distillation column. It contains a higher amount of protein per total mass as the polysaccharides are extracted to produce more ethanol, resulting in a reduced amount of co-product at the end of the process. The total wet solids sent to the drier are decreased, thus decreasing the drier load.

23.4 Ethanol Production from Corn Cob and Corn Stover

The agricultural residues from maize production are potential sources of sugar for ethanol production, in addition to starch and by-products. When maize is harvested in the field, the corn grain is separated from the cobs, stalks, and leaves. While the grain is transported for storing and processing, the stover is currently not widely collected. However, this biomass could be used for lignocellulosic ethanol production. Corn stover includes stalks, leaves, and corn cobs. Unlike the corn grains, of which the major component is starch, the main components of corn stover are cellulose, hemicellulose, and lignin (Table 23.1). Corn stover production is 75 million dry tons per year in the USA, which represents about 38% of all biomass from agricultural

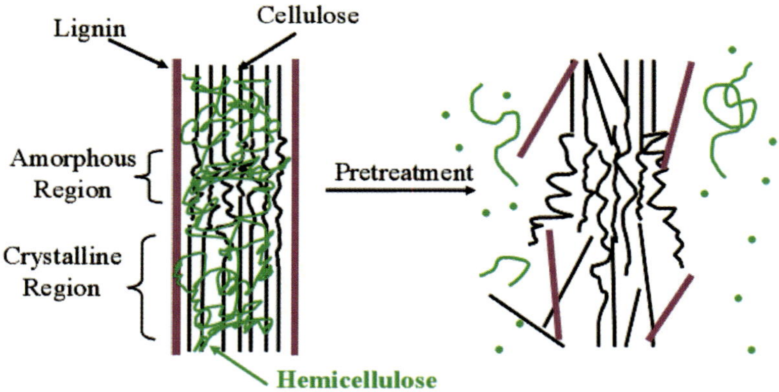

Fig. 23.5 Pretreatment effect on lignocellulosic material. (Reproduced from Mosier et al. (2005a))

lands (US DOE 2005). However, the return of some of this material to the soil is also important for maintaining the organic matter content and fertility of the soil (Blanco and Lal 2007; Kim and Dale 2005; Mann et al. 2002).

Lignocellulosic material is characterized by its strength and complexity due to a network formed between hemicellulose and cellulose in close association with lignin. A number of processing steps is required to overcome this complex structure to make it suitable for fermentation. The first step in producing cellulosic ethanol is biomass handling where the size of the lignocellulose is reduced to make handling easier and ethanol production more efficient. During pretreatment, cellulose structure is disrupted, the lignin seal is broken, and the hemicellulose is partially removed. This increases the specific surface area that is accessible to enzymes. Pretreatment is one of the many steps in the cellulose-to-ethanol process, but represents a currently critical step for hydrolysis. An effective pretreatment is performed at conditions that avoid degradation of pentose from hemicellulose, or glucose from cellulose, and limit formation of degradation products that inhibit the growth of fermentative microorganisms. Pretreatments should also limit energy, chemical, and/or enzyme usage in order to limit the cost of the pretreatment process itself (Mosier et al. 2005a). The effect of pretreatment on the structure of lignocellulose is depicted in Fig. 23.5.

After pretreatment, the cellulose and hemicellulose chains are more accessible to enzymes in the form of polymers and oligomers. Hydrolysis by enzymes breaks the chains into monomers. Enzymatic hydrolysis has the potential to make ethanol, derived from cellulose biomass, competitive when compared to other liquid fuels on a large scale (Wyman 1999). The monomers are then fermented by natural yeast and by genetically engineered bacteria or yeast. *Saccharomyces cerevisiae* (yeast) is currently used to produce ethanol for corn. *Saccharomyces* yeasts are currently applied in large-scale corn-to-ethanol or sugarcane-to-ethanol industry.

Ethanol production from lignocellulose requires not only fermentation of glucose but also fermentation of pentose sugars. Wild-type *Saccharomyces* is not able to ferment pentose. Utilization of genetically modified yeasts, specifically engineered for the purpose of co-fermenting xylose and glucose, address this need (Ho et al. 2000). The result of fermentation is a mixture of water, ethanol, and residues, with CO_2 being formed and removed as a gas from the fermentation.

Starch is a storage compound consisting of glucose linked via α-1,4 and α-1,6 glucosidic linkages (amylose and amylopectin). It is present in plants as an energy source. Cellulose, on the other hand, is a structural cell wall component with a function to provide rigidity to the plant. It is a highly crystalline and compact substrate, composed of glucose linked via β-1,4 glycosidic bonds, which makes it resistant to enzymatic hydrolysis (Gray et al. 2006). In this sense, it is not surprising that lignocellulosic substrates are more resistant to biological attack than starch. It is estimated, on a protein weight basis, that 40–100 times more enzyme is required to degrade cellulose to glucose than is used to hydrolyze starch in order to produce equivalent amounts of ethanol. Since the cost of enzyme production is not substantially different (Merino and Cherry 2007), the overall cost of enzyme use is higher for lignocellulosic ethanol conversion.

23.4.1 Cellulolytic Microorganisms

Several organisms exist in nature as complex consortia (fungi, bacteria, and protozoa) that work synergistically to deconstruct the plant cell wall. They rely on biomass degradation for their survival. Although all of these organisms can be considered as potential sources of biomass-degrading enzymes, fungi are predominantly used for the enzymatic cellulose hydrolysis to glucose, e.g., *Trichoderma*, *Penicillium*, and *Aspergillus* species (Galbe and Zacchi 2002). These microorganisms produce (at high productivity) a complex mix of enzymes with high catalytic efficiency in hydrolyzing cellulose to soluble sugars and glucose. These features are required for low-cost enzyme supply. Additionally, enzymes from fungi are generally secreted in the growth medium, facilitating the separation of active enzymes in a liquid form suitable for using in a hydrolysis reactor (Merino and Cherry 2007).

The saprophytic mesophilic fungus *Trichoderma reesei* is currently used for the commercial manufacturing of cellulase and hemicellulase products and has been proposed as the most promising organism for production of enzymes for lignocellulose conversion to fermentable sugars. In addition to a long history of safe commercial use, this fungus secretes significant quantities of enzymes and has well-developed genetic systems (Potera 2006; Li et al. 2007).

23.4.2 Cellulolytic Enzymes

The cellulolytic system produced by *T. reesei* is formed by exoglucanases (two cellobiohydrolases, CBHI and II), a number of endoglucanases (EG), and β-glucosidases. CBH I and CBH II constitute approximately 60% and 20%, respectively, of the protein secreted by this fungus, and they have a critical role in the efficient hydrolysis of cellulose (Hazell et al. 2000). EG randomly attack the cellulose chain, creating new reducing and non-reducing ends for the cellobiohydrolases. A third type of enzyme, β-glucosidase, which hydrolyzes cellobiose and some other short-chain cellodextrins into glucose, is also necessary. In the absence of β-glucosidase, end-product inhibition from cellobiose will occur for the other enzymes (Hahn-Haegerdal et al. 2006). β-glucosidase can also be inhibited by the accumulation of glucose. The sensitivity of the enzyme to inhibition by a specific end-product is reflected by experimentally measured inhibition constant (K_i). When a β-glucosidase (K_i 3.6 mM, competitively inhibited by glucose) from a different microorganism source (anaerobic fungus *Orpinomyces* strain PC-2) was used in combination with *T. reesei* cellulases, the conversion of cellulosic materials into glucose by *T. reesei* enzymes was significantly enhanced, demonstrating the potential of this specific enzyme for use in biofuel and feedstock chemical production (Li et al. 2007).

The maximum cellulase activity of most fungal cellulases and β-glucosidases is generally at 50°C and a pH of 4.0–5.0. Optimal conditions may vary with the hydrolysis time and are also dependent on the source of the enzymes (Hahn-Haegerdal et al. 2006). It is well known that efficient cellulose hydrolysis requires a complex enzyme mix. In order to decrease the amount of enzymes required for this purpose it is important that enzymes with superior specific activities be identified, or additional enzymes be added, reducing the total enzyme loading. It is also important to determine the synergism among the enzymes. A significant contribution to the degradation of lignocellulosic substrates can also be seen by the presence in certain hydrolytic enzymes of carbohydrate-binding modules connected by linkers to the catalytic modules.

Hydrolysis performed separately from fermentation is known as separate hydrolysis and fermentation (SHF). Cellulose hydrolysis carried out in the presence of the fermentative organism is referred to as simultaneous saccharification and fermentation (SSF). Simultaneous saccharification of both cellulose (to glucose) and hemicellulose (to pentose) and co-fermentation of both glucose and pentose (SSCF) is realized by combining enzymes with genetically engineered microbes (Mosier et al. 2005a). The ethanol fermentation of glucose, mannose, and galactose by baker's yeast, *Saccharomyces cerevisiae*, is well established on a large scale. For the conversion of pentose sugars (xylose and arabinose) to ethanol, most of the work has been focused so far on xylose (Hahn-Hagerdal et al. 2006). Pentose-fermenting *Escherichia coli* (Ingram et al. 1987) and *Klebsiella oxytoca* (Burchhardt and Ingram 1992) have been generated by introducing ethanologenic genes from *Zymomonas mobilis*. The use of genetically modified yeasts specifically engineered for the purpose of co-fermenting xylose and glucose has also been described (Ho et al. 2000).

Ethanol is recovered from the fermentation broth by distillation (Ladisch et al. 1984). The residual lignin, unreacted cellulose and hemicellulose, ash, enzyme, organisms, and other components end up in the bottom of the distillation column. These materials may be concentrated, and burned as fuel to power the process,

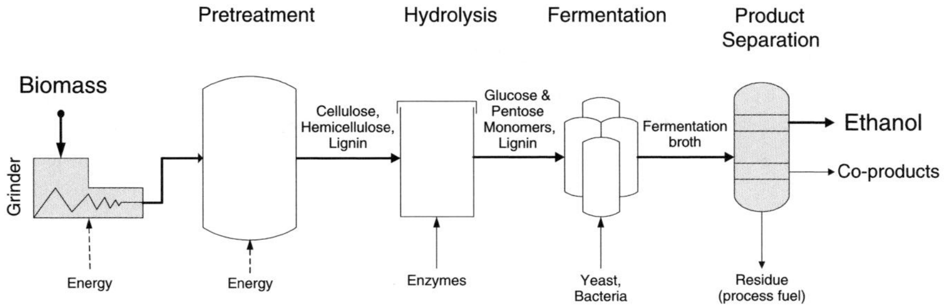

Fig. 23.6 Process flow diagram for ethanol production from lignocellulose

or converted to various co-products (Wyman 1994). CO_2 is recycled into plant matter through production agriculture.

Figure 23.6 shows a process sequence for ethanol production from lignocellulose. The key steps are preparation (size reduction) of biomass, pretreatment to soften up and disrupt the structure of the cellulose (Fig. 23.5), hydrolysis to break the cellulose down into sugars, and then fermentation of the sugars to ethanol. Product separation is the final step and consists of distillation followed by drying using either molecular sieve or a corn-based adsorbent (Ladisch et al. 1984; Gulati et al. 1996).

23.5 Comparison of Ethanol Yields for Conversion of Starch, By-products, and Corn Stover

It is widely recognized that feedstock costs are a critical factor in the production of fuel ethanol. One way of reducing these costs is to increase the feedstock yields per acre that are harvested. Today, only the starch from maize is used for ethanol conversion. The yield would be increased if a larger portion of the maize plant were to serve as a feedstock. Lignocellulosic ethanol technology is the key to enable the conversion of plant matter other than starch into renewable fuel.

A comparison of four different best case scenarios is given in Table 23.3, where compositions are as shown in Table 23.4. In the base scenario, ethanol is produced using only the starch portion of the plant. Scenario 2 includes ethanol production from both starch and by-products (fiber, DG, and DDGS). In scenarios 3a and 3b, ethanol is derived from starch, by-products, and corn stover. The two different scenarios including corn stover account for different harvesting techniques. In scenario 3a (normal cut), the maize is cut at 40 cm above the ground. This has been evaluated as the optimal height for harvesting (Hoskinson et al. 2007). In the study by Hoskinson et al., different harvest scenarios have been assessed in terms of sustainability, economics, nutrient value, and suitability for ethanol production. In scenario 3b (low cut), a larger portion of the corn stover is collected when compared to scenario 3a (10 cm cutting height).

Table 23.3 Theoretical ethanol production using starch, cellulose, and hemicellulose as feedstocks, reported in dry kilograms per acre

Component	Base scenario (starch)	Scenario 2 (starch and by-products)	Scenario 3a "normal cut"[a] (starch, by-products, and corn stover)	Scenario 3b "low cut"[b] (starch, by-products, and corn stover)
Feedstock				
Corn grain	4000	4000	4000	4000
Corn stover	0	0	2060	2703
Sugar				
Glucan from starch	2868	2868	2868	2868
Glucan from cellulose	0	96	96 + 750	96 + 984
Xylan	0	220	220 + 505	220 + 662
Ethanol				
Ethanol from starch	1628	1628	1628	1628
Ethanol from cellulose	0	55	55 + 426	55 + 559
Ethanol from hemicellulose	0	128	128 + 293	128 + 384
Total ethanol (kg)	1628	1811	2530	2754

[a]Normal cut is 40 cm above the ground
[b]Low cut is 10 cm above the ground

The following assumptions were made for estimating the total ethanol yield in Table 23.3. Calculations are based on maximum theoretical ethanol yield per acre of maize using current practices (starch ethanol). This means that all available sugar polymers (according to the different scenarios) are converted into monomeric sugars and that all sugars are fermented to ethanol without losses. The data of the compositional analyses of corn grain and corn stover for calculation of the sugar yields are shown in Table 23.4. The data in the last two columns represent the compositions of

Table 23.4 Compositions of corn grain (scenarios 1 and 2) and corn stover (scenarios 3a and 3b). n/m indicates not measured

Type of material	Corn grain[a]	Corn stover "normal cut"[b]	Corn stover "low cut"[b]
Starch	71.7	n/m	n/m
Cellulose	2.4	36.4	38.2
Hemicellulose	5.5	24.5	23.4
Protein	10.3	4.1	4.5
Oil	4.3	n/m	n/m
Lignin	0.2	12.7	15.5
Ash	1.4	2.5	2.7
Other	4.2	19.8	15.7
Total	100.0	100.0	100.0

[a]Gulati et al. (1996)
[b]Hoskinson et al. (2007). Normal cut is 40 cm above the ground; low cut is 10 cm above the ground

corn stover that incorporate the compositional differences between corn cobs, corn stalks, and leaves.

Data for both corn grain and corn stover yields have been adopted from Hoskinson et al. (2007), which are representative values for Story County, Iowa, and as reported by the National Agricultural Statistics Service (NASS) (USDA National Agricultural Statistics Service 2005). The average corn grain yield per acre and per year in the USA is assumed to be 150 bushels (equal to 3225 kg) (USDA 2006), which is lower than the data provided by NASS (4000 kg). However, both sources give the same weight ratio of 1:1 for corn grain yield to corn stover yield, which is an important parameter in differences in ethanol yield when comparing the four scenarios.

In scenario 2, glucan and xylan yields from cellulose and hemicellulose, respectively, are derived from the corn hull. In scenario 3, there are additional glucan and xylan yields when compared to scenario 3 which account for the sugars available from the corn stover. A number of formulae were used to calculate the ethanol yields for the three scenarios. These encompass hydrolysis of glucan to glucose and xylan to xylose (Eqs. 23.1 and 23.2), and fermentation of glucose and xylose to ethanol (Eqs. 23.3 and 23.4):

$$Glucose\,(\text{kg}) = Glucan\,(\text{kg}) \times \frac{180}{162} \tag{23.1}$$

$$Xylose\,(\text{kg}) = Xylan\,(\text{kg}) \times \frac{150}{132} \tag{23.2}$$

$$EtOH\,(\text{kg}) = Glucose\,(\text{kg}) \times 0.511 \tag{23.3}$$

$$EtOH\,(\text{kg}) = Xylose\,(\text{kg}) \times 0.511 \tag{23.4}$$

The results are summarized in Table 23.3. In the base scenario, 1628 kg of ethanol is produced for each acre of harvested corn when only starch is used as a feedstock. The ethanol yield is 11.2% (183 kg) higher compared to the base scenario when co-products are converted into monomeric sugars and subsequently fermented. The increase in ethanol yield is due to the additional sugar yield which originates from both cellulose and hemicellulose in the hull of the corn kernel. Kim et al. (2008c) report a 14% increase in ethanol yield when by-products are used for additional ethanol production. The 2.8% difference can be explained by the different modeling approaches. While this book chapter separates ethanol yields from starch and lignocellulose, Kim et al. distinguish between starch ethanol production and by-product ethanol production. In the latter case, ethanol produced from by-products includes ethanol from the unutilized residual starch in the by-products. Therefore, the number in Kim et al. (2008c) is slightly higher.

A 55% (902-kg) higher ethanol yield (compared to base scenario) per acre is possible when the corn stover is also harvested and processed (scenario 3a). If agricultural practices were improved so that the maize could be low cut (scenario 3b), the total ethanol yield would rise to 2754 kg of ethanol. This equals an increase of 69% compared to the base scenario. However, as mentioned previously, continuous complete removal of the stover is probably not desirable due to adverse effects on the soil and hence only about half of the available stover will be harvested.

23.6 Conclusion

The potential of maize to provide feedstock for alcohol production in excess of 15 billion gallons of ethanol per year in the USA will depend on utilization of cellulose and hemicellulose portions of the plant. Consequently, cellulose conversion will play a role in increasing yields. This chapter has discussed both the challenges and potential of converting the pericarp, stalks, cobs, and leaves to obtain additional fermentable sugars and ethanol. If sustainable, cost-effective, and environmentally compatible agricultural practices are developed and coupled to cellulose conversion technology, maize has the potential to provide 20–24 billion gallons of ethanol per year through a combination of starch processing and cellulose conversion.

Acknowledgements This work was supported by the US Department of Energy (Contract: DE-FG36-04GO14220) in cooperation with The Midwest Consortium for Biobased Products and Bioenergy and also by Purdue Agricultural Research Programs.

References

Belyea RL, Rausch KD, Tumbleson ME (2004) Composition of corn and distillers dried grains with solubles from dry grind ethanol processing. Bioresource Technol 94(3):293–298

Blanco H, Lal R (2007) Soil and crop response to harvesting corn residues for biofuel production. Geoderma 141(3–4):355–362

Burchhardt G, Ingram LO (1992) Conversion of xylan to ethanol by ethanologenic strains of *Escherichia coli* and *Klebsiella oxytoca*. Appl Environ Microbiol 58:1128–1133

Dock C, Hess M, Antranikian G (2008) A thermoactive glucoamylase with biotechnological relevance from the thermoacidophilic Euryarchaeon *Thermoplasma acidophilum*. Appl Microbiol Biotechnol 78:105–114

Eggeman T, Elander RT (2005) Process and economic analysis of pretreatment technologies. Bioresource Technol 96(18):2019–2025

Energy Information Administration (2007) US fuel ethanol oxygenate production. US Department of Energy, Washington, DC

Galbe M, Zacchi G (2002) A review of the production of ethanol from softwod. Appl Microbiol Biotechnol 59:618–628

Graboski MS (2002) Fossil energy use in the manufacture of corn ethanol. Colorado School of Mines. Prepared for the National Corn Growers Association, Chesterfield

Gray KA, Zhao L, Emptage M (2006) Bioethanol. Curr Opin Chem Biol 10:1–6

Gulati M, Kohlmann K, Ladisch MR, Hespell R, Bothast RJ (1996) Assessment of ethanol production options for corn products. Bioresource Technol 58(3):253–264

Hahn-Haegerdal B, Galbe M, Gorwa-Grauslund MF, Liden G, Zacchi G (2006) Bioethanol – the fuel of tomorrow from the residues of today. Trends Biotechnol 24(12):549–556

Hazell BW, Te'o VSJ, Bradner JR, Bergquist PL, Nevalainen KMH (2000) Rapid transformation of high cellulase-producing mutant strains of *Trichoderma reesei* by microprojectile bombardment. Lett Appl Microbiol 30:282–286

Ho N, Chen Z, Brainard A, Sedlak M (2000) Genetically engineered *Saccharomyces* yeasts for conversion of cellulosic biomass to environmentally friendly transportation fuel ethanol. American Chemical Society Symposium Series, Washington, DC

Hoskinson RL, Karlen DL, Birrell SJ, Radtke CW, Wilhelm WW (2007) Engineering, nutrient removal, and feedstock conversion evaluations of four corn stover harvest scenarios. Biomass Bioenergy 31(2–3):126–136

Ingram LO, Conway T, Clark DP, Sewell GM, Preston JF (1987) Genetic engineering of ethanol production in *Escherichia coli*. Appl Environ Microbiol 53:2420–2425

Kim S, Dale BE (2005) Life cycle assessment of various cropping systems utilized for producing biofuels: bioethanol and biodiesel. Biomass Bioenergy 29(6):426–439

Kim Y, Hendrickson R, Mosier N, Bals B, Balan V, Dale B, Ladisch M (2008a) Enzyme hydrolysis and ethanol fermentation of liquid hot water and AFEX pretreated distillers' grains at high solids loadings. Bioresource Technol 99(12):5206–5215

Kim Y, Mosier N, Hendrickson R, Brewer M, Ho NWY, Sedlak M, Dreshel R, Welch G, Dien BS, Aden A, et al. (2008b) Composition of corn dry-grind ethanol by-products: DDGS, wet cake, and thin stillage. Bioresource Technol 99(12):5165–5176

Kim Y, Mosier N, Ladisch M (2008c) Process simulation of modified dry grind ethanol plant with recycle of pretreated and enzymatically hydrolyzed distiller's grains. Bioresource Technol 99(12):5177–5192

Ladisch MR, Voloch M, Hong J, Bienkowski P, Tsao GT (1984) Cornmeal adsorber for dehydrating ethanol vapors. Ind Eng Chem Proc Des Dev 23(3):437–443

Li X-L, Skory CD, Ximenes EA, Jordan DB, Dien BE, Hughes SR, Cotta MA (2007) Expression of an AT-rich xylanase gene from the anaerobic fungus *Orpinomyces* sp. strain PC-2 in and secretion of the heterologous enzyme by *Hypocrea jecorina*. Appl Biochem Biotechnol 74(6):1264–1275

Mann L, Tolbert V, Cushman J (2002) Potential environment effects of corn (*Zea mays* L.) stover removal with emphasis on soil organic matter and erosion. Agric Ecosyst Environ 89(3):149–166

Merino ST, Cherry J (2007) Progress and challenges in enzyme development for biomass utilization. Adv Biochem Engin Biotechnol 108:95–120

Mosier N, Wyman C, Dale B, Elander R, Lee Y, Holtzapple M, Ladisch M (2005a) Features of promising technologies for pretreatment of lignocellulosic biomass. Bioresource Technol 96:673–686

Mosier NS, Hendrickson R, Brewer M, Ho N, Sedlak M, Dreshel R, Welch G, Dien BS, Aden A, Ladisch MR (2005b) Industrial scale-up of pH-controlled liquid hot water pretreatment of corn fiber for fuel ethanol production. Appl Biochem Biotechnol 125(2):77–97

Pandey A (1995) Glucoamylase research – an overview. Starch 47:439–445

Pordesimo LO, Hames BR, Sokhansanj S, Edens WC (2005) Variation in corn stover composition and energy content with crop maturity. Biomass Bioenergy 28(4):366–374

Potera C (2006) Progress with biofuels depends on drive microbiology research. Microbe 1:317–322

Turner P, Mamo G, Karlsson EN (2007) Potential and utilization of thermopiles and thermostable enzymes in biorefining. Microb Cell Factories 6(9):1–23

USDA (2006) Agricultural statistics. Table 1–36. http://www.nass.usda.gov/publications/ag_statistics/index.asp. Accessed 14 December 2007

USDA National Agricultural Statistics Service (2005) Quick stats, Iowa county data – crops. http://www.nass.usda.gov. Accessed 18 December 2007

US DOE (2005) Biomass as feedstock for a bioenergy and bioproducts industry: the technical feasibility of a billion-ton annual supply. US Department of Energy, Office of Science and Office of Energy Efficiency and Renewable Energy, Washington, DC

US DOE (2007) Biomass feedstock composition and property database. US Department of Energy, Office of Science and Office of Energy Efficiency and Renewable Energy, Washington, DC. http://www1.eere.energy.gov/biomass/feedstock_databases.html. Accessed 13 December 2007

Van der Veen ME, Veelaert S, Van der Goot AJ, Boom RM (2006) Starch hydrolysis under low water conditions: a conceptual design. J Food Eng 75:178–186

Wyman CE (1994) Ethanol from lignocellulosic biomass: technology, economics, and opportunities. Bioresource Technol 50(1):3–15

Wyman CE (1999) Biomass ethanol: technical progress, opportunities, and commercial challenges. Annu Rev Energy Environ 24:189–226

Index

Printing: Krips bv, Meppel, The Netherlands
Binding: Stürtz, Würzburg, Germany